PHYSICS OF DENSE MATTER

INTERNATIONAL ASTRONOMICAL UNION
UNION ASTRONOMIQUE INTERNATIONALE

SYMPOSIUM No. 53
HELD IN BOULDER, COLORADO, U.S.A., 21–25 AUGUST, 1972

PHYSICS OF DENSE MATTER

EDITED BY

CARL J. HANSEN

*Department of Physics and Astrophysics
and The Joint Institute for Laboratory Astrophysics,
University of Colorado, Boulder, Colo., U.S.A.*

ASSISTED BY

LORRAINE H. VOLSKY

D. REIDEL PUBLISHING COMPANY

DORDRECHT-HOLLAND / BOSTON-U.S.A.

1974

Published on behalf of
the International Astronomical Union and
the International Union of Pure and Applied Physics
by
D. Reidel Publishing Company, P. O. Box 17, Dordrecht, Holland

All Rights Reserved
Copyright © 1974 by the International Astronomical Union

Sold and distributed in the U.S.A., Canada, and Mexico
by D. Reidel Publishing Company, Inc.
306 Dartmouth Street, Boston,
Mass. 02116, U.S.A.

Library of Congress Catalog Card Number 73-91431

Cloth edition: ISBN 90 277 0406 6
Paperback edition: ISBN 90 277 0407 4

No part of this book may be reproduced in any form, by print, photoprint, microfilm,
or any other means, without written permission from the publisher

Printed in The Netherlands by D. Reidel, Dordrecht

TABLE OF CONTENTS

INTRODUCTORY REMARKS	VII
THE ORGANIZING COMMITTEES	VIII
LIST OF PARTICIPANTS	IX
J. W. NEGELE / The Equation of State of Matter at Sub-Nuclear Density	1
H. A. BETHE / Equation of State at Densities Greater than Nuclear Density	27
V. R. PANDHARIPANDE / Variational Method for Dense Systems	47
J. R. BUCHLER / Properties and Synthesis of Heavy Nuclei and Properties of Neutron Star Matter	67
J. C. WHEELER / 'Statistical Bootstrap' Equation of State for Cold Ultra-Dense Matter	77
Y. C. LEUNG and C. G. WANG / A Simple Equation of State of Matter at Super-Nuclear Densities	93
R. F. SAWYER / Pions in Neutron Star Matter	105
Y. NE'EMAN / Hypercollapsed Nuclear Matter	111
M. RUDERMAN / Matter in Superstrong Magnetic Fields	117
V. CANUTO and S. M. CHITRE / Quantum Crystals in Neutron Stars	133
G. GREENSTEIN / Superfluidity in Neutron Stars	151
J. P. HANSEN, B. JANCOVICI, and D. SCHIFF / Phase Diagram of a Charged Bose Gas (Abstract Only)	167
G. KALMAN and S. T. LAI / Superluminal Sound and Ferromagnetic Transition in the Zeldovich Model	169
S. A. COLGATE / Supernovae and Neutron Stars	183
D. PINES, J. SHAHAM, and M. A. RUDERMAN / Neutron Star Structure from Pulsar Observations	189
S. TSURUTA / Cooling of Dense Stars	209
G. BÖRNER and J. M. COHEN / Pulsar Observations and Neutron Star Models	227
J. M. COHEN and G. BÖRNER / Hadron Star Models	237
H. M. VAN HORN / Differential Rotation in Degenerate Stars	251
F. K. LAMB and P. G. SUTHERLAND / Continuum Polarization in Magnetic White Dwarfs	265

R. F. O'CONNELL / Polarized Radiation from White Dwarfs and Atoms in Strong Magnetic Fields 287

R. OMNÈS / Matter-Antimatter Separation and the Antimatter Problem in Cosmology 301

Y. NE'EMAN / Quasar Counts and the Lagging Core Model ('White Holes') 319

A. G. W. CAMERON / Concluding Remarks 321

INTRODUCTORY REMARKS

The subject of this Symposium is indicative of the breadth of modern astronomy in that we are to mainly consider here the nature of physical environments whose character is similar to, or more extreme than, nuclear matter. The reason for this consideration has been the discovery of those observed astronomical objects, the pulsars, and their associates, neutron stars. Here are phenomena whose explanation must ultimately be derived from the work of both astronomers and physicists of many disciplines. To quote from the concluding remarks of A. G. W. Cameron:

> It is a measure of the newness of this field of dense matter calculations that probably the majority of people presenting papers are not members of the IAU. A large number of the participants are physicists who have entered the field very recently because of their interest in the properties of neutron stars. Probably Commission 35 should co-opt many of you in order to give dense matter a greater representation within the deliberations of the IAU.

On behalf of the Local Organizing Committee we should like to thank the participants in the Symposium for an enlightening and stimulating week of good science and sociality and hope that Boulder will see you again. We also thank the Graduate School, the Departments of Physics and Astrophysics and Aerospace Engineering, and The Joint Institute for Laboratory Astrophysics, all of the University of Colorado, for support both moral and financial.

Part of the time spent in editing this work was supported by NSF Grant GP-36245 through the University of Colorado.

C. J. HANSEN
W. E. BRITTIN

SCIENTIFIC ORGANIZING COMMITTEE

E. Schatzman (Chairman), V. A. Ambartsumian, J. N. Bahcall,
H. Bethe, A. G. W. Cameron, R. Dashen, S. Frautschi, D. Pines,
E. E. Salpeter, Y. Zeldovich

LOCAL ORGANIZING COMMITTEE

W. E. Brittin (Chairman), J. P. Cox, C. J. Hansen, M. Uberoi

LIST OF PARTICIPANTS

N. Ashby, Boulder, Colo., U.S.A.
Z. Barkat, Ithaca, N.Y., U.S.A.
B. K. Berger, Boulder, Colo., U.S.A.
H. A. Bethe, Ithaca, N.Y., U.S.A.
M. S. Bhatia, Bowie, Md., U.S.A.
G. Borner, Greenbelt, Md., U.S.A.
W. E. Brittin, Boulder, Colo., U.S.A.
J. R. Buchler, New York, N.Y., U.S.A.
A. G. W. Cameron, New York, N.Y., U.S.A.
E. Canel, Birmingham, Great Britain
V. Canuto, New York, N.Y., U.S.A.
T. R. Carson, New York, N.Y., U.S.A.
L. M. Celnikier, Meudon, France
D. M. Chitre, Salt Lake City, Utah, U.S.A.
S. M. Chitre, New York, N.Y., U.S.A.
H.-Y. Chiu, Stony Brook, N.Y., U.S.A.
J. M. Cohen, Greenbelt, Md., U.S.A.
S. A. Colgate, Socorro, N. M., U.S.A.
R. Cover, Ann Arbor, Mich., U.S.A.
J. P. Cox, Boulder, Colo., U.S.A.
A. E. Glassgold, New York, N.Y., U.S.A.
G. Greenstein, Amherst, Mass., U.S.A.
C. J. Hansen, Boulder, Colo., U.S.A.
G. Horwitz, Jerusalem, Israel
I. Iben, Urbana, Ill., U.S.A.
N. Itoh, Urbana, Ill., U.S.A.
B. Jancovici, Orsay, France
G. Kalman, Chestnut Hill, Mass., U.S.A.
S. T. Lai, Chestnut Hill, Mass., U.S.A.
P. Ledoux, Seattle, Wash., U.S.A.
L. Mestel, Manchester, Great Britain
M. Nauenberg, Santa Cruz, Calif., U.S.A.
U. Nauenberg, Boulder, Colo., U.S.A.
Y. Ne'eman, Tel-Aviv, Israel
J. Negele, Cambridge, Mass., U.S.A.
R. F. O'Connell, Baton Rouge, La., U.S.A.
R. Omnès, Orsay, France

V. R. Pandharipande, Ithaca, N.Y., U.S.A.
J. M. Pearson, Montreal, Canada
D. Pines, Urbana, Ill., U.S.A.
P. E. Reichley, Pasadena, Calif. U.S.A.
W. Rense, Boulder, Colo., U.S.A.
A. Rich, Ann Arbor, Mich., U.S.A.
R. W. Richardson, New York, N.Y., U.S.A.
R. Richtmyer, Boulder, Colo., U.S.A.
N. Roughton, Denver, Colo., U.S.A.
J. P. Rozelot, Bagnères-de-Bigorre, France
M. Ruderman, New York, N.Y., U.S.A.
I. Rudnick, Los Angeles, Calif., U.S.A.
R. F. Sawyer, Santa Barbara, Calif., U.S.A.
J. Shaham, Urbana, Ill., U.S.A.
S. Shore, Stony Brook, N.Y., U.S.A.
P. Sutherland, New York, N.Y., U.S.A.
T. Tsuneto, New Brunswick, N. J., U.S.A.
S. Tsuruta, Greenbelt, Md., U.S.A.
M. Uberoi, Boulder, Colo. U.S.A.
H. M. Van Horn, Rochester, N. Y., U.S.A.
G. Vauclair, Meudon, France
J. Ventura, New York, N.Y., U.S.A.
C. G. Wang, Cambridge, Mass., U.S.A.
J. C. Wheeler, Cambridge, Mass., U.S.A.

THE EQUATION OF STATE OF MATTER AT SUB-NUCLEAR DENSITY

J. W. NEGELE

Laboratory for Nuclear Science and Department of Physics,
Massachusetts Institute of Technology, Cambridge, Mass. 02139, U.S.A.

Abstract. An extremely simple form for the energy density of a nuclear many-body system is derived from the two-body nucleon-nucleon interaction. This theory, which yields excellent results for energies and density distributions of finite nuclei, is used to determine the ground state configuration of matter at sub-nuclear density. As the baryon density is increased, nuclei become progressively more neutron rich until neutrons eventually escape, yielding a Coulomb lattice of bound neutron and proton clusters surrounded by a dilute neutron gas. The clusters enlarge and the lattice constant decreases with increasing density, approaching a completely uniform state near nuclear density.

1. Introduction

There exists sufficient astrophysical evidence that pulsars are in fact neutron stars that it is worthwhile to seriously investigate the equation of state of dense baryonic matter utilizing one's current understanding of theoretical nuclear physics. In the present work, we shall restrict our attention to that limited regime in which the two tacit assumptions of nuclear physics are satisfied, namely that the temperature be low compared with characteristic nuclear excitation energies and that the meson degrees of freedom may be legitimately suppressed and replaced by a phenomenological two-body potential determined by scattering data and deuteron properties.

The first condition appears to be satisfied for observed pulsars, since the surface temperature may be inferred to be less than 10^8 K, corresponding to 10 keV and the thermal conductivity is expected to be very high. Thus, if we ignore the early stages of formation, on the relevant temperature scale of nuclear energies, neutron stars are exceedingly cold and may be treated in first approximation as being at a temperature of absolute zero. Hence it is legitimate to use the same zero-temperature perturbation expansions which are used in nuclear many-body theory, and the equation of state is completely specified by calculating the binding energy per baryon as a function of baryon density.

The suppression of meson degrees of freedom in nuclear physics has been extremely successful. At a density of 0.17 nucleons fm^{-3}, corresponding to the interior density of a heavy nucleus, the average distance between nuclei is 2.2 fm, whereas the one pion exchange force has a range of 1.4 fm. Thus, it appears quite reasonable to first approximate the interaction between two nucleons in nuclear matter as the interaction which would occur in free space, and then treat as a small perturbation the explicit meson many-body effects. The lowest order example of such an effect would be the three-body force arising from the interaction of a pion in transit between two nucleons with yet a third nucleon. Such processes have been investigated in nuclear matter and yield a correction to the binding energy per particle on the order of 2 MeV out of a total potential energy contribution of 40 MeV (Brown and Green, 1969). Thus, present evidence indicates that if we restrict our attention to nuclear configurations with densities less than or equal to that occurring in the interior of large nuclei, the structure should be dominated by the two-body nucleon-nucleon interaction and explicit meson many-body effects should produce only very small corrections amenable to perturbation theory.

Having determined the upper limit on densities to be considered to be that occurring in the interior of large nuclei, hereafter referred to simply as nuclear density and having the value 0.17 nucleon $fm^{-3} = 2.8 \times 10^{14}$ g cm^{-3}, we now consider the lower limit.

At densities above 10^7 g cm^{-3} for temperatures below 10^8 K, matter is expected to be a solid. This is because the Coulomb interaction between ions is only weakly screened and at sufficiently low temperatures, the Coulomb energy is minimized by a bcc lattice. Applying the Lindemann melting criterion which requires the mean fluctuations of ions to be small compared with the average ion spacing, the ions are expected to form a solid below the temperature $T_m \approx Z^2 e^2/(100\, r_Z)$ where r_Z is the average spacing between ions (Pines, 1970). For ^{56}Fe, this solidification occurs by the time one reaches a density of 10^7 g cm^{-3}, and as the density increases, r_Z decreases, assuring that the condition will be maintained at all subsequent densities. At extremely high densities, one might worry the Coulomb lattice breaking up solely due to the zero point motion of the ions, but for the highest density configuration treated in this work, $_{32}$Ge at 1.4×10^{14} g cm^{-3}, the zero point fluctuation is still very small compared with the ion spacing.

For Coulomb lattices of ions at densities between 10^7 and 10^9 g cm^{-3}, the equation of state is determined directly from the experimentally observed mass table using straightforward corrections for the electrons and lattice Coulomb energy (Baym et al., 1971a). At $\varrho = 6.2 \times 10^9$ g cm^{-3}, the Fermi energy of the relativistic electrons shifts the energy balance favoring neutron rich nuclei so strongly that ^{84}Se, which is unstable but experimentally observable, gives way to ^{82}Ge which has not been observed but may be reliably extrapolated (Myers and Swiatecki, 1965). Extrapolations from the mass table continue to be reliable up to a density of roughly 4.3×10^{11} g cm^{-3}, at which point the neutrons in ^{118}Kr are just barely bound and any further increase in density causes the last few neutrons to 'drip' out of the nuclei and form a low density neutron gas in the intervening space between the nuclear clusters. Although the theory presented in the present work should also be valid in the entire pre-drip sequence of extrapolated nuclei, we believe the mass table extrapolations are satisfactory well below the drip point and have concentrated our attention on the last sequence of pre-drip nuclei with 82 neutrons: ^{124}Mo, ^{122}Zr, ^{120}Sr and ^{118}Kr, and on the free neutron regime after drip occurs, between 4×10^{11} g cm^{-3} and 2×10^{14} g cm^{-3}.

In order to assess the significance of this region of densities, calculations of stellar density distributions by Baym et al. (1971a) using an equation of state by Baym et al. (1971b) which is very similar to that obtained in this work, are shown in Figure 1. From this graph, it is observed that the very lightest neutron stars do not exceed nuclear density and that the region from 4×10^{11} to 2×10^{14} g cm^{-3} is the most crucial in determining the structure and stability of these stars. In particular, one may observe qualitatively the effect of neutron drip and the approach to nuclear density on the equation of state. At 4×10^{11} g cm^{-3} where neutron drip occurs, the sudden increase in slope of the density distribution indicates the appreciable softening of the equation of state when neutrons begin to drip out of nuclei. Similarly the flatness of the curve near nuclear density reflects the stiffness of the equation of state once nuclei begin to touch and the gravitational pressure attempts to compress the nucleons to a density

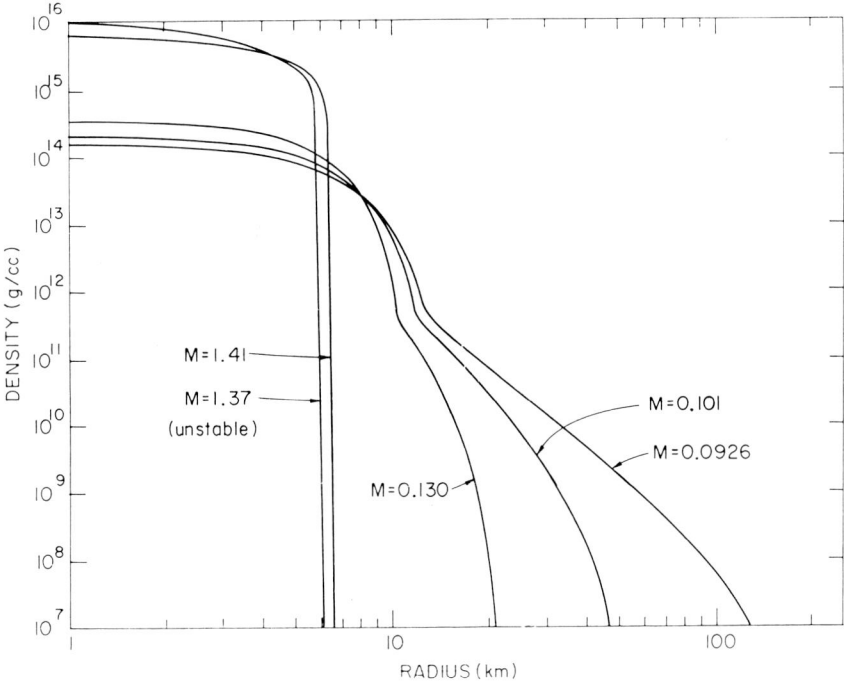

Fig. 1. Density profiles of neutron stars, calculated by Baym et al. (1971a).

higher than that occurring in a free nucleus. The need for quantitative precision in describing both the softness at the neutron drip point and stiffness near nuclear density motivates the present application of a microscopic theory of nuclear structure to the examination in somewhat greater detail of the equation of state in this regime.

2. Review of Previous Investigations

It is natural that the first attempts to treat the free neutron regime should be based on extrapolations from the semi-empirical mass formula (Bethe et al., 1970; Langer et al., 1969). The limitations of such an approach arise, obviously, from the fact that the mass formula parameters are determined only by a very restricted region of nuclear configurations: the ratio of protons to neutrons is greater than ~ 0.6, the chemical potential (minus the removal energy) for neutrons and protons is on the order of -8 MeV and there is zero external pressure on the nucleus. In the free neutron regime, however, the configurations are very far from satisfying these conditions. The ratio of protons to neutrons in nuclei is 0.1 to 0.3; the neutron chemical potential in the nucleus must equal that of the free gas and thus cover a range from 0 to $+20$ MeV; the pressure from the exterior neutron gas becomes significant, and the surface becomes much more diffuse and instead of approaching zero density outside the nucleus, approaches the neutron gas density. For all these reasons, extrapolations based on the semi-empirical mass formula are extremely unreliable, and one is forced to undertake a more fundamental theory.

Baym et al. (1971b, hereafter denoted BBP) introduced the information which is unobtainable from the mass formula by means of the theory of uniform nuclear matter. Based on nuclear matter calculations with the Reid (1968) soft core potential by Siemens (1970) in the region of roughly equal neutron and proton densities and by Siemens and Pandharipande (1971) in the region of almost pure neutron matter, BBP obtained the nuclear binding energy per nucleon surface shown in Figure 2. One should note that both Coulomb and gravitational interactions are necessarily omitted from these infinite nuclear matter calculations. In addition, since only two-body reaction matrix diagrams are included, a phenomenological correction was applied

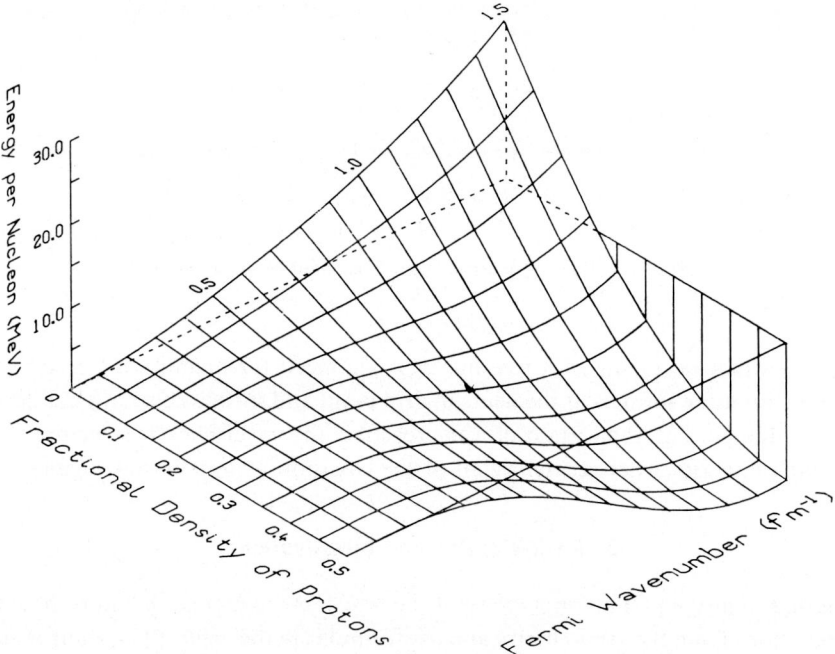

Fig. 2. Energy per nucleon of uniform nuclear matter ignoring Coulomb and gravitational forces, taken from BBP.

to include the effects of all higher order diagrams, three-body forces, and relativistic corrections. Since such corrections enter most strongly in the $^3S_1-^3D_1$ coupled partial waves which are connected by the tensor force, the phenomenological correction was defined to occur only in isospin $T=0$ states and thus acts only between unlike particles. By means of this correction, the minimum of the saturation curve was shifted from 11 MeV to 16.5 MeV at a density corresponding to $k_F = 1.43$ fm^{-1}.

Having determined the binding energy surface for all densities and ratios of protons and neutrons, BBP treat the Coulomb lattice of nuclei surrounded by a neutron gas in the Wigner-Seitz approximation. The unit cell is divided into a uniform density interior nucleus comprised of neutrons at high density and protons of low density and a uniform neutron gas exterior, comprised of neutrons at low density. A phenomenological surface energy expression motivated by Thomas-Fermi theory corrects for the presence of a finite surface and the Coulomb energy is calculated assuming a constant density of electrons filling the entire cell to yield charge neutrality. Then at a specified baryon density, the ground state configuration is determined by requiring that the system be stable against β-decay, that the pressure inside the nucleus equal the pressure in the outside gas, that the neutron chemical potential inside the nucleus equal that of the outside gas and by minimizing the total binding energy with respect to the radius of the unit cell.

Whereas this approach constitutes a clear improvement over previous work, it is still subject to two limitations. The first is an incomplete theory of the nuclear surface, with the general form being specified by Thomas-Fermi theory, but with the actual surface thickness being determined from the neutron Fermi wavelength instead of variationally. The second is the complete absence of nuclear shell effects, which are already observed to play an important role in determining the composition of the ground state matter prior to neutron drip.

Similar calculations by Arponen (1971), Barkat *et al.* (1972) and Buchler and Barkat (1971a, b) utilize slightly different parameterizations of the nuclear energy surface and determine ground state density distributions variationally in a Thomas-Fermi theory. The surface energy is included by introducing a gradient term in the expression for the energy as a functional of the density with a phenomenological coefficient adjusted to fit ordinary nuclei.

The semi-classical Thomas-Fermi approximation has been compared with the quantum mechanical Hartree-Fock method for a semi-infinite surface by Ravenhall *et al.* (1972). Using the Skyrme (1959) interaction, they demonstrate significant differences between the density distributions and equilibrium configurations with the two methods (see also Vautherin and Brink, 1972). The final result they obtain in Hartree-Fock theory for the charge of the nucleus as a function of matter density lies between the results of BBP and Barkat *et al.* (1972) and constitutes the most reliable calculation discussed thus far. We will subsequently compare the calculations of this present work with those of Ravenhall *et al.* (1972) and show how the differences arise from curvature terms in the surface energy and nuclear shell effects, both of which are necessarily omitted in a semi-infinite matter calculation.

3. Theory of Finite Nuclei

Present evidence indicates that a strongly repulsive core in the nucleon-nucleon interaction plays an essential role in producing finite nuclei which obtain the proper binding energy without collapsing to unphysically high densities. Once one adopts a potential with a strongly repulsive core, such as the Reid soft core potential used in this work, perturbation theory can no longer be ordered simply in terms of numbers of interactions of the bare potential, but rather must be re-expressed in terms of the reaction matrix. The reaction matrix, G, sums all orders of ladder diagrams containing the bare interaction and unoccupied intermediate states and may be written

$$G(w) = v - v \sum_{ab} \frac{|ab\rangle \langle ab|}{e_a + e_b - W} G(w),$$

where the projector labels a, b are summed over all unoccupied states and W indicates an energy specified by the single particle energies of the two interacting particles and the particular Goldstone diagram under consideration.

Physically, by permitting two interacting particles to interact any number of times when they approach each other, one is allowing the two-body wave function to respond to the presence of the potential by generating correlations which strongly decrease the probability of the particles penetrating into the repulsive core region. This may be visualized by defining the correlated wave function ψ by the relation $G\phi \equiv v\psi$, where ϕ is the uncorrelated wave function which is a plane wave in nuclear matter. The correlated wave function in the $^1S_0$ partial wave channel obtained in this way in nuclear matter is shown in Figure 3.

The field of nuclear matter theory is treated in great detail in two recent extensive review articles by Bethe (1971) and Sprung (1972). For our present purposes, however, it is sufficient to concentrate our attention on the general feature of the two-body correlations. The difference between ϕ and ψ, which we shall refer to as the defect function and denote by χ, contains all the information concerning two-body correlations and will play a central role in our theory of finite nuclei.

Whereas Figure 3 only shows the defect function in a single partial wave, one can more effectively visualize the total two-body correlation by plotting the sum of the squares of the defect functions in each partial wave as shown in Figure 4 for several densities of nuclear matter. The integral of the sum of the squares of the defect functions, multiplied by the density and appropriate statistical factors, yields the total probability of exciting a particle out of the uncorrelated Fermi sea into some excited state, which turns out to be approximately 13% at nuclear density. The most striking feature displayed in Figure 4 is that in addition to the maximum at 0.5 fm expected from the fact that the correlated wave function does not penetrate the hard core significantly, there is a second maximum at 1.2 fm. This correlation arises from second and higher order processes in which the strong, long range tensor force coupling the $^3S_1$–$^3D_1$ channels introduces deuteron-like spatial correlations in the two-body wave function. A particularly significant feature of this correlation is its strong density dependence. Recalling the definition of the reaction matrix, it is clear that at high density there are fewer intermediate states available for second order contributions to ψ, and Figure 4 shows how strongly these correlations are diminished at high density.

Having investigated the two-body correlations occurring between nucleons in nuclear matter, we now apply this knowledge to finite nuclei *via* the local density approximation. Physically, one argues that the short range correlations between two nucleons in the interior of ^{208}Pb, which is roughly 13 fm in diameter, should be the same as for two nucleons in infinite uniform matter at the same neutron and proton density. For nucleons near the nucleon surface, which is 2 fm thick, the argument is still quite good for the 0.4 fm correlations due to the core, and begins to break down only for the longer range tensor force correlations.

The most convenient technique for using the correlations calculated in nuclear matter in finite nuclei is the construction of an effective interaction. Considering, for simplicity, relative matrix elements in a finite nucleus in a specific relative partial wave one may introduce a complete set of plane waves and define a non-local effective interaction as follows:

$$\langle \phi_{nl} | G | \phi_{nl} \rangle = \sum_k \langle \phi_{nl} | G | \phi_k \rangle \langle \phi_k | \phi_{n'l} \rangle =$$

$$= \int\int \langle \phi_{nl} | r \rangle \langle r | \sum_k v | \Psi_k \rangle \langle \phi_k | r' \rangle \langle r' | \phi_{n'l} \rangle \, d^3r \, d^3r' \equiv$$

$$\equiv \int\int \phi_{nl}(r) \, v_{\text{eff}}(\varrho, W, r, r') \, \phi_{n'l}(r') \, d^3r \, d^3r'.$$

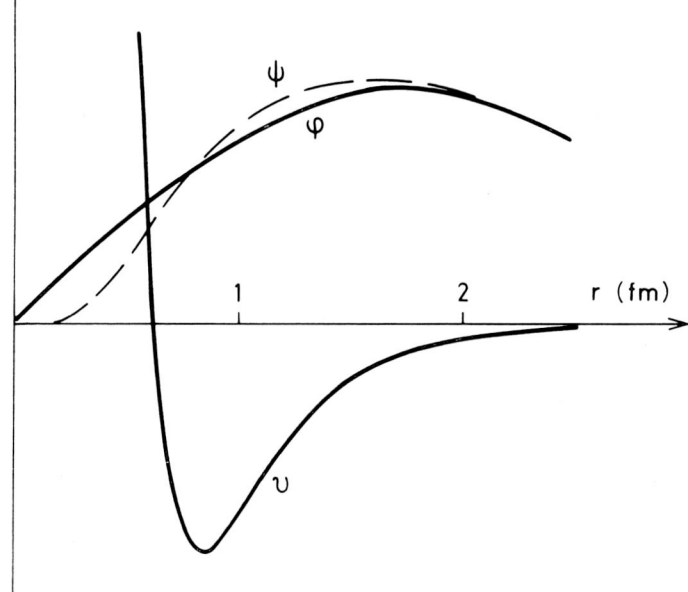

Fig. 3. Reid potential and uncorrelated and correlated nuclear matter wave functions for the $^1S_0$ partial wave.

For a practical calculation, it is necessary to make a number of additional simplifying approximations to v_{eff} yielding an effective interaction which is local in relative coordinate and is averaged over relative angular momentum states in each spin and isospin channel. These approximations are justified in detail elsewhere (Negele, 1970) and are immaterial to the conceptual basis of the theory. The final result is an effective interaction to be used with a Slater determinant wave function which depends not only on the relative coordinate of two interacting particles, but also on the density at the location where the two particles are interacting and on the energy of the two interacting particles. The fact that both the density dependence and energy dependence

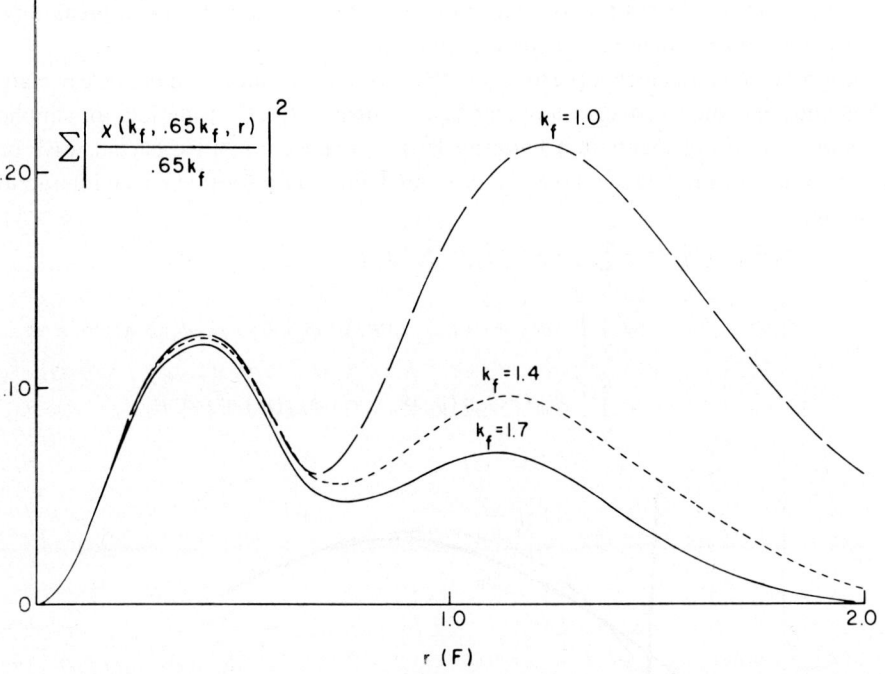

Fig. 4. Sum of the squares of the defect functions in nuclear matter taken from Negele (1970).

contribute to saturation, that is to making the interaction less attractive at high density, is evident from examining the second order term in the reaction matrix expansion, $v(Q/e)v$. As the density increases, the projector onto unoccupied states, Q, excludes more and more phase space from the second order sum, which is always attractive, thereby decreasing the attraction. In addition, the single particle energies become larger in magnitude with increasing density, thereby increasing the magnitude of the energy denominator e and decreasing the attraction.

Once one has accepted an effective interaction which is density and energy dependent, it is straightforward to write out the expression for the energy of a Slater determinant and to functionally differentiate with respect to each single particle wave function. In addition to the usual Hartree-Fock terms obtained from varying $\psi_m^*(r_1)$

and $\psi_n^*(r_2)$ appearing explicitly in

$$\sum_{mn} \int\int \Psi_m^*(r_1) \Psi_n^*(r_2) g\left[|r_1 - r_2|, \varrho\left(\frac{r_1 - r_2}{2}\right), e_m + e_n\right] \times$$
$$\times [\Psi_m(r_1) \Psi_n(r_2) - \Psi_n(r_1) \Psi_m(r_2)]$$

one also obtains terms of the form $(\partial G/\partial \varrho)(\partial \varrho/\partial \psi_s^*)$ and $(\partial G/\partial e_m)(\partial e_m/\partial \psi_s^*)$. These terms express the fact that by increasing the probability of finding particle s at some region in the nucleus, the total energy contribution from that region is not only increased because of the pair interactions of particle s with all other nucleons, but it is also slightly decreased by the fact that two other particles, say m and n interacting in that region, now find their interactions less attractive due to the fact that particle s has now made some phase space unavailable for intermediate scattering states and has slightly increased their energy denominators. These extra terms in the resulting density-dependent Hartree-Fock (DDHF) equations, loosely referred to as 'rearrangement terms' are crucial to obtaining saturation in finite nuclei and in obtaining agreement with experimental single particle energies and binding energies.

The effective interaction described above is actually defined in detail to reproduce exactly the same nuclear matter saturation curve as Siemens' (1970) original G matrix when evaluated in a Fermi gas of plane wave states. Thus, in order to obtain the proper binding energy per particle, it is necessary to introduce a phenomenological correction to include all the higher order terms in nuclear matter theory. In the same spirit as BBP, this was chosen to be a short range force acting only between unlike particles and for the results presented in this work, the two parameters were adjusted to give 16.53 MeV binding energy per particle at a saturation density of $k_F = 1.33$ fm^{-1}.

For spherical nuclei, it is straightforward, although numerically cumbersome, to directly solve the non-local integro-differential DDHF equations and by iteration obtain self-consistent wave functions. The binding energies per particle obtained in this way (Bethe, 1971) agree with experimental energies for O, Ca, Zr and Pb to within 0.5 MeV, and subsequent calculations are in even closer agreement (Negele, 1970) and (Campi and Sprung, 1972). The single particle energies agree with experimental energies to within several MeV and in particular the spin-orbit splittings in light nuclei are correct.

Having shown the necessity of having an accurate quantum mechanical theory of the nuclear surface, we wish to emphasize the agreement of the DDHF results with experimental evidence concerning the surface. The agreement of binding energies throughout the periodic table mentioned previously indicates the semi-empirical surface energy parameter is accurately reproduced. In addition, it is possible to check the detailed spatial distribution of protons by comparison with elastic electron scatterings results, as shown in Figures 5 and 6 for ^{40}Ca and ^{208}Pb, respectively. Although an accurate fit for a single nucleus might be deemed fortuitous, systematic agreement throughout the periodic table yields strong evidence that the delicate balance between Coulomb energy, symmetry energy, surface energy and bulk volume energy is being very accurately reproduced.

The DDHF theory of finite nuclei as described above is unfortunately too computationally time-consuming to apply to the equation of state in the free neutron regime. Self-consistent calculations for 208 particles in ^{208}Pb are already lengthy and expensive and in the free neutron regime, one desires to search over a variety of configurations with up to 5000 particles in a unit cell.

The primary computational complication in the DDHF theory is the presence of the non-local exchange term, which necessitates both the Legendre expansion of the exchange potential and the explicit solution of a non-local Schrödinger equation. For this reason, it is useful to examine the structure of the exchange term and develop a systematic expansion which treats it in a much more convenient manner. Although

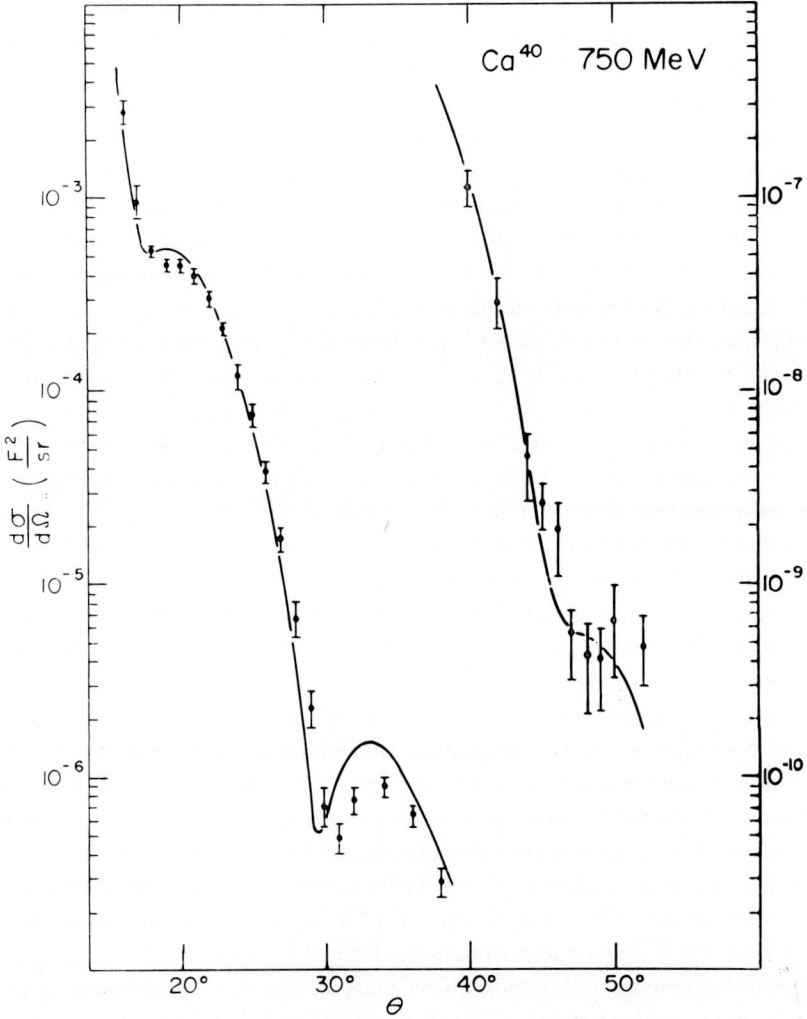

Fig. 5. Elastic electron scattering cross sections calculated with DDHF wave functions, compared with experimental results for ^{40}Ca.

this is a crucial practical development, it should be emphasized that it introduces no new conceptual assumptions since one can construct the exact exchange term and compare it with the subsequent expansion and insist on any specified degree of accuracy.

The density matrix expansion (DME) is a systematic expansion of the two-body density matrix $\varrho(r_1, r_2) = \sum_n \psi_n^*(r_1) \psi_n(r_2)$ in coordinate space (Negele and Vautherin, 1972). One begins by writing the angle-averaged density matrix as a formal expansion about the center of mass of the two interacting particles,

$$\hat{\varrho}(\mathbf{R}, r) = \int \frac{d\Omega}{4\pi} \sum_a \psi_a^*\left(\mathbf{R} + \frac{\mathbf{r}}{2}\right) \psi_a\left(\mathbf{R} - \frac{\mathbf{r}}{2}\right) =$$

$$= \int \frac{d\Omega}{4\pi} \exp\left\{\frac{\mathbf{r} \cdot (\nabla_1 - \nabla_2)}{2}\right\} \sum_a \psi_a^*(R_1) \psi_a(R_2)$$

$$= \frac{1}{2} \int d\cos\theta \exp\left\{\cos\theta r\left(\frac{\nabla_1 - \nabla_2}{2}\right)\right\} \varrho(R_1, R_2)$$

$$\equiv F\left[\left(\frac{\nabla_1 - \nabla_2}{2}\right)^2\right] \varrho(R_1, R_2)$$

$$F(k^2) = \begin{cases} \dfrac{\sinh(kr)}{kr} & k^2 > 0 \\ \dfrac{\sin(kr)}{kr} & k^2 < 0. \end{cases}$$

Expanding this around some value $-k^2$, one obtains

$$F\left[\left(\frac{\nabla_1 - \nabla_2}{2}\right)^2\right] = F(-k^2) + F'(-k^2)\left[\left(\frac{\nabla_1 - \nabla_2}{2}\right)^2 + k^2\right] + \cdots$$

$$= j_0(kr) + \frac{r}{2k} j_1(kr) \left[\left(\frac{\nabla_1 - \nabla_2}{2}\right)^2 + k^2\right] + \cdots.$$

Physically, the operator $(\nabla_1 - \nabla_2)/2$ is just the relative momentum operator. Hence it is reasonable to average both sides of this expression over the values of relative momentum k which would appear in nuclear matter at the same density. Averaging with the appropriate phase space factors and suitably rearranging the series, one obtains

$$\hat{\varrho}(\mathbf{R}, r) = \varrho_{SL}(rk_F) \varrho(\mathbf{R}) + r^2 g(rk_F)\left[\tfrac{1}{4}\nabla^2 \varrho(\mathbf{R}) - \tau(\mathbf{R}) + \tfrac{3}{5} k_F^2 \varrho(\mathbf{R})\right]$$

$$\varrho_{SL}(rk_F) = \frac{3 j_1(rk_F)}{rk_F} \qquad \varrho(R) = \sum_a |\psi_a(\mathbf{R})|^2$$

$$g(rk_F) = \frac{35 j_3(rk_F)}{2(rk_F)} \qquad \tau(R) = \sum_a |\nabla \psi_a(\mathbf{R})|^2.$$

The first term is just the Slater mixed density for a Fermi gas and the subsequent terms systematically expand the deviation of the finite nucleus density matrix from the nuclear matter result.

The accuracy of the first two terms is demonstrated in Figure 7. The exact density matrix is calculated for neutron wave functions in ^{208}Pb as a function of relative coordinate, s, at several values of the center of mass coordinate, R. Since the exact density matrix is not rigorously isotropic in $\mathbf{s}$, the extremal values are indicated on the graph by the error bars. The Slater approximation, denoted by the short dashed line, is observed to be in excellent agreement with the exact calculation in the interior and to systematically overestimate the mixed density in the surface. The sum of the first two terms is indicated by the solid line and yields an excellent approximation through-

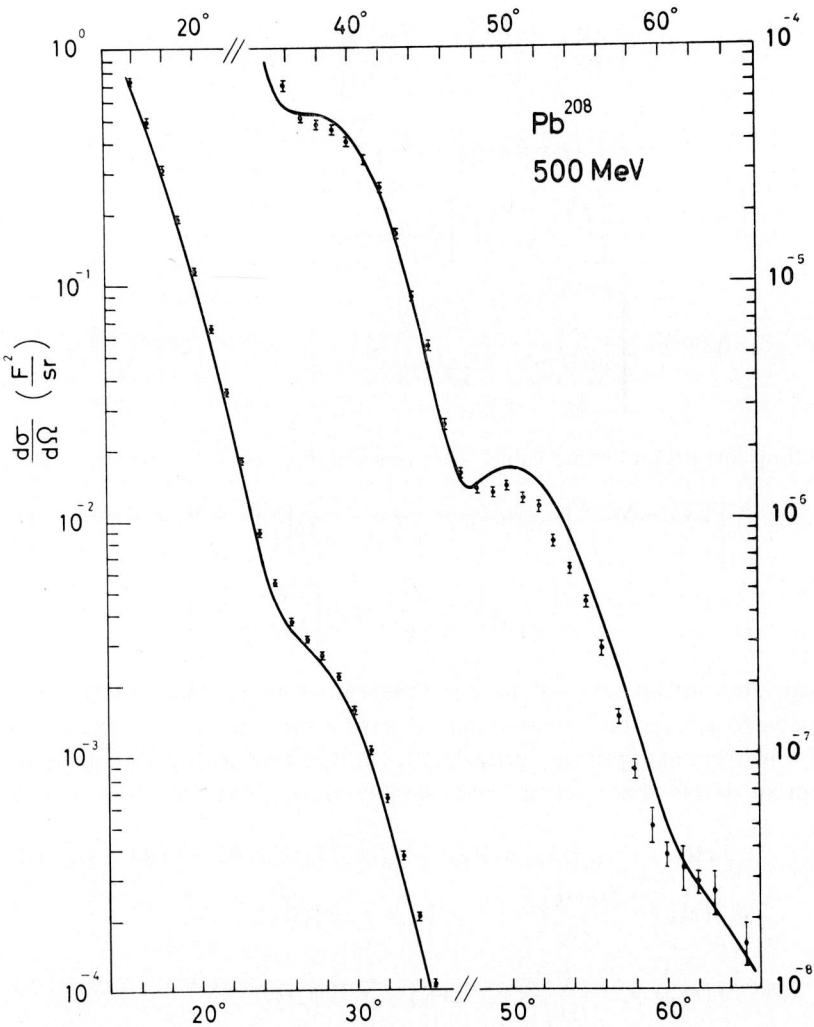

Fig. 6. Elastic electron scattering cross sections calculated with DDHF wave functions, compared with experimental results for ^{208}Pb.

out the region of relative coordinate in which the two-body potential is significant.

Truncation of the expansion of the density matrix at second order yields an extremely simple factorized expression of specified functions of relative coordinate multiplying $\varrho(R)$, derivatives of $\varrho(R)$ and the kinetic energy density, $\tau(R)$. The total Hamiltonian density may then be obtained by performing the integral over relative coordinates of the density matrix squared times the effective interaction. Distinguishing between proton and neutron densities and kinetic energies, the final form of the Hamiltonian density is

$$H(R) = \frac{\hbar^2}{2M}(\tau_n + \tau_p) + A(\varrho_n, \varrho_p) + B(\varrho_p, \varrho_n)\tau_p + B(\varrho_n, \varrho_p)\tau_n +$$
$$+ C(\varrho_p, \varrho_n)|\nabla_{\varrho_p}|^2 + C(\varrho_n, \varrho_p)|\nabla_{\varrho_n}|^2 + D(\varrho_n, \varrho_p)\nabla_{\varrho_n}\cdot\nabla_{\varrho_p},$$

where the functions A, B, C, and D are specified integrals involving the effective interaction at the local neutron and proton density, $\varrho_{\rm SL}(rk_{\rm F})$ and $g(rk_{\rm F})$.

One useful conceptual feature of this expansion is the fact that $H(\mathbf{R})$ may be separated into a term which is precisely the nucleon matter potential energy per particle, the kinetic energy, and terms which depend only on integrals of the long range part of $v_{\rm eff}$ times τ and gradients of ϱ. The long range part of $v_{\rm eff}$ is very close to the bare potential, which in turn is the most unambiguously determined part of the nuclear potential. Thus, the finite nucleus corrections are virtually independent of the

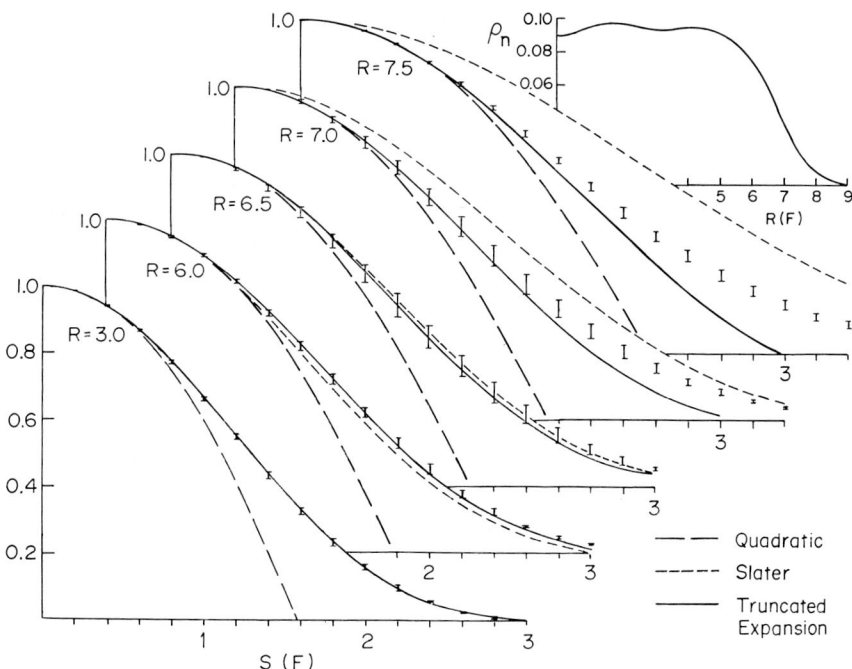

Fig. 7. Comparison of the square of the exact neutron density matrix in ^{208}Pb with the truncated expansion, taken from Negele and Vautherin (1972).

model of the nuclear force, and all ambiguities concerning the short range repulsion, off energy shell behavior, and higher order corrections enter only through the nuclear matter energy surface as a function of ϱ_n and ϱ_p as graphed in Figure 2.

Technically, the DME is much simpler than the DDHF theory, involving only a local Schrödinger equation with a position dependent effective mass. The results of this simplified theory have been compared in detail with the DDHF results (Negele and Vautherin, 1972) and are in excellent agreement. Because of the computational simplicity of the DME, it was possible to search more extensively over the parameters of the higher order correction adjustment and thereby obtain binding energies per particle which agree with experimental values within 0.1 MeV. The only additional change required to apply this theory of finite nuclei to the free neutron regime was to introduce a more careful parameterization of $A(\varrho_n, \varrho_p)$ in the region of very low neutron density so as to accurately reproduce the uniform nuclear matter calculations of Siemens and Pandharipande (1971).

TABLE I

BE/A for pre-drip nuclei

Nucleus	Extrapolated value (MeV)	DME (MeV)
$_{40}^{122}$Zr	7.67	7.54
$_{38}^{120}$Sr	7.45	7.32
$_{36}^{118}$Kr	7.20	7.05

Although our primary emphasis is on the free neutron regime, the DME theory of finite nuclei has been used to check the extrapolation by Myers and Swiatecki (1965) of the last pre-drip sequence of nuclei.

As shown in Table I, the DME results are in good agreement with the extrapolated values, and in particular, the differences between neighboring nuclei are in excellent agreement. Thus, we corroborate the results of Baym et al. (1971a) for the composition of matter prior to neutron drip. The spatial distribution of protons in ^{118}Kr is shown in Figure 11 for subsequent reference.

4. The Neutron Drip Regime

The equation of state in the neutron drip regime is obtained by minimizing the energy per nucleon of spherically symmetric configurations of nucleons in a Wigner-Seitz unit cell.

For charge neutrality, a cell with Z protons contains an equal number of electrons. Because of the large Fermi-Thomas screening length,

$$1/k_{FT} = \left(\frac{4}{\pi}\frac{e^2}{\hbar c}\right)\frac{1}{k_e},$$

where k_e is the electron Fermi wave number, the electrons are approximated by a uniform gas, and the small screening correction is neglected (BBP). The kinetic energy per electron is then

$$\frac{T}{Z} = m_e c^2 \left\{ \frac{3}{8x^3} \left[x(1+2x^2)(1+x^2)^{1/2} - \ln(x + (1+x^2)^{1/2}) \right] - 1 \right\},$$

where $x = \hbar k_e / m_e c$. The Coulomb exchange energy for the electrons is of the order of $e^2/\hbar c$ times the mean electron kinetic energy and is also neglected. With these approximations, the electron chemical potential is

$$\mu_e = (k_e^2 \hbar^2 c^2 + m_e^2 c^4)^{1/2} - m_e c^2 + \hbar c \times$$

$$\times \int d^3 r_1 \, d^3 r_2 \left[\varrho_p(r_1) + \varrho_e(r_1) \right] \frac{e^2}{|r_{12}|} \varrho_e(r_2).$$

The nuclear energy for the unit cell is expressed in terms of the nucleon wave functions using the DME Hamiltonian density functional. The direct Coulomb energy is calculated straightforwardly from the electron and proton densities, and the Slater

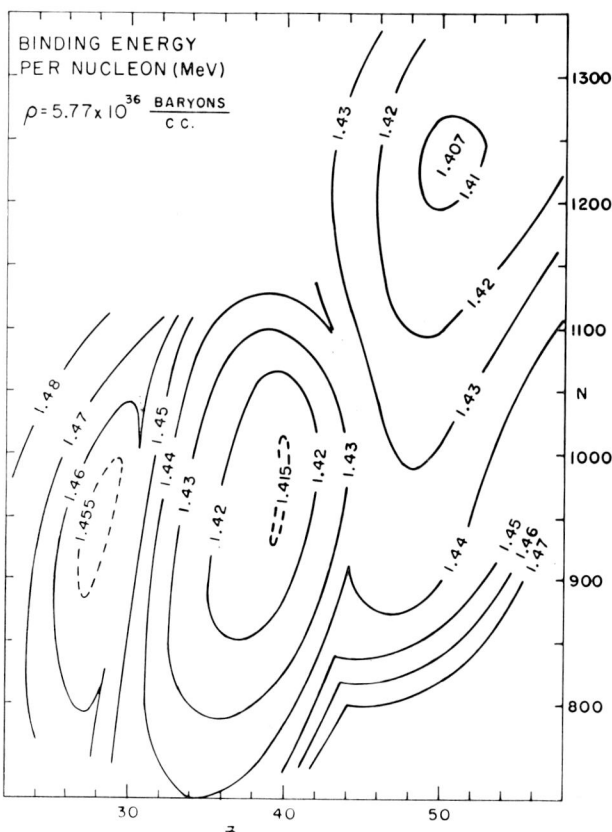

Fig. 8. Contour plot of energy per particle as a function of number of protons and neutrons in a unit cell.

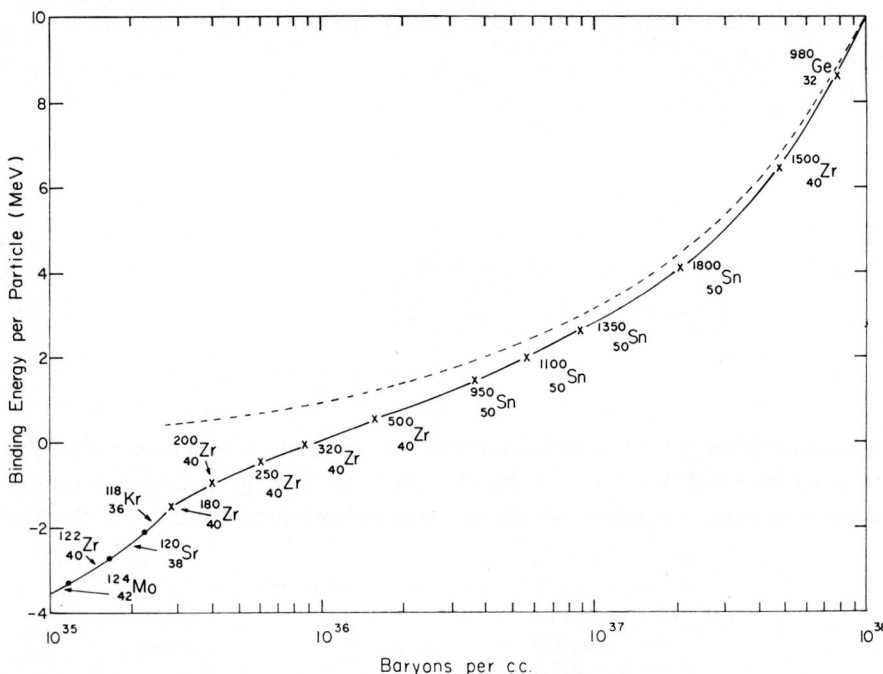

Fig. 9. Energy per particle versus baryon density.

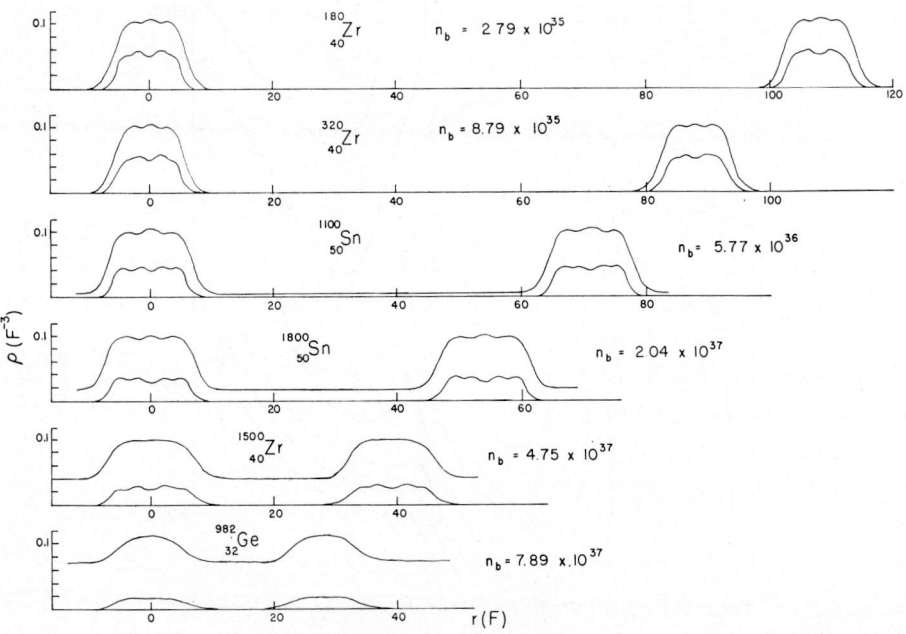

Fig. 10. Proton and neutron density distributions occurring along an axis joining the centers of two adjacent unit cells.

approximation is used for the nuclear exchange Coulomb energy. Since the total energy expression is variational in the nuclear wave functions, the proton and neutron chemical potentials, μ_P and μ_N, are given by the eigenvalues of the last occupied orbitals.

In addition to the approximations discussed previously in connection with the DME,

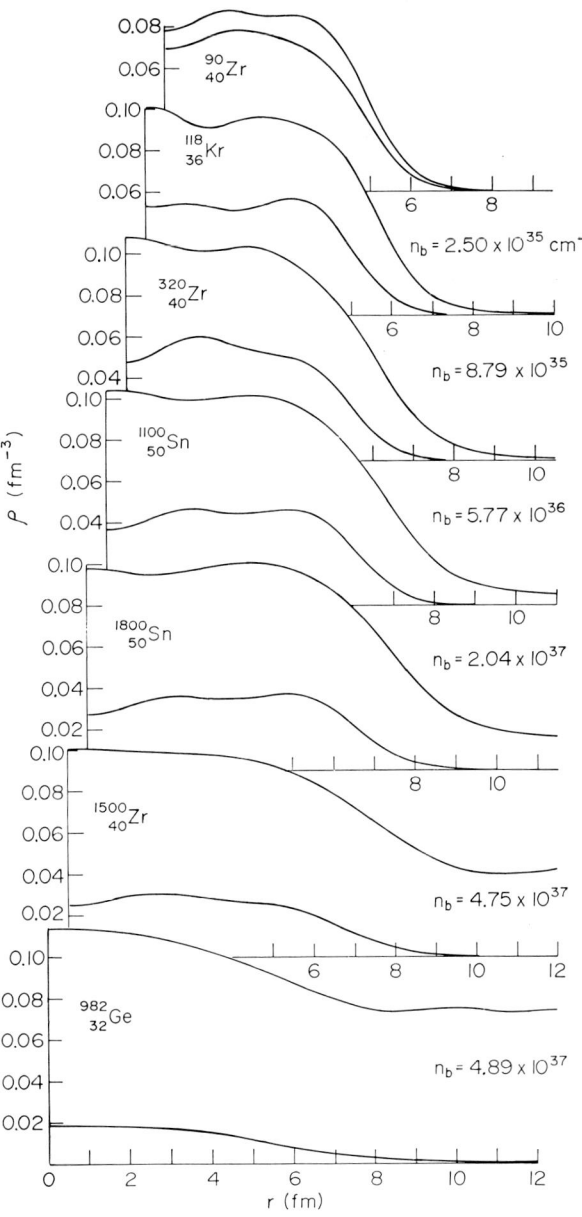

Fig. 11. Proton and neutron densities versus distance from the center of a unit cell.

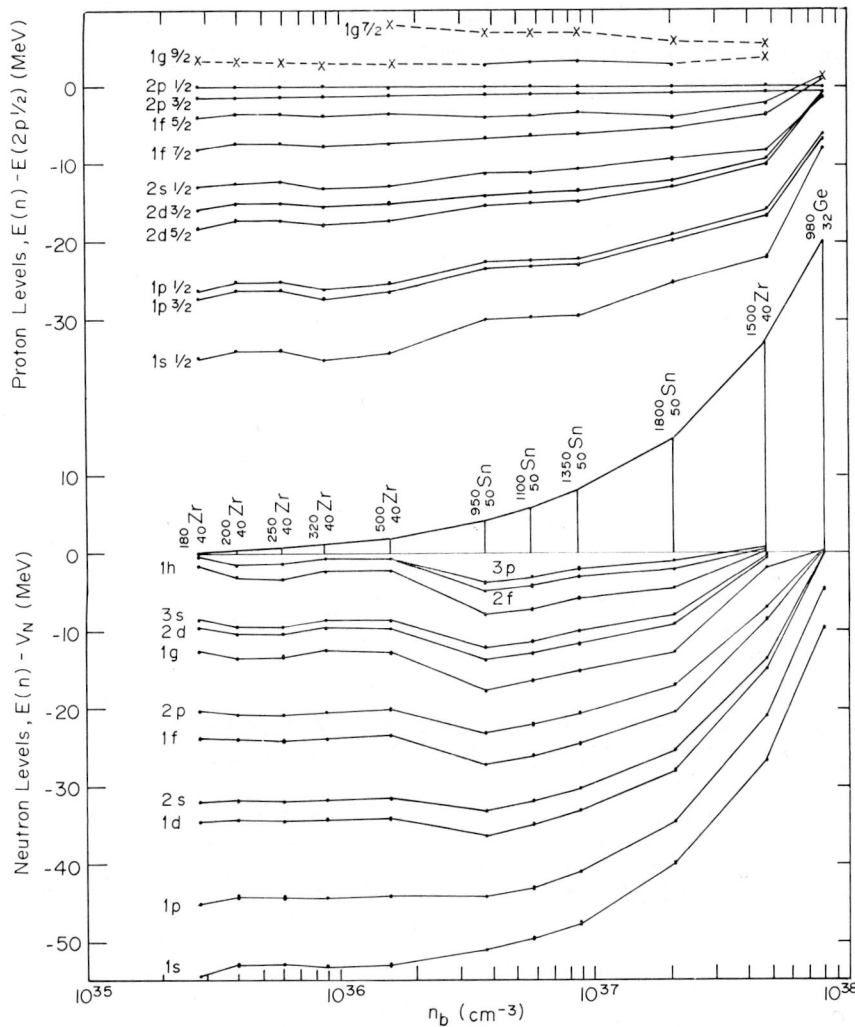

Fig. 12. Single particle spectrum of protons and neutrons.

it is extremely useful to omit the spin orbit splittings for neutron states, thereby reducing the total number of wave functions to be considered by almost one half. Since the one-body spin orbit potential is proportional to $(1/r)(d\varrho/dr)$ and the spectrum of the continuum neutron states is essentially determined by the uniform neutron gas region, the effect of the spin orbit force on the neutron level ordering is negligible. Thus the density and energy density obtained by filling the barycentric states representing the average of the levels $j=l-\frac{1}{2}$ and $j=l+\frac{1}{2}$ with $2l+1$ particles should be an excellent approximation. For protons, however, the spin orbit splitting is crucial in determining the level ordering and thus which orbitals are actually occupied. In order to retain a theory which is completely variational, the appropriate proton spin orbit term to be added to the energy density is (Negele and Vautherin, 1972) $H_{SO} = W/2[\nabla(\varrho_N + 2\varrho_P) \cdot \mathbf{J}_P]$ where $\mathbf{J}_P$ is the proton spin density $\mathbf{J}_P = (\mathbf{r}/r^4)\sum_i j_i[(j_i + 1) - l_i(l_i+1) - \frac{3}{4}]u_i^2(r)$ and W is a constant determined from the two-body spin orbit force.

With the approximations specified above, it is straightforward to perform a self-consistent Hartree-Fock calculation for the nuclear wave functions in a unit cell, for a given cell radius and number of neutrons and protons. An initial guess is made for the single particle potentials and effective masses and the radial wave functions and eigenvalues are calculated subject to the boundary condition that the wave function or its derivative must vanish at the cell radius, depending on parity. The lowest N neutron states and Z proton states are occupied, ϱ and τ are calculated for neutrons and protons and new potentials and effective masses are obtained. The lowest N and Z states for the new potentials are then filled and the process is iterated until self-consistency is achieved. Because of the multiplicity of local minima, it is necessary to try several dissimilar initial guesses for the potentials to assure that the absolute minimum energy configuration has been obtained.

To determine the minimum binding energy per nucleon at a specified baryon density,

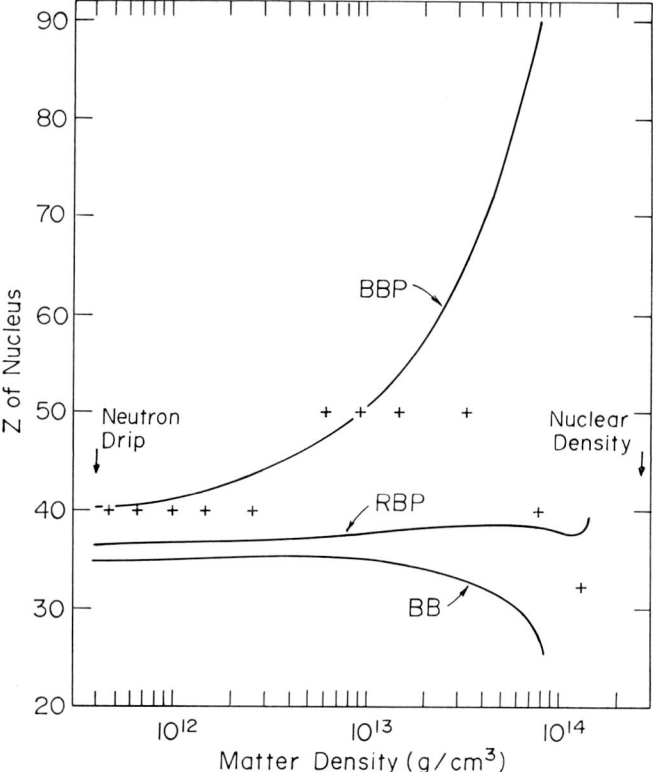

Fig. 13. Number of protons per nucleus, denoted by crosses, compared with previous predictions taken from Ravenhall et al. (1972).

TABLE II

Numerical results in the free neutron regime. – The quantities $\tilde{\varrho}_G$ and $\tilde{x}$ are the approximate density of the exterior neutron gas and the approximate ratio of protons to neutrons in nuclei respectively, and cannot be defined uniquely due to density fluctuations. All other quantities are defined in the text.

n_b (cm^{-3})	N	Z	μ_N (MeV)	μ_P (MeV)	$\tilde{\varrho}_G$ (fm^{-3})	$\tilde{x}$	$\dfrac{E}{A} - m_n$ (MeV)	$\dfrac{E_{\text{gas}}}{A} - m_n$ (MeV)
2.79×10^{35}	140	40	0.2	-26.8	4×10^{-5}	0.53	-1.425	0.436
4.00×10^{35}	160	40	0.3	-29.4	9.7×10^{-5}	0.53	-0.962	0.543
6.00×10^{35}	210	40	0.6	-29.5	2.6×10^{-4}	0.53	-0.462	0.692
8.79×10^{35}	280	40	1.0	-28.5	4.8×10^{-4}	0.53	-0.050	0.865
1.59×10^{36}	460	40	1.4	-29.4	1.2×10^{-3}	0.52	0.541	1.214
3.73×10^{36}	900	50	2.6	-33.6	3.0×10^{-3}	0.46	1.465	1.926
5.77×10^{36}	1050	50	3.3	-34.5	4.7×10^{-3}	0.45	1.966	2.408
8.91×10^{36}	1300	50	4.2	-35.8	7.8×10^{-3}	0.44	2.610	2.981
2.04×10^{37}	1750	50	6.5	-43.6	1.84×10^{-2}	0.35	4.097	4.422
4.75×10^{37}	1460	40	10.9	-54.0	4.36×10^{-2}	0.28	6.428	6.660
7.89×10^{37}	950	32	15.0	-68.3	7.37×10^{-2}	0.16	8.611	8.657

one must search over the number of nucleons in the unit cell, thus specifying the cell radius, and the ratio of neutrons to protons, as well as the spatial distribution of nucleons described above. The results of such a search for a preliminary version of this theory are shown as a contour plot in Figure 8. One notes several local minima generated by the energy fluctuations arising from the shell closures for 28, 40, and 50 protons, and in this case the absolute minimum occurs at $Z = 50$. In practice, the search is greatly expedited by considering only β-stable configurations. At each iteration one computes $E(Z+1, N-1) \sim \mu_e + \mu_p - \mu_n + m_e + m_p - m_n$ and converts neutrons into protons plus electrons or vice-versa. The resulting β-stable self consistent solution corresponds to the minimum along a line of constant $N+Z$ in Figure 8, and one has only to perform a single parameter search over the number of particles in a cell.

The resulting ground state solutions for 11 densities are presented in Table II and Figure 9. Since it is impossible to distinguish neutrons in nuclei from those in the gas, we adopt the unusual convention of labeling nuclei by the total number of nucleons in the unit cell. In Figure 9, the energies per particle for the ground state configurations, denoted by the crosses, are compared with the energies of a β-stable uniform gas of electrons, protons, and neutrons, thus demonstrating the significant gain in binding energy obtained by forming a Coulomb lattice of nuclei surrounded by a low density neutron gas.

The spatial distribution of neutrons and protons at various densities is presented in Figures 10 and 11. Figure 10 shows the densities obtained along a line joining the centers of two adjacent unit cells. As the baryon density increases, one observes a smooth, systematic progression of configurations. The cell radius decreases, the neutron gas density increases and the density of protons in nuclei decreases. By the time one reaches ^{982}Ge, the difference in energies of various local minima are sufficiently small that it is not meaningful to proceed to higher density. The energy per baryon is already very close to the uniform gas result, and it seems quite reasonable to assume that the density distributions at higher density continue the systematic behavior and smoothly approach a uniform density.

In Figure 11, the density distributions of the nuclei at the center of the unit cells are shown in greater detail. For comparison, ^{90}Zr, a naturally occurring isotope, and ^{118}Kr, the last pre-drip nucleus, are also presented. In addition to the systematic

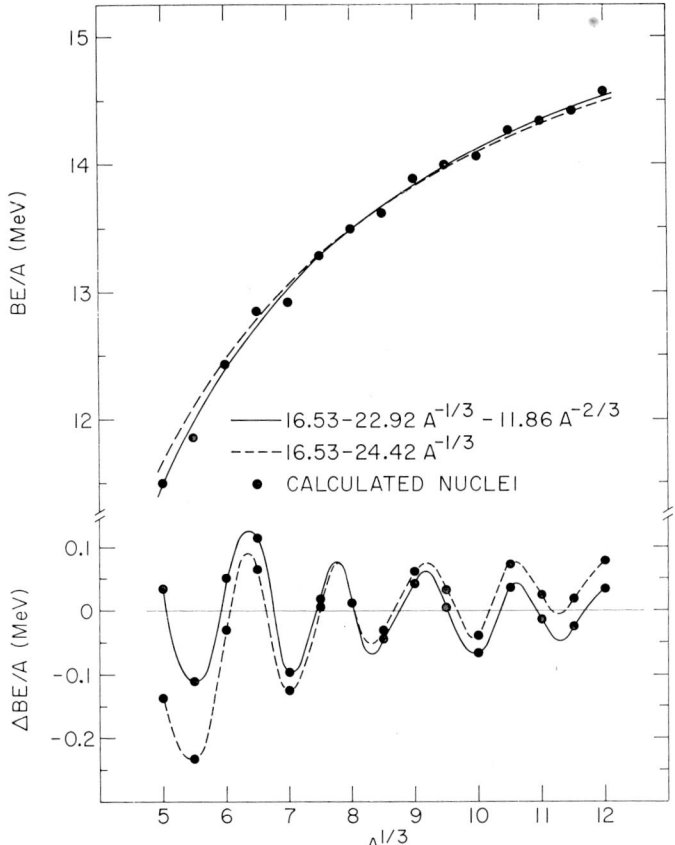

Fig. 14. Least squares fit to the masses of large, mirror nuclei to determine the curvature coefficient.

effects mentioned previously, one also observes a strong systematic increase of the nuclear surface thickness and the diminishing of neutron density fluctuations as the number of neutrons becomes sufficiently large and one approaches a statistical regime.

One of the most striking features of Figures 10 and 11 is the degree to which the nuclei in the free neutron regime resemble ordinary nuclei. This similarity is also manifested in the behavior of the single particle energies, as shown in Figure 12. In order to eliminate irrelevant constant shifts in the absolute single particle energies, relative energy spacings are plotted with the proton energies shown relative to the $2p_{1/2}$ state and the neutron energies shown relative to the continuum. The fact that the usual shell model level sequence is maintained throughout the free neutron regime and that the interior nuclear density does not deviate significantly from the density of ordinary nuclei strongly suggest that the DDHF theory should be just as reliable in this regime as in observable nuclei where it has been experimentally tested.

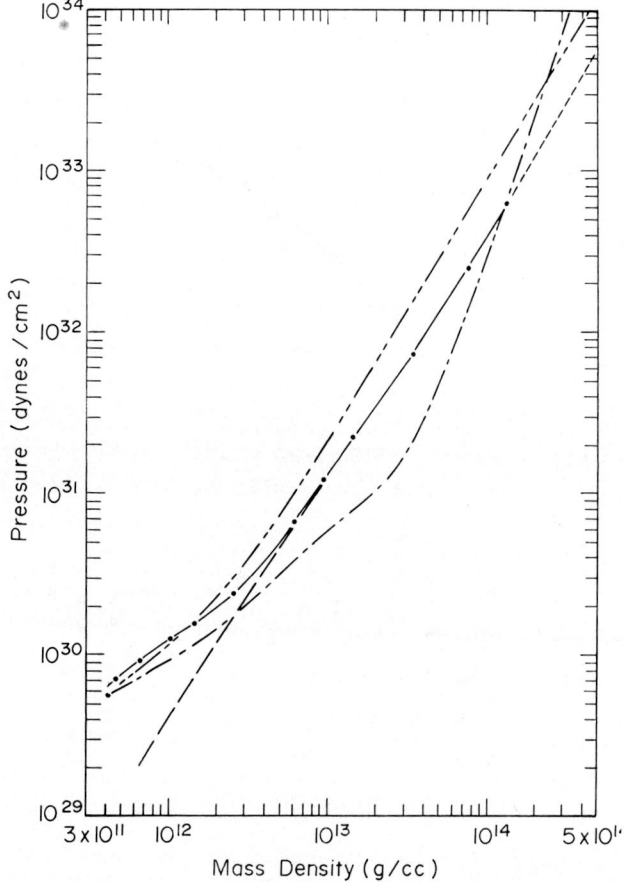

Fig. 15. Equation of state compared with previous predictions taken from BBP.

The composition of the resulting nuclei in this theory are compared with the predictions of BBP, Buchler and Barkat (1971a, b) (BB), and Ravenhall et al. (1972) (RBP) in Figure 13. We have already argued that RBP is the most reliable of the three previous theories. The equilibrium conditions derived by BBP show that if the nuclear surface energy is expressed in the form $BE \sim W_{SURF} A^{2/3}$, then the equilibrium size of the nucleus is given by $W_{SURF} A^{2/3} = 2 E_{COUL}$, where E_{COUL} is the total Coulomb energy

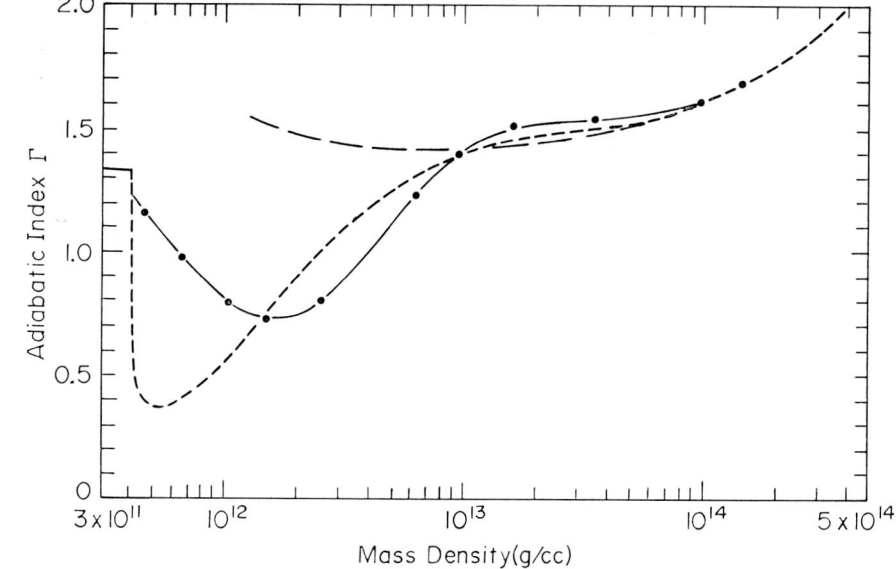

Fig. 16. Adiabatic index compared with the BBP result.

per nucleus, including the lattice energy. However, in the spherical geometry of a Wigner–Seitz unit cell, in contrast to a plane surface, there are additional contributions to the surface energy proportional to $A^{1/3}$ arising from the curvature of the surface as well as a small pressure term which we will ignore for the present argument. The curvature term may be evaluated by calculating a sequence of unphysically large, mirror nuclei with the Coulomb force turned off. The resulting binding energies are least squares fit with a mass formula of the form $BE = W_{\rm VOL} A + W_{\rm SURF} A^{2/3} + W_{\rm CURV} A^{1/3}$ as shown in Figure 14. If one repeats the argument of BBP with this expression, the new equilibrium condition is $(W_{\rm SURF} + 2 W_{\rm CURV} A^{-1/3}) A^{2/3} = 2 E_{\rm COUL}$. Thus, substituting the values from Figure 14, although the curvature term in the binding energy is only about 11% of the surface term for $A \sim 200$, it shifts the equilibrium size by 22%. Taking an average value of Z from RBP to be 37, this effect would be expected to shift the

TABLE III

Coefficients for the equation of state

I	Ground state C_I	Uniform gas C_I
0	-4.0	-4.0
1	2.8822899×10^{-1}	1.4821424
2	5.9150523×10^{-1}	$-4.0373482 \times 10^{-2}$
3	9.0185940×10^{-2}	6.0455728×10^{-2}
4	$-1.1025614 \times 10^{-1}$	$-1.5307639 \times 10^{-2}$
5	2.9377479×10^{-2}	3.4774416×10^{-3}
6	$-3.2618465 \times 10^{-3}$	$-4.3627154 \times 10^{-4}$
7	1.3543555×10^{-4}	2.3383473×10^{-5}

charge to 45. Because of shell effects, as demonstrated in Figure 8, the equilibrium point in our present theory is shifted away from 45, either to the shell closure at 40 or to the closure at 50, but it is significant that the average shift is accurately explained by this curvature effect.

Given the total ground state energy per particle, including masses, E_T, as a function of baryon density, n_b, the pressure, adiabatic index and mass density are determined as follows: $P = n_b^2 (\partial E_T/\partial n_b)$, $\Gamma = (n_b/P)(\partial P/\partial n_b)$ and $\varrho = n_b E_T/c^2$. Since searches were carried out only at 11 discrete densities, it was necessary to curve fit the resulting energies shown in Figure 9 with a smooth function. The total energies per baryon, E_T, for both the ground state and uniform gas configurations were fit by the following function:

$$E_T = m_n + c_0 + \exp\left\{\sum_{I=1}^{7} c_I x^{(I-1)}\right\},$$

where E_T is in MeV, m_n is the neutron mass, $x = ln(n_b \times 10^{-35})$, n_b is the baryon density in baryons per cm^3, and c_0 through c_7 are tabulated in Table III.

The equation of state obtained by fitting the 11 ground state nuclear configurations in the free neutron regime is shown in Figure 15 by the solid line. At high density, it joins smoothly onto the BBP curve, denoted by the short dashed line. The equation of state for a uniform gas is indicated by the long dashed curve, and becomes indistinguishable from the ground state curve at intermediate densities. For comparison, the Harrison-Wheeler equation of state (Hartle and Thorne, 1968) and the V_γ equation of state of Langer *et al.* (1969) are also plotted as taken from BBP.

The adiabatic index calculated with the present theory is shown in Figure 16 by the solid line. One should note that taking the second derivative of the curve fit to the 11 calculated energies strongly amplifies noise in the fit, and therefore undue significance should not be attached to the fine details of this curve. For comparison, the adiabatic index obtained for a uniform gas is shown by the long dashes and the BBP result is shown by the short dashes.

5. Conclusion

We have attempted to show in this work that it is possible to construct a reliable theory of a nucleon many-body system derived from the two-body nucleon-nucleon interaction. The relevant two-body correlations are incorporated in a two-body effective interaction, and the energy density is expressed as an extremely simple functional of the density and kinetic energy density *via* the density matrix expansion. This theory yields excellent agreement with experimentally observable properties of finite nuclei and should provide a reliable extrapolation to the nucleus-like clusters occurring in the free neutron regime of dense matter.

The primary uncertainty affecting the equation of state we have calculated in the free neutron regime is the small uncertainty in the function specifying the uniform matter potential energy for a small ratio of protons to neutrons. The adjustment for the higher order corrections, the ambiguity in off-energy-shell behavior of the nuclear force, and the omission of pairing correlations (Yang and Clark, 1971), all contribute to the uncertainty in the energy of a neutron gas. Fortunately, throughout most of the free neutron regime, the nuclear clusters occurring in the center of the unit cell are insensitive to small changes in the energy functional for a neutron gas, so that the composition and energy of the cluster is essentially unaffected. Since most of the baryons in the cell are neutrons in the gas, their energy change may be computed directly from the change in the neutron gas energy and thus the change in the total energy and therefore the equation of state may be obtained from the data in Table II without repeating the lengthy self-consistent calculations reported in this work.

Finally, the substantial agreement between the final equation of state obtained in this work and that of BBP suggests that calculations of Baym *et al.* (1971a) based on the BBP results should be essentially unchanged by the present work.

Acknowledgments

It is a pleasure to acknowledge the contribution of D. Vautherin who collaborated in the development of the density matrix expansion and in the writing of the program to calculate wave functions in a Wigner Seitz cell. In addition, this work has benefitted greatly at various stages from discussions with H. A. Bethe, G. Baym and C. J. Pethick. The hospitality of the Niels Bohr Institute during the spring of 1970, where the author's interest in this problem was first stimulated, and of Brookhaven National Laboratory during the summer of 1972, where this manuscript was prepared, is gratefully acknowledged. This work was supported in part through funds provided by the U.S. Atomic Energy Commission under Contract No. AT(11-1)-3069.

References

Arponen, J.: 1971, University of Helsinki (preprint).
Barkat, Z., Buchler, J. R., and Ingber, L.: 1972 (to be published).
Baym, G. A., Pethick, C., and Sutherland, P.: 1971a, *Astrophys. J.* **170**, 299.
Baym, G. A., Bethe, H. A., and Pethick, C. J.: 1971b, *Nucl. Phys.* **A175**, 225.
Bethe, H. A.: 1971, *Ann. Rev. Nucl. Sci.* **21**, 93.
Bethe, H. A., Börner, G., and Sato, K.: 1970, *Astron. Astrophys.* **7**, 279.
Brown, G. E. and Green, A. M.: 1969, *Nucl. Phys.* **A173**, 1.
Buchler, J. R. and Barkat, Z.: 1971a, *Phys. Rev. Letters* **27**, 48.
Buchler, J. R. and Barkat, Z.: 1971b, *Astrophys. Letters* **7**, 167.
Campi, X. and Sprung, D. W.: 1972, *Nucl. Phys.* **A194**, 401.
Hartle, J. B. and Thorne, K. S.: 1968, *Astrophys. J.* **153**, 807.
Langer, W. D., Rosen, L. C., Cohen, J. M., and Cameron, A. G. W.: 1969, *Astrophys. Space Sci.* **5**, 259.
Myers, W. D. and Swiatecki, W. J.: 1965, UCRL Report 11980.
Negele, J. W.: 1970, *Phys. Rev.* **C1**, 1260.
Negele, J. W. and Vautherin, D.: 1972, *Phys. Rev.* **C5**, 1472.
Pines, D.: 1970, Proc. of XVII International Conf. on Low Temperature Physics.
Ravenhall, D. G., Bennett, C. D., and Pethick, C. J.: 1972, *Phys. Rev. Letters* **28**, 978.
Reid, R. V.: 1968, *Ann. Phys. N.Y.* **50**, 411.
Siemens, P. J.: 1970, *Nucl. Phys.* **A141**, 225.
Siemens, P. J. and Pandharipande, V. R.: 1971, *Nucl. Phys.* **A173**, 561.
Skyrme, T. H. R.: 1959, *Nucl. Phys.* **9**, 615.
Sprung, D. W. L.: 1972, *Advances in Nuclear Physics* (to be published).
Vautherin, D. and Brink, D. M.: 1972, *Phys. Rev.* **C5**, 626.
Yang, C. H. and Clark, J. W.: 1971, *Nucl. Phys.* **A174**, 49.

EQUATION OF STATE AT DENSITIES GREATER THAN NUCLEAR DENSITY

H. A. BETHE

Laboratory of Nuclear Studies Cornell University, Ithaca, N.Y. 14850, U.S.A.

Abstract. An equation of state is developed for densities from nuclear density (3×10^{14} g cm^{-3}) to about 10^{16} g cm^{-3}. The repulsive interaction between baryons dominates and empirical arguments for its existence are given. This interaction is attributed to vector meson exchange, and is derived from classical field theory whereupon a Yukawa potential results. The potential actually assumed is a modification of the Reid potential. Arguments are given that the baryons will not form a crystal lattice. The actual calculations were done using Pandharipande's method. The particles present at high density certainly include nucleons, Λ and Σ. The presence of Δ is questionable but that of π is likely. Results are given for the concentration of various species. With the more likely assumption about interactions, the concentration of each permissible species of particle is about equal at $\varrho = 10^{16}$ g cm^{-3}. The relation between energy and density is nearly independent of the assumptions on the species permitted and the energy is about 3 GeV particle^{-1} at $\varrho = 10^{16}$ g cm^{-3}. The relation between pressure and energy density is given, which yields a sound velocity equal to c at a few times 10^{15} g cm^{-3}. Results for the structure of neutron stars are given. The maximum mass is about 2 solar masses and the maximum moment of inertia 10^{45} g cm^2.

1. Density Ranges of Interest

We may distinguish the following ranges of density:

(a) $\varrho < 3 \times 10^{11}$ g cm^{-3}. In this density range we have neutron-rich nuclei and degenerate electrons.

(b) $3 \times 10^{11} < \varrho < 2 \times 10^{14}$ g cm^{-3}, which means less than 0.12 particles fm^{-3}. In this range, we have free neutrons, and some neutron-rich nuclei. Electrons compensate for the charge of the nuclei. With increasing density, the nuclei occupy an increasing fraction of the space.

(c) $2 \times 10^{14} < \varrho < 10^{15}$ g cm^{-3}, i.e. between 0.12 and 0.6 nucleons fm^{-3}. In this range, there are neutrons, a few percent protons, and an equal number of electrons as protons.

(d) $\varrho > 10^{15}$ g cm^{-3}, i.e. more than 0.6 nucleons fm^{-3}. In this density range, we have neutrons, protons, and excited baryons, especially Σ. There are few electrons. There may be some π^- at the higher densities.

Density range (a) has been well treated by Baym *et al.* (1971b). Density range (b) is dominated by the free neutrons whose equation of state has been calculated by Siemens and Pandharipande (1971) and is reported in some detail by Baym *et al.* (1971a). The behavior of the nuclei in this range is reported in this volume by John Negele. The pairing of neutrons gives a very important contribution to the energy; this has been calculated by Yang and Clark (1971), and has been combined with Siemens' equation of state by Bethe (1971).

In density range (c), the equation of state can be obtained directly from nuclear matter theory; this has been done by Siemens and Pandharipande (1971).

In this paper we are concerned with range (d). We shall restrict this range at the upper end by requiring $\varrho < 10^{16}$ g cm^{-3} because this is the highest density which is likely to occur in neutron stars. By doing this we avoid complicated problems regarding higher excited states of baryons, and perhaps some others.

2. Repulsive Interaction

2.1. Empirical

At high densities, the baryons will come very close to each other; the radius of the sphere containing one baryon is about 0.35 fm at $\varrho = 10^{16}$ g cm^{-3}. It is very difficult to get direct experimental evidence on the interaction of nucleons at such short distances. Experiments at energies $E < 400$ MeV involve de Broglie wavelengths > 0.5 fm, and are therefore not able to give fine details of the interaction at shorter distances. Experiments at higher energy involve inelastic scattering of nucleons in which pions or heavier mesons are produced; it is impossible to deduce a potential from these experiments.

It must be recognized that the high density problem is not similar to the high energy problem. In fact, we know that neutron stars are at very low temperature, perhaps 10 keV, so that the baryons take the lowest energy configuration compatible with the density. There is no excess energy. The momenta of the baryons may be quite high especially when two of them come close to each other, but their energy is not. The situation is therefore entirely analogous to that in normal atomic nuclei, and in low energy scattering experiments, and we should get our guidance from these.

All potentials that have been proposed to explain low energy scattering include a strong repulsive core. The existence of this repulsion is shown directly by the phase shift of the s-state scattering at energies above 300 MeV. This core is also essential to explain the saturation of nuclear forces. Calogero and Simonov (1969, 1970a, b, 1972) have shown that it is necessary to have an ordinary repulsive force at short distances which can overcome other forces such as tensor, spin orbit, and exchange. If such a force does not exist, heavy nuclei will collapse.

It has been suggested that perhaps the repulsion derived from scattering data exists only at intermediate distances like 0.3–0.7 fm, and is followed by an attraction at smaller distances. It can easily be shown that such an assumption would lead to collapse of nuclei if the volume integral of the potential for $r < 0.7$ fm is zero or negative. In view of all this, we shall assume that the observed repulsion persists to small r.

2.2. Vector Meson Exchange

A natural explanation of the repulsion is given by the exchange of vector mesons, chiefly the ω meson which has isospin 0, and is neutral. The interaction due to a neutral vector meson field is very similar to the familiar interaction due to the electromagnetic field, except that the mesons have finite mass while light quanta have zero mass. It was shown in the early days of meson theory that such a field has a static limit (Wentzel, 1949). It is possible to separate the effect of stationary, point nucleons from any effects due to the motion of the nucleons. This is similar to the separation of the Coulomb force from magnetic interactions in electrodynamics. The analog of the Coulomb force is the Yukawa force,

$$Y = g^2 e^{-\mu r}/r, \tag{1}$$

where g is the coupling constant and μ the reciprocal Compton wavelength of the meson.

Corrections to (1) are proportional to v^2/c^2, where v is the velocity of the nucleons. In our theory it will turn out that v/c is generally less than $1/2$ so that these corrections are small. They can be calculated. Renormalization terms are expected to be of the relative order, $\hbar c/g^2$ which in our case is a small number, about 0.1 or less, and they also contain at least a factor of v^2/c^2. We therefore claim that the major interaction at small distances is given by a sum of Yukawa terms (1).

The great advantage of this classical treatment compared to the usual treatments of high energy theory is that the latter require the separate consideration of the exchange of one, two, or more mesons. As more mesons are exchanged, the theory becomes very complicated with cuts, many Riemann sheets, etc. In our case it will turn out that $g^2/\hbar c$ is about 20 so that 20 or more mesons can easily be exchanged. The classical treatment is precisely the formalism adapted to this situation.

In addition to the repulsion, there is a somewhat weaker attraction at intermediate range as is shown clearly by low energy scattering experiments. This is generally attributed to the simultaneous exchange of two pions which together behave much as a scalar particle, $s=0$ and isospin$=0$. This quasi-particle is often called σ.

The most important exchanged particles, ω and σ, are both isoscalar, $t=0$. Therefore they should interact equally with any baryon. This will be assumed in our calculations, and the coupling constant g will be derived from nucleon interaction (see Section 4).

3. Method of Calculation

3.1. Nuclear Matter Theory

In our work, we first tried to use nuclear matter theory. This however is not applicable when the parameter κ of that theory becomes greater than about $1/2$. This is the case in density range (d). Our attempts to use nuclear matter theory lead to very peculiar results which are clearly false.

3.2. Crystal Lattice

Next we assumed that the neutrons (or other baryons) form a regular crystal arrangement at high density. In fact, this has been suggested by Anderson and Palmer (1971), in analogy with condensed helium. Some calculations of this type will be reported in this volume by Canuto.

We do not believe that neutron and baryon matter at high density is crystalline. At very high density the interaction (1) goes over into the Coulomb interaction. It is well known to solid state physicists that particles repelling each other with Coulomb forces, and having strong quantum effects, do not form a crystalline lattice at high density but only at low density. Therefore there should not be crystal structure for, say, $\mu r_0 < 1$. If μ is taken to correspond to ω mesons this means $r_0 < 0.26$ fm.

There is still the possibility of crystal structure at intermediate densities. However, the interaction (1) is rather soft. Near a given r it will behave as r^{-n} where

$$n = 1 + \mu r. \tag{2}$$

Now clearly r must be chosen small enough that the repulsive potential dominates over the attractive potential. If we assume it to be twice as large as the attractive potential, then for interaction II of Table I, we get $n = 2.6$; for interaction III n is hardly greater than unity. A soft potential is completely different from the $n = 12$ Lennard-Jones potential which is customary for condensed helium. Therefore the analogy between helium and nuclear matter is not valid, and there is no reason to assume crystal structure.

This agrees with our own experience. Coldwell (1972) has tried to calculate dense neutron matter using a crystal model. His energy, derived from a variational calculation, turned out much greater than ours, calculated from a liquid model, indicating that there is no crystallization. Pandharipande in 1970 found no significant difference between the crystal and the liquid energy. Johnson and I, using the Guyer-Zane method for treating quantum crystals, found an energy somewhat lower than Coldwell but still much higher than our liquid calculations. We therefore do not believe that there is a crystal arrangement at any of the densities we have considered.

3.3. Actual method used

We use the method developed by Pandharipande which is a special case of the Jastrow method. He will himself report on his method in this volume. Johnson and I have used the simplest form of Pandharipande's method which he calls "lowest order constrained variation." In his paper, Pandharipande will show that this method agrees with a more sophisticated cluster expansion to within 5 or 10% for dense nuclear matter. He will also show that his method gives excellent results for liquid ^{4}He and ^{3}He. The case of nuclear matter is simpler than that of liquid helium, (i) because the repulsion is softer, and (ii) because both potential and kinetic energy are positive so there is no cancellation.

4. Details of the Potential

4.1. The Reid potential and modifications

It is clear that we need a soft core repulsion to make the calculations succeed. If a hard core was assumed, as in the Hamada-Johnston potential, then the highest possible density would be one in which all the hard cores touch and form a close packed lattice. Beyond this density, nuclear matter could not exist. This is clearly absurd, as is also the physical concept of a potential which goes to infinity at some distance.

The first choice is therefore the Reid (1968) potential. This is the only available

potential which has a core of Yukawa shape (1) which we have argued to be the reasonable behavior of the interaction at short distances. The Reid potential also has the proper behavior at large r, namely one-pion exchange. In fact, the potential is of the form
$$V = \sum c_n e^{-nx}/x,$$
$$x = \mu_\pi r,$$
(3)

where μ_π is the reciprocal Compton wavelength of the pion, $\mu_\pi = 0.7$ fm^{-1}. The c_n are coefficients which are determined by a phenomenological analysis of nucleon scattering up to 400 MeV. There is one c_n, usually $n=7$, which is repulsive and is interpreted as the exchange of ω mesons. A term $n=4$ or 3 has negative c_n and represents the attraction due to exchange of σ mesons. Finally, $n=1$ represents one-pion exchange and is a small contribution for our purposes.

The trouble with the Reid potential for our purpose is that it has very different form for states of different angular momenta L, S and J. This was acceptable for the purpose for which Reid constructed his potential, namely to form a basis for calculations of nuclear matter at normal density. However, in our case we want especially the repulsive term to have the same n for all LSJ. In Reid's potential, the repulsion is $n=7$ for even-l states, $n=6$ for $^3P_2$, but $n=3$ for $^3P_1$ and ^1P. The $n=3$ makes no sense in connection with ω-mesons.

Since we need uniformity in the repulsive force, Pandharipande took all odd-l states to have $n=6$. Mikkel Johnson and I took all states to be $n=7$ because the repulsion in the ^1S state is best determined, and this was chosen to be $n=7$ by Reid. We also took the coefficient c_7 to be the same for all states LSJ. This will be further discussed in sub-Section 4.4. In our most recent calculations, we changed to $n=5.5$ (see sub-Section 4.2) (Interaction III).

The intermediate range attraction for odd-l states was taken by Pandharipande to be the same as $^3P_2$. Johnson and I assumed $n=4$ in our older calculations; in our newest calculation Johnson found $n=3.5$ to be the best fit to scattering data. Table I shows the potential constants c_n used in various calculations; I, II and III are successive models used by Johnson and Bethe.

The table shows that generally odd-l states have less attraction than even-l states. Our odd-l attraction is less than that chosen by Pandharipande. This will be important in the following: The attraction in the ^1D state is less than in $^1S_1$ – this is definitely established by the scattering data. We assumed the ^1D interaction to persist for all higher even-l states, and likewise the Pandharipande interaction to be valid for all odd-l states.

4.2. MODIFIED RANGE OF REPULSION

The mass of the ω meson is 5.5 times that of π; therefore the chief repulsion should really be $n=5.5$. Johnson is now making new fits using $n=5.5$ for even-l. For odd-l, we still use $n=6$. The corresponding coefficients are listed in Table I. The coefficients are of course smaller if n is smaller. The effect of the repulsion both on nuclear scattering data and on high density nuclear matter is approximately proportional to the

TABLE I

The coefficients c_n and 'masses' n in the Reid-type interactions of Equation (3). 'Pand' is the interaction used by Pandharipande (1971a, b). The others are 3 interactions used by Johnson and Bethe. The c_n are in MeV

Interaction		n	c_n	n	c_n
Pand.	$l=0$	4	-1650	7	6484
	l even $\neq 0$	4	-1113	7	6484
	l odd	4	-933	6	4152
I	l even	4	-1113	7	6484
	l odd	4	-370	7	6484
II	^1S	4	-1650	7	6484
	^3S	4	-934	7	6484
	l even $\neq 0$	4	-1113	7	6484
	l odd	4	-774	7	6484
III	^1S	3.5	-1240	5.5	3120
	^3S	3.5	-650	5.5	3120
	even $l \neq 0$	3.5	-800	5.5	3120
	odd l, $S=1$	4	-808	6	3750
	odd l, $S=0$	3	-600	6	12140

volume integral of the interaction which is proportional to

$$\int V \, d\tau \approx c_n/n^2. \tag{4}$$

This quantity is nearly preserved when going from $n=7$ to the new data with $n=5.5$.

4.3. Comparison with High Energy Data

High energy experiments give results for $g^2/\hbar c$. For ω, the best value is about 8 to 10. With our form (3) of the interaction we have simply.

$$g_n^2/\hbar c = c_n/m_\pi. \tag{5}$$

Taking the mass of the pion $m_\pi = 140$ MeV, Johnson's coefficient $c_{5.5} = 3120$ gives

$$g_\omega^2/\hbar c = 22. \tag{6}$$

This is more than twice the high energy experimental value. The discrepancy is unexplained. According to a conversation with J. Hamilton there seems to be some extra repulsion of short range between nucleons and some mesons. The cause for this repulsion is unknown. Some such repulsion may also act in our case, and contribute to our repulsive core. We believe that the nucleon scattering experiments are more reliable evidence than the high energy experiments.

The medium range attraction has the correct range for an exchange of two pions, taking into account that these pions will have some relative kinetic energy. Theoretical attempts (Chemtob et al., 1971) to explain the medium range attraction have so far given too small a coefficient for the attraction. G. Brown (private communication) believes that the attraction between the two pions has to be taken into account to increase this coefficient. In the meantime, the phenomenological value is the only reliable one.

4.4. INTERACTION BETWEEN VARIOUS LSJ STATES OF THE NUCLEONS

Two vector mesons are known to exist, ω and ϱ, both of about the same mass, i.e. 770 MeV. The ω meson has the same kind of coupling as the electromagnetic field, that is the interaction energy is proportional to $\mathbf{j} \cdot \mathbf{A}$. The ϱ meson has negligible $\mathbf{j} \cdot \mathbf{A}$ coupling but appreciable Pauli coupling proportional to $\boldsymbol{\sigma} \cdot \mathbf{B}$. The resulting central force can then be shown to be proportional to

$$g_{\text{rep}}^2 = g_\omega^2 + g_\varrho^2 \, \boldsymbol{\sigma}_1 \cdot \boldsymbol{\sigma}_2 \, \boldsymbol{\tau}_1 \cdot \boldsymbol{\tau}_2. \tag{7}$$

The operator multiplying g_ϱ^2 has the values

$$\boldsymbol{\sigma}_1 \cdot \boldsymbol{\sigma}_2 \, \boldsymbol{\tau}_1 \cdot \boldsymbol{\tau}_2 = \begin{cases} -3 \text{ for } L \text{ even} \\ +1 \text{ for } L \text{ odd}, S = 1 \\ +9 \text{ for } L \text{ odd}, S = 0. \end{cases} \tag{8}$$

Qualitatively therefore the repulsion should obey the inequality

$$g^2 (^1S) = g^2 (^3S) \leqslant g^2 (^3P) \leqslant g^2 (^1P). \tag{9}$$

We are constructing our potentials so as to obey this inequality. The potential II used by Johnson and me in most of our calculations puts all g^2 equal; this amounts to assuming no coupling to ϱ. If we take (8) seriously, we should have

$$g^2 (^3P) = \tfrac{2}{3} g^2 (S) + \tfrac{1}{3} g^2 (^1P). \tag{10}$$

There is evidence that $g^2 (^1P)$ is indeed very large.

The attraction is purely empirical, as was mentioned in sub-Section 4.3. The attraction for ^3P is less than ^1S. There is good evidence from scattering data that the central attraction for ^3S is also smaller than ^1S. For simplicity in calculation, we have chosen the ^3S attraction such that the mean potential for the np interaction is equal to the potential for the ^1D state. This means

$$\tfrac{1}{4} g_\sigma^2 (^1S) + \tfrac{3}{4} g_\sigma^2 (^3S) = g_\sigma^2 (^1D). \tag{11}$$

At low density, the ^3S state is made very attractive by the action of the tensor force. At densities higher than nuclear matter, this stops being true because the tensor forces saturate (Bethe, 1971). We have assumed that for our high densities, the tensor force has no effect at all. Since the central force in the S state is less attractive for unlike particles than for like ones, this means that it is more advantageous to have like particles in high density matter. The effect of this will be apparent in the results, Section 6.

4.5. OTHER BARYONS

Using SU_3 symmetry, and remembering that both the ω and the σ mesons have zero isospin, it is reasonable to assume that the interaction of excited baryons, such as

Λ, Σ and Δ, are the same as for nucleons. We have therefore made this assumption. Suitable averages are taken over spin and isospin of these particles when they interact.

Experimental information exists essentially only on the ΛN interaction. This is not enough for any firm answer, especially because it is restricted to low energy. However, the information is compatible with our assumption, taking into account that there is no spin exchange between nucleons and Λ.

5. Particles Present

5.1. Particles Possible

The particles that could in principle be present in nuclear matter are listed in Table II, with their masses. We have omitted particles heavier than Δ because their effect on the equation of state is likely to be small. Even the effect of Δ is not very large. At densities $\varrho > 10^{16}$ g cm^{-3}, this situation may change.

TABLE II

Masses of the lowest-energy states of the baryon, and of 3 important mesons, in MeV

Particle	MeV
Neutron, proton	939
Λ	1116
$\Sigma(+, 0, -)$	1193
$\Delta(+, 0, -)$	1236
$\Xi(0, -)$	1317
π	140
ϱ	765
ω	784

The reason for the appearance of heavier baryons is that thereby the Fermi momentum (k_F) may be reduced. Both kinetic and potential energy (Section 7) are decreased by this fact. Once a pure neutron gas would have a very high k_F, it pays to spend the extra energy involved in the higher rest mass of the heavier baryons, in exchange for smaller k_F.

Another important point is the charge. Protons have the same mass as neutrons, but their charge must be compensated by negative charges. If electrons are used for this purpose, their Fermi energy is extremely high for a given momentum because they move essentially with velocity of light. Muons offer no advantage. At some high density, possible π^- may be useful (see sub-Section 5.4). But the cheapest way to provide negative charge is by using a negative baryon, such as Σ^- or Δ^-. Then k_F is reduced at the same time.

The Δ particles have spin 3/2, therefore each of them has twice the statistical weight of either neutron or proton. The relative statistical weights of the particles (NP), ($\Lambda\Sigma$), (Δ^{0+-}) are therefore 2, 4 and 6.

5.2. Problems concerning the Δ

5.2.1. *The Size*

The Δ may be considered as a π bound to a nucleon N. In this view, the Δ will have a fairly large size. One may then expect that at high pressure, the pion is squeezed out, just like an atom suffers pressure ionization. Like the electrons in the atomic case, the π may then move freely through the baryon matter, and the Δ ceases to exist.

5.2.2. *Self-Energy*

Sawyer (1972a) has pointed out that Δ decays rapidly into nucleon plus π which gives the Δ state a width of about 100 MeV. Connected with the decay, by dispersion theory, is also a real correction to the mass (mass renormalization). Now the Pauli principle inhibits decay to $N+\pi$, because many neutron states are occupied. Therefore it changes the mass correction, and it can easily be shown that the mass of Δ in a sea of nucleons will increase, by one to a few hundred MeV.

Sawyer's argument is almost certainly correct in a theory which only enumerates the particles, and will in this case hold *a fortiori* for heavier baryon states with larger width. However, if the *interaction* between baryons is taken into account as in our theory, M. Johnson has questioned Sawyer's argument because the Goldstone formalism of nuclear matter theory states the Pauli principle should not be taken into account in intermediate states. At present, I am inclined to believe self-energy is different from nuclear interactions, and that Sawyer's argument applies even in our theory. This, of course, reduces the concentration of Δ at any given density.

5.2.3. *Pauli Principle*

The Pauli principle might also act as follows: A π may be transmitted from baryon 1 to baryon 2, thus,

$$\Delta_1 + N_2 \rightarrow N_1 + \Delta_2, \tag{12}$$

i.e. the particle 1 transforms from a Δ into a nucleon while particle 2 makes the reverse transformation. In this case we are not talking about an intermediate state but about the fact that we cannot assign to a given baryon the property of being Δ.

When the transformation (12) occurs, the new nucleon N_1 should show anti-symmetry with all the nucleons already existing. Likewise the old nucleon N_2 should be part of the anti-symmetric wave function. This would indicate that Δ's and nucleons together should satisfy antisymmetry. Therefore we would not increase the available momentum space or statistical weight when we introduce Δ.

However, the correct answer seems to be that the one-pion exchange (12) is simply forbidden if the nucleon state N_1 and/or the Δ state Δ_2 is in the Fermi sea. We therefore believe that Δ and nucleon may occupy the *same* momentum states, and that the introduction of Δ *does* increase the available phase space.

The status of Δ must be considered moot, especially because of the size effect (argument I). In our calculations we have included Δ at the observed mass, but its exclusion

5.3. π^- MESONS

If Δ^- is not included among the particles, the π^- probably should be. At low energy, π^- is repelled by a neutron because π must be in a relative S state. This increases the effective mass of π^- to

$$m_{\text{eff}}(\pi^-) = 140 + 220(\varrho_n - \varrho_p), \tag{13}$$

(see Bethe, 1971, p. 158) where ϱ_n and ϱ_p are the densities of neutrons and protons in particles fm^{-3}. At high density however, π may be in a P state in which case it is attracted to the nucleon.

5.3.1. Sawyer's Model

Sawyer (1972b) suggested that there be a close coupling between a neutron of momentum **p** with a proton of momentum **p**−**q** plus a pion of momentum **q**. He assumes that all π^- are in the same quantum state of momentum **q**. This coupling is similar to the 'phase locking' of two atomic states by a strong laser beam in resonance. This possibility certainly exists, and was further elaborated upon by Scalapino (1972). The total energy will be decreased by the close coupling, but we have calculated that this decrease is only about 1% of the energy due to the repulsive potential (1) (at high density).

5.3.2. Scattering

We thought at one time that an even greater decrease of energy could be obtained by forward scattering of π by the nucleons. However, it appears that the mechanism for forward scattering essentially leads back to Sawyer's theory.

We have not investigated the relation of this model to the assumption of Δ. It is clear that the use of a negative potential energy for π is closely related to the existence of Δ, and that the Δ in this model is only transient.

5.4. CONCENTRATION OF SPECIES

This is obtained by minimizing the total energy E as a function of the concentrations c_k of the various species. Equivalent to this is the requirement that the chemical potential

$$\mu_k = \frac{\partial E}{\partial c_k} \tag{14}$$

be the same for all species of the same charge.

At the same time one has to satisfy the condition of charge neutrality. This requires among other things that

$$\mu_e = \mu_\mu = \mu_0 - \mu_+ = \mu_- - \mu_0, \tag{15}$$

where μ_0 is the chemical potential for neutral baryons, μ_+ and μ_- that of positively and negatively charged ones.

6. Results

6.1. Concentration of Species

The calculation was made by Mikkel Johnson, using interaction I of Table I. The result is shown in Figure 1. The concentration of every species approaches

$$\varrho_k \approx \tfrac{1}{2} \text{ baryon fm}^{-3} \tag{16}$$

at high density. The concentration for each species of Δ is twice this amount because of the spin of 3/2. The fact that ϱ_k remains small keeps k_F down to about 2.5 fm^{-1}, corresponding to a velocity $v_F = 1/2\ c$. This is the justification for our use of non-relativistic Schrödinger theory in all our calculations, and also for the neglect of magnetic-like forces, cf. Section 2. In Section 7 we shall discuss the effect of this small k_F on the total energy.

The concentration of Σ^- begins to increase at rather low total density, about 0.3 fm^{-3}. Δ^-, because of its higher mass, follows later. The neutral species, even Λ,

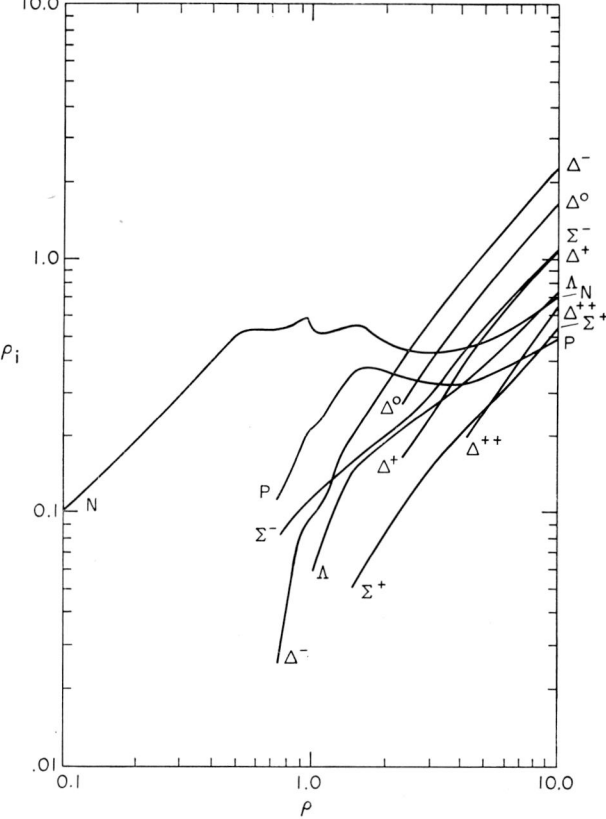

Fig. 1. Density of various baryon species, ϱ_i, versus total baryon density, in baryons fm^{-3} (1 baryon fm$^{-3} = 1.66 \times 10^{15}$ g cm^{-3}). Interaction I of Table I was assumed.

come in substantially later because they do not contribute to neutralizing the protons. Σ^+ and Δ^+ are even less important, and Δ^{++} has minimal concentration even at high density.

Similar results had been obtained earlier by Pandharipande (1971a). He also calculated a pure neutron gas (Pandharipande, 1971b). His interaction is also given in Table I; it is more attractive especially in odd states.

A second calculation was done by Johnson using Interaction II (Figure 2). This interaction has been described in Section 4.4. The ^3S attraction has been reduced, and the ^1P repulsion increased. Both of these facts make the interaction between unlike particles more repulsive. This favors a pure neutron gas. Indeed, the neutron partial

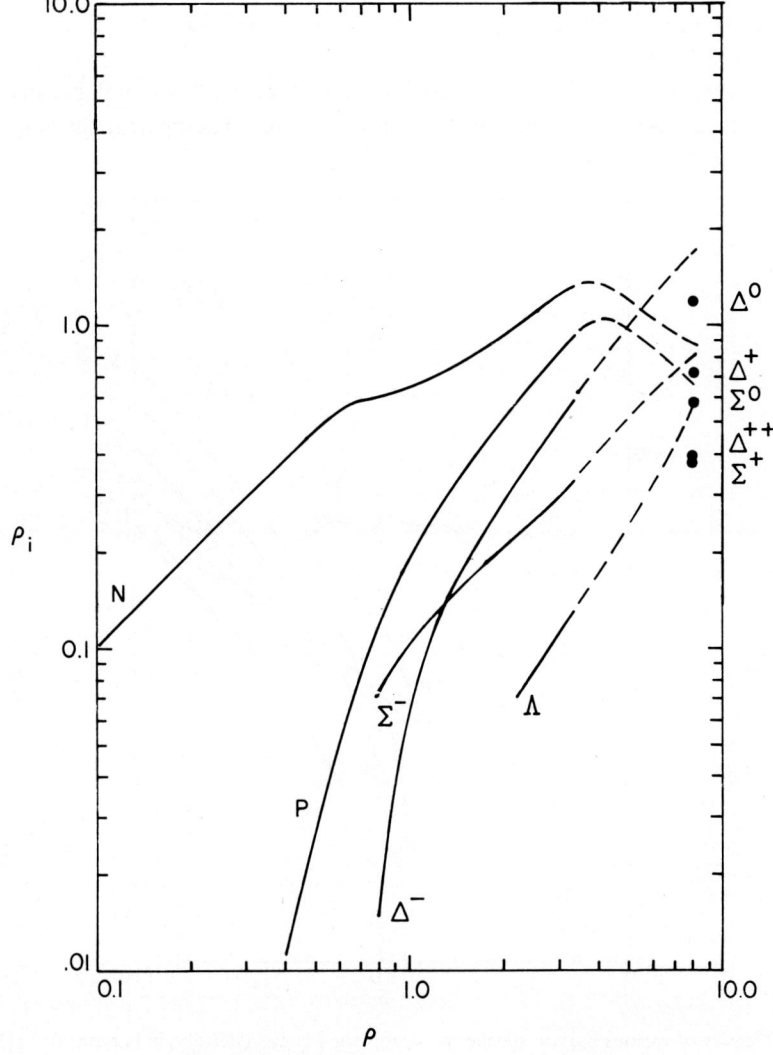

Fig. 2. Same as in Figure 1, but with interaction II of Table I.

density goes up to about 1 fm^{-3}, and the other particles stay lower. In this calculation, as with Interaction I, the existance of Δ has been assumed. The qualitative features discussed with Figure 1 persist to an even greater extent.

Pandharipande, using somewhat more radical assumptions about the difference in potential between like and unlike nucleons, found that a pure neutron gas persists to the highest density. All this shows that the concentration of species is rather sensitive to the assumed interaction between baryons.

6.2. ENERGY VERSUS DENSITY

Having obtained the concentrations we can now calculate the total energy per baryon in high density matter. This energy excludes the mass energy Mc^2.

Figure 3 compares three calculations. The upper solid curve was obtained by Johnson and Bethe, using interaction I and assuming all excited baryons. It is rather

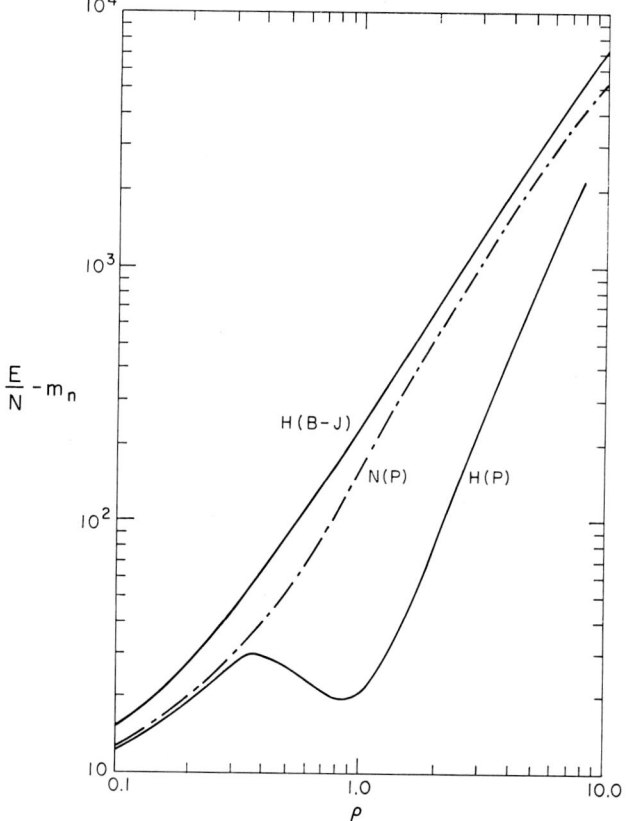

Fig. 3. Energy per baryon (excluding the rest energy of the neutron, $M_n c^2$) in MeV, versus density in baryons fm^{-3}. N(P) = pure neutron matter, according to Pandharipande (interaction Pand. of Table I). H(P) same interaction, but including excited baryons (hyperons and Δ) in the calculation. This is interaction A of Pandharipande (1971b). H(B-J) interaction I of Table I, including excited baryons.

close to the dashed curve which is Pandharipande's result for pure neutron matter, using his interaction as given in Table I. The lower solid curve is Pandharipande's (1971a) result for a gas including excited baryons; it is substantially lower than that of Johnson and Bethe. The reason for this is of course that Pandharipande used a smaller repulsive, and greater attractive force for odd states than Johnson and Bethe.

The Pandharipande excited curve has the peculiarity that it shows a decrease between densities 0.4 and 1 fm^{-3}. This would mean negative pressure. In reality, one must make the Maxwell construction which then shows a phase transition, going from about 0.2 to 1 fm^{-3} at constant pressure. If this equation of state were correct, such a phase transition would have to occur. I have thought briefly what this might mean to neutron stars, and it does not seem to be useful to explain any of their properties, especially because it would occur at rather low pressure. Actually we do not think that Pandharipande's choice of force constants is justified; therefore his solid curve in Figure 3 should be disregarded.

Figure 4 shows calculations using interaction II (solid curve). The dashed curve would result if only neutrons existed in the medium. It is seen that the effect of excited baryons on the energy is quite small, viz., a reduction of energy by about 20%. Therefore, the resulting equation of state is not sensitive to the assumed baryon interaction. Comparison between Figures 3 and 4 shows also rather small differences between interactions I and II.

The lower solid curve in Figure 4 was obtained by Pandharipande assuming excited baryons, but reducing all attractive potentials by 10% as compared to his lowest curve in Figure 3. It is seen that this small change of interaction has eliminated the peculiar phase transition of Figure 3.

Figure 4 shows that the energy at a density of 10^{16} g cm^{-3} (6 baryons per fm^3) is about 3 GeV per particle or three times the rest energy.

As has already been stated, Pandharipande finds that a consistent cluster expansion will reduce the energy of high density nuclear matter by about 5–10%.

6.3. Pressure

Figure 5 plots the ratio

$$p/\varepsilon \tag{17}$$

against ε. Here ε is the energy density while p is the pressure. These two quantities have the same dimension. We plot p/ε because a plot of p itself would go through so many decades that it would not be very revealing. The energy density ε includes the rest mass energy Mc^2. Figure 5 shows that as ε increases from 10^2 to 10^4 MeV fm^{-3}, the pressure goes from about 1% of ε to 100%. At still higher energy densities, $p > \varepsilon$. An ε of 10^4 corresponds to about 4 baryons fm^{-3}.

The fact that p exceeds ε is slightly disturbing. We should expect that the sound velocity

$$c \, dp/d\varepsilon \tag{18}$$

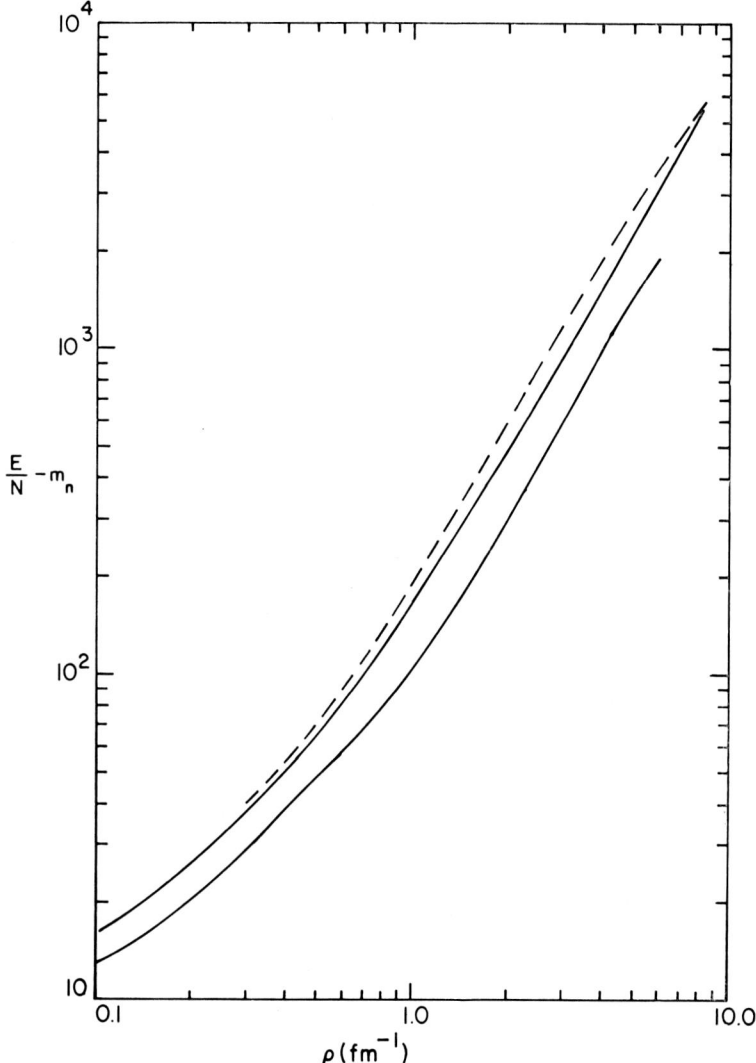

Fig. 4. Same as Figure 3, but upper solid curve is interaction II of Table I, including excited baryons. Dashed curve same with neutrons only. Lowest curve: Pandharipande (1971b), interaction C (with excited baryons).

should be less than the velocity of light, c. The reason that our theory violates this is due to the neglect of special relativity. It has been shown from general principles by Aichelburg *et al.* (1971) that in fact the sound velocity will never exceed the velocity of light. As a practical prescription one might use our theory up to the point where (18) becomes c, and thereafter replace (18) simply by c.

The two curves in Figure 5 refer to interactions I and II.

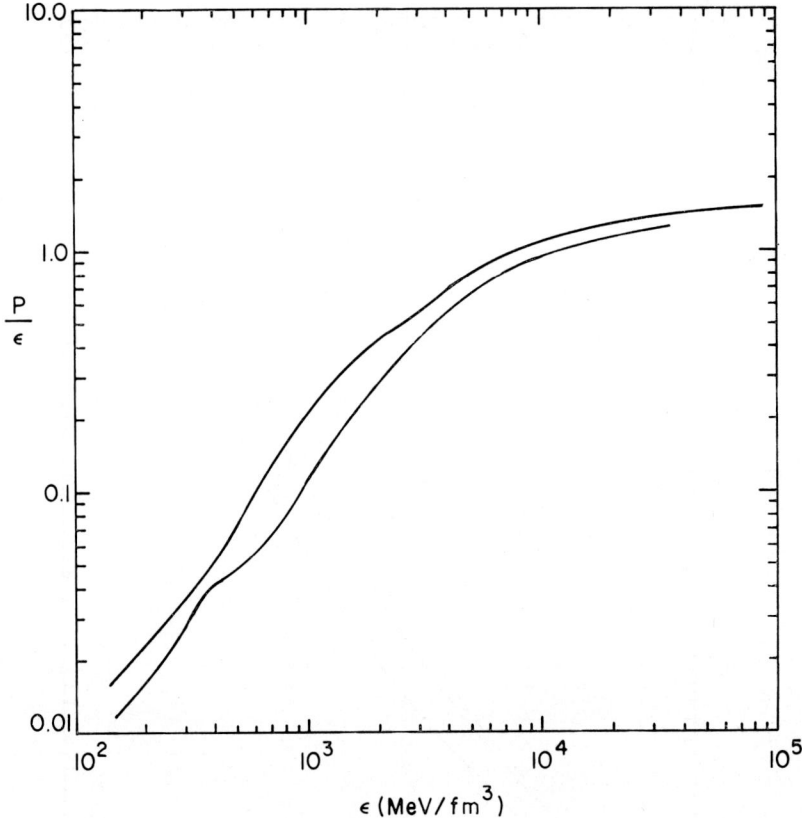

Fig. 5. The ratio of pressure P to energy density ε, a dimensionless quantity, is plotted versus ε itself. The ε includes the rest energy of the neutrons, and is given in MeV fm^{-3}; one of these units equals 1.6×10^{33} erg cm^{-3}.

7. Some Features of the Results

7.1. Contribution of various parts of the interaction

Our potential is more attractive in even than in odd states. Therefore the potential energy will be smaller if the relative momentum of two interacting baryons is small, because in this case $L=0$ states are favored.

It is convenient and possible in our density range to expand the potential energy is powers of the relative momentum, and hence of k_F. Of course only even powers will contribute. Table III lists these contributions for various densities. The contribution proportional to k^2 is always positive (repulsive), corrected by a negative contribution proportional to k^4. The sum of these momentum-dependent contributions to the potential energy is equal to the kinetic energy at $\varrho = 0.5$ fm^{-3}; at higher density, the potential energy contribution is much larger. Therefore the advantage of having low k_F comes chiefly from the potential energy and not, as might be suspected, from the kinetic.

TABLE III

Contributions to the energy in MeV per particle, from various parts of the interaction (see Section 7 of text). ϱ is the density in baryons fm^{-3}

ϱ	k^0	k^2	k^4	Kin
10	5340	1950	−520	83
4.5	1480	717	−175	50
1.5	120	308	−102	54
1.0	14	186	−54	54
0.5	−49	101	−25	74
0.25	−33	32	−6	47
0.1	−16	7	−0.5	26

The contribution independent of momentum is negative up to $\varrho = 0.9$ fm^{-3} because for these low densities the attraction dominates over the repulsion (disregarding the k^2 terms). At higher density the repulsion dominates. At $\varrho = 4.5$ fm^{-3}, the momentum-independent part is much greater than the momentum-dependent part of the potential energy; from this point on it does not make very much difference whether we have even or odd states.

7.2. Mass of the Nucleons

It may seem somewhat alarming that the potential energy is as much as 3 Mc^2 at high density. Is it then justified to use the normal rest mass of the nucleon?

The answer is affirmative because the quantity which occurs in the Klein-Gordon equation is

$$(E - V(r))^2 - M^2 c^4. \tag{19}$$

Now the total energy arises primarily from the average potential energy, and therefore $E - V(r)$ is very close to Mc^2 for nearly all r, except when two nucleons come very close together – but in that case the wave function is very small. The potential is very uniform at high density, and E simply reflects the existence of a large positive average potential. Since (19) is likely to be small, nonrelativistic theory is justified, and the ordinary nucleon rest mass may be retained.

For some time we thought that possibly the gravitational potential energy should also be included. We have been assured by C. Møller in a private communication that this is not the case but that we should calculate the equation of state in an inertial system regardless of gravitation. The result will then be inserted into the gravitational equations.

8. Results for Neutron Stars

Baym *et al.* (1971b) have used the equation of state of Pandharipande in a calculation of the hydrostatic equilibrium of a neutron star. Malone (unpublished) has done the same for our interaction II. The equations of general relativity have been used. Since our equation of state with excited baryons is similar to Pandharipande's neutron equation the results are similar.

According to Baym *et al.* (1971b) as well as Malone, neutron stars are stable with masses between about 0.1 and 2 solar masses. Table IV gives the results of Malone's calculations. The second column gives the gravitational mass M_g of the star, in units of the mass of the Sun, and the first gives the 'conserved' mass M_c, i.e. the total mass which all the baryons would have if they had no binding energy. The central density is listed, in units of 10^{15} g cm^{-3}; it increases with the mass which is a condition for stability of the star. The radius for most neutron stars is close to 10 km. The binding energy, which is the sum of nuclear and gravitational energies, varies from 4 to over 100 MeV per baryon; probably only neutron stars with binding of more than about 10 MeV can actually be formed.

The most interesting quantity is the moment of inertia which is seen to be of the order of 10^{44} g cm^2 units with the maximum occurring at a mass of 1.8 solar masses and having the value

$$I_{max} = 11 \times 10^{44} \text{ g cm}^2. \tag{20}$$

By contrast if a soft equation of state is used, such as that of Leung and Wang (1971), the maximum moment of inertia is 1 or 0.15 units (of 10^{44} g cm^2) according to the nuclear interaction they use at low density.

Useful observations exist only on the Crab Nebula. For this pulsar, the period and its time rate of change have been very accurately observed. If one attributes all the luminosity of the outer part of the nebula to energy transfer from the pulsar, and if one assumes that no pulsar energy escapes from the nebula in an invisible form, then

TABLE IV

Properties of neutron stars of various masses, using the interaction II of Table I, and the equation of state given by the upper solid line in Figure 4. Mass of neutron stars in units of the mass of the Sun, other details are described in Section 8 of text

$M_c/M_\odot$	$M_g/M_\odot$	ϱ_c (10^{15} g cm^{-3})	R km	B.E. MeV	I 10^{44} g cm^2
0.0940	0.0936	0.16	138	4.4	1.13
0.1495	0.148	0.30	19 0	8.4	0.66
0.229	0.226	0.40	13.7	15	1.05
0.374	0.364	0.50	11.8	27	1.9
0.567	0.544	0.70	11.1	38	3.2
0.817	0.770	0.95	10.7	53	4.9
1.13	1.04	1.2	10.4	73	7.3
1.47	1.32	1.6	10.0	97	9.6
1.76	1.54	2.3	9.5	118	11.0
1.95	1.67	4.1	8.5	134	10.7
	LW 0.43		6.8	a	1.05
	LW 0.22		3.9	b	0.15

[a] Leung and Wang, with Reid potential at low density.
[b] Leung and Wang, with Lomon and Feshbach potential.

the total energy emitted by the pulsar can be deduced from the observations. Together with Ω and $\dot{\Omega}$ this leads to a moment of inertia

$$I_{\text{obs}} = 2.5 \times 10^{44} \text{ g cm}^2. \tag{21}$$

With our equation of state, this corresponds to a mass of about 0.5 $M_\odot$ which is very satisfactory. With the soft equation of state this moment of inertia could not be obtained at all. We consider this some confirmation of our equation of state, but not a very strong one because the interpretation of the Crab energy is uncertain. The main point is that the order of magnitude of the 'observed' moment of inertia is that which can reasonably be expected from a neutron star.

I am much indebted to Dr M. Johnson for doing the calculations of the equation of state, Dr Pandharipande for devising the method, and Mr R. Malone for calculating the properties of neutron stars. I am also grateful to the many persons mentioned in this report for interesting suggestions and discussions.

References

Aichelburg, P. C., Ecker, G., and Sexl, R. U.: 1971, *Nuovo Cimento* **2B**, 63.
Anderson, P. W. and Palmer, R. G.: 1971, *Nature Phys. Sci.* **231**, 145.
Baym, G., Bethe, H. A., and Pethick, C. J.: 1971a, *Nucl. Phys.* **A175**, 225.
Baym, G., Pethick, C. J., and Sutherland, P.: 1971b, *Astrophys. J.* **170**, 299.
Bethe, H. A.: 1971, *Ann. Rev. Nucl. Sci.* **21**, 93.
Calogero, F. and Simonov, Yu. A.: 1969, *Nuovo Cimento* **64B**, 337.
Calogero, F. and Simonov, Yu. A.: 1970a, *Phys. Rev. Letters* **25**, 881.
Calogero, F. and Simonov, Yu. A.: 1970b, *Nuovo Cimento Letters* **4**, 219.
Calogero, F., Simonov, Yu. A., and Surkov, E. L.: 1972, *Phys. Rev.* **C5**, 1493.
Chemtob, M., Durso, J. W., and Riska, D. O.: 1971, *Phys. Letters* **36B**, 313.
Coldwell, R. L.: 1972, *Phys. Rev.* **D5**, 1273.
Leung, Y. C. and Wang, C. G.: 1971, *Astrophys. J.* **170**, 499.
Pandharipande, V. R.: 1971a, *Nucl. Phys.* **A178**, 123.
Pandharipande, V. R.: 1971b, *Nucl. Phys.* **A174**, 641.
Reid, R. V., Jr.: 1968, *Ann. Phys. N.Y.* **50**, 411.
Sawyer, R. F.: 1972a, *Astrophys. J.* **176**, 531.
Sawyer, R. F.: 1972b, *Phys. Rev. Letters* **29**, 382.
Scalapino, D. J.: 1972, *Phys. Rev. Letters* **29**, 386.
Siemens, P. and Pandharipande, V. R.: 1971, *Nucl. Phys.* **A173**, 561.
Wentzel, G.: 1949, *Quantum Theory of Fields*, Interscience Publ., New York.
Yang, C. H. and Clark, J. W.: 1971, *Nucl. Phys.* **A174**, 49.

DISCUSSION

Ruderman: The 'disease' of some of the highest density calculations, that $p > \varrho c^2$ or $c_s^2 = \mathrm{d}p/\mathrm{d}\varrho > c^2$, may not be cured by relativity. Some years ago (*Phys. Rev.* (1968) **170**, 1176 and **172**, 1286) Sidney Bludman and I explored the fully relativistic theory of many classical sources repelling each other *via* a neutral vector meson field. The same disease arises there also, as soon as one replaces the positive self energy of the source (infinite for point sources) by a smaller 'renormalized' source mass. In the quantum field theoretic treatment of such models (*Phys. Rev.* (1970) **1D**, 3243) instability against pair production of a source particle-antiparticle pair would apparently occur before the $c_s^2 > c^2$ disease could, so that the latter would never be expected. This suggests a test that should be made in the very dense neutron matter calculations ($\varrho \gtrsim 10^{16}$ g cm^{-3}) whenever $p > \varrho c^2$. Because the forces are described as coming from a vector meson (ω) exchange and a dipion scalar (σ) exchange an antineutron is as strongly attracted in such matter as a neutron is repelled. By appropriately anti-correlating an additional neutron and correlating an anti-neutron wavefunction with respect to the positions of the already present neutrons, the very strong interactions will always lower the total energy to make the pair below $2\ Mc^2$. If the total ≤ 0 the matter is, of course, unstable.

Bethe: It is comforting to know a mechanism by which the sound velocity will always be kept below the velocity of light. The actual calculations of the correlation function of an additional neutron and antineutron at high density may prove rather difficult.

Itoh: I would like to ask about the shape of the nuclear soft-core potential. Could you exclude experimentally the Gaussian soft-core potential proposed by Tamagaki?

Bethe: It is difficult to exclude it purely from experiment. However, I think one should combine experimental evidence with theoretical arguments. In our problem, this means assuming the exchange of ω mesons which gives a Yukawa potential.

Ruderman: David Pines and I wish to make a joint comment on the question of the nature of the high density limit of a Fermi gas with inverse square law repulsions among the fermions. It is indeed true as Professor Bethe remarked that the high density limit of a degenerate electron gas (in a uniform positive background) is a gas and not a solid. This is expected because as the average separation (b) between electrons decreases, their Fermi energy ($\sim \hbar^2/mb^2$) becomes much larger than their repulsive energy (e^2/b) so in the ground state it is more important to minimize the former than the latter. However this is no longer true when $e^2/\hbar c \sim 1/137$ is replaced by $g^2/\hbar c \sim 10$. For non-relativistic fermions the ratio $(g^2/b)(\hbar^2/mb^2)^{-1} > 10$. Its minimum is reached for relativistic baryons when the Fermi energy $\sim \hbar c/b$. Then the ratio of repulsive energy to Fermi energy becomes just $g^2/\hbar c \sim 10$. Thus this ratio is always sufficiently large at all densities that minimizing the repulsive energy by forming a crystal might be expected.

Bethe: The comment by Dr Ruderman is certainly correct. Therefore one cannot, on general principles, prove that neutron matter must remain liquid at high density. Whether it *actually* becomes solid, can only be shown by an explicit calculation with the actual forces between baryons. Our calculations have given negative answers to this question. This was true of Pandharipande's calculation a year ago with a method similar to the one he will describe in his contribution to this conference. About the same time Mikkel Johnson and I used the method of Guyer and Zane, and also found that the energy of the solid was much higher than that of the liquid. Canuto and Chitre, in their contribution to this conference, do find a positive result, viz. that neutron matter does crystallize at high density. It will be necessary to examine very carefully whether their method converges, or whether clusters of more than two bodies will change the result. For the present, the question whether baryon matter actually becomes solid must remain open.

VARIATIONAL METHOD FOR DENSE SYSTEMS

V. R. PANDHARIPANDE

Laboratory of Nuclear Studies, Cornell University, Ithaca, New York, U.S.A. and Tata Institute of Fundamental Research, Colaba, Bombay-5, India

Abstract. The variational method for calculating energy of quantum fluids, and its applications to the Bose liquid ^{4}He, Fermi neutron gas, and liquid ^{3}He are discussed. The correlation functions are parameterized by their healing distance, and can depend on the states occupied by the correlated particles in the model wave function. They are calculated by constrained variation of lowest order contributions. The healing distance has a prescribed value in lowest order calculations, whereas it is sufficiently large in hopefully exact energy calculations. The direct many-body cluster diagrams are summed with successive approximations of an integral equation. The contribution of exchange diagrams is shown to decrease rapidly with the number of exchanges, and their sums are truncated after the energy has converged to within a few percent.

1. Introduction

An equilibrium mixture of hyperons, interacting through two-body potentials is the simplest model of dense matter, and its region of validity has been discussed by H. Bethe in his paper in this volume. (We will refer to it, hereafter, as Paper I.) In the non-relativistic limit of this model, the many-body Schrödinger equation

$$\left\{\sum_i -\frac{\hbar^2}{2m_i}\nabla_i^2 + \tfrac{1}{2}\sum_{i<j} v_{ij}\right\} \Psi(1\ldots N) = E\Psi(1\ldots N) \tag{1}$$

should be solved for the variationally determined ground state composition to calculate the zero temperature energy as a function of density.

For practical reasons the form of the trial wavefunction used to solve (1) variationally is restricted to

$$\Psi(1\ldots N) = \prod_{i<j} f_{ij} \Phi(1\ldots N), \tag{2}$$

where Φ is a model fluid state wavefunction, and the correlation function f_{ij} is determined by minimizing the energy. We discuss two methods: in the first, the variation is constrained to enable us to make energy calculations in lowest order; in the second, the energy calculation is, we hope, exact and the variation is unconstrained.

Since there are many distinguishable hyperons of similar mass, hyperonic matter can be anywhere inbetween a Fermi and a Bose system when its composition is heterogeneous. At typical maximum densities in neutron stars (Baym *et al.*, 1971) the unit radius r_0,

$$\frac{4\pi}{3} r_0^3 \varrho = 1 \tag{3}$$

almost equals r_c, the radius of the repulsive core in the n-n potential. In liquid ^{3}He and ^{4}He also the r_0 at equilibrium density is a few percent less than the He-He atomic potential core radius. The binding energies of both types of liquid helium, Fermi ^{3}He and Bose ^{4}He are known, and have been extensively studied by variational methods with wavefunction (2) (as in Schiff and Verlet, 1967; and Murphy and Watts, 1970). These liquids form a suitable testing ground for many-body techniques used in hyperonic matter, and hence calculation of their energies is discussed in considerable detail. The He-He atomic potential core is very hard, and has r^{-12} behavior at small r. It tends to induce much stronger correlations than the n-n soft core potential with its r^{-2} to r^{-3} behavior as pointed out in Paper I. Thus liquid He is probably too severe a test for the theoretical methods.

2. The Radial Distribution Function

The radial distribution function of a Bose fluid is defined as

$$g_{mn} = \Omega \int \prod_{i<j} f_{ij}^2 \, d\tau' \bigg/ \int \prod_{i<j} f_{ij}^2 \, d\tau, \tag{4}$$

where $d\tau'$ omits integration over $\mathbf{r}_{mn}$, and Ω is the normalization volume. The subscripts denote coordinate variables of the functions, thus $g_{mn} = g(\mathbf{r}_{mn})$.

By convention we antisymmetrize only the left-hand side Ψ^* in calculations of expectation values for Fermi fluids. Thus initially the particles 1, 2, ... i respectively occupy plane wave states $\phi_1, \phi_2 \ldots \phi_i$ with momenta $\mathbf{k}_1, \mathbf{k}_2 \ldots \mathbf{k}_i$, and summation over particles is changed to summation over states. The radial distribution function for particles initially in states $\mathbf{k}_m$ and $\mathbf{k}_n$ can now be defined as

$$g(\mathbf{k}_m, \mathbf{k}_n) = \frac{\Omega \int \left(A \prod_i \phi_{q,i}^* \right) \left(\prod_{i<j} f_{ij}^2 \right) \left(\prod_i \phi_{i,i} \right) d\tau'}{\int \left(A \prod_i \phi_{q,i}^* \right) \left(\prod_{i<j} f_{ij}^2 \right) \left(\prod_i \phi_{i,i} \right) d\tau}, \tag{5}$$

where $\phi_{q,i}$ is the abbreviation of $\phi_q(\mathbf{r}_i)$. The true g is obviously

$$g(\mathbf{r}) = \frac{1}{N(N-1)} \sum_{i \neq j} g(\mathbf{k}_i, \mathbf{k}_j). \tag{6}$$

The f_{ij} approaches unity at large r_{ij}, and is small or zero at $r=0$. It is then convenient to substitute

$$f_{ij}^2 = 1 + F_{ij} \tag{7}$$

for all pairs other than mn in (4) or (5). The F_{ij} is a short-range function with absolute value generally less than unity, and products of $(1 + F_{ij})$ can be expanded in powers of F as follows:

$$\prod_{i<j} (1 + F_{ij}) = 1 + \sum_{i<j} F_{ij} + \prod_{i<j,k<l \, (ij \neq kl)} F_{ij} F_{kl} \cdots . \tag{8}$$

Integrals over various terms in (8) are represented by diagrams of the type shown in

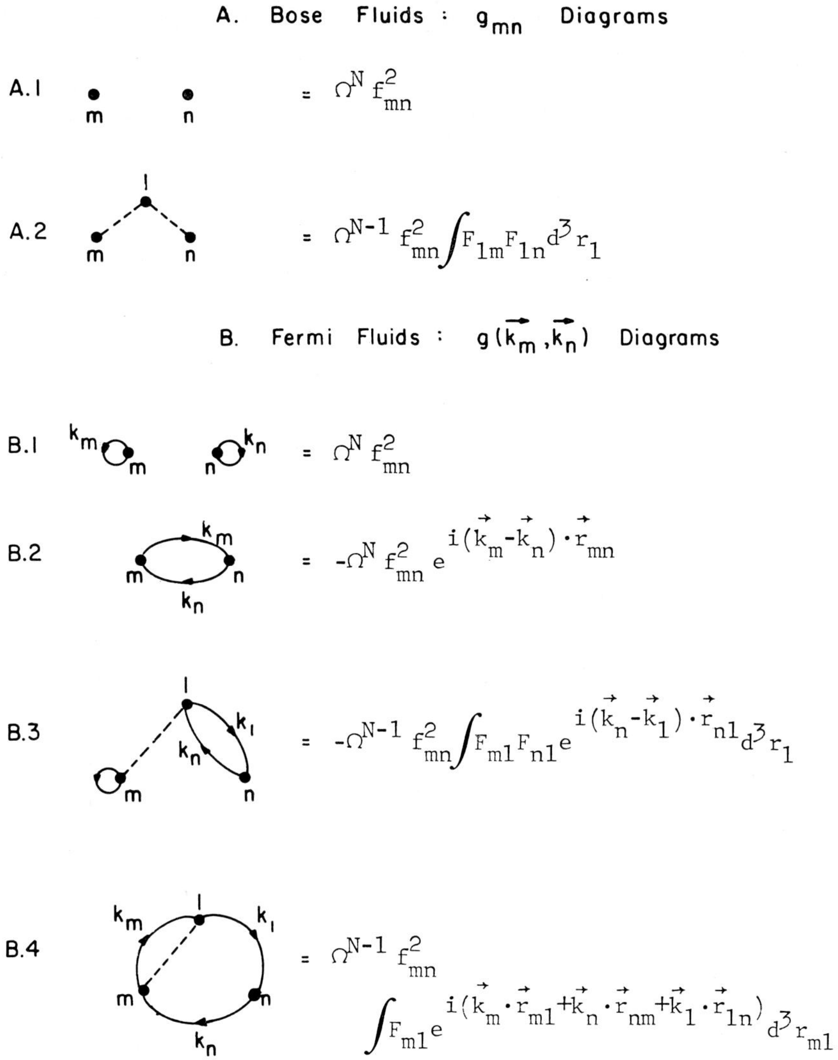

Fig. 1. Radial distribution function diagrams.

Figure 1. The points in these diagrams represent the particle coordinates, broken lines represent F functions, and numerator diagrams must always contain points m and n with f_{mn}^2. In Fermi fluid diagrams the full line k_q entering r_i represents the final state ϕ_q occupied by the particle in the left-hand side Ψ^*. The lines $k_1 \ldots k_i$ must originate from points $1 \ldots i$ respectively by our convention. The g_{mn} is simply the sum of all irreducible numerator diagrams divided by Ω^N (Pandharipande and Bethe, 1972).

3. The Energy

The potential, and the part of the kinetic energy obtained by collecting terms in which

∇^2 operates on a single f_{ij} can be included in an effective potential V_{ij}, and 'potential energy' W defined as

$$V_{ij} = v_{ij} - \frac{\hbar^2}{m} \frac{\nabla^2 f_{ij}}{f_{ij}}, \tag{9}$$

$$W = \frac{1}{2\Omega} \sum_{i,j} \int V_{ij} g_{ij} \, d^3 r_{ij}. \tag{10}$$

The ∇_i^2 operating on f_{ij} and f_{ik} gives additional kinetic energy U,

$$U = -\frac{1}{\Omega^2} \frac{\hbar^2}{2m} \sum_{i,j,k} \int g_3(r_{ij}, r_{ik}) \frac{\nabla_i f_{ij} \cdot \nabla_i f_{ik}}{f_{ij} f_{ik}} d^3 r_{ij} \, d^3 r_{ik}, \tag{11}$$

where $g_3(r_{ij}, r_{ik})$ is a three-particle distribution function defined analogous to g_{ij}. The total energy of the Bose fluid is simply

$$E = W + U. \tag{12}$$

In Fermi fluids the terms in which ∇^2 operates on Φ give exactly the Fermi gas energy T,

$$T = \sum_i \frac{\hbar^2}{2m_i} k_i^2, \tag{13}$$

because Φ is an eigenfunction of ∇^2. A term W_F is obtained in addition to W from $(\nabla_i^2 + \nabla_j^2)$ operating on ϕ_i or ϕ_j, and f_{ij}, that is,

$$W_F = -\frac{1}{\Omega} \sum_{k_i, k_j} \frac{\hbar^2}{m_{ij}} \frac{\nabla f_{ij} \nabla \phi_{ij}}{f_{ij} \phi_{ij}} g(\mathbf{k}_i, \mathbf{k}_j) d^3 r_{ij}, \tag{14}$$

where m_{ij} is the reduced mass for particles i and j, and the relative model wave function is

$$\phi_{ij} = \exp\left[i \tfrac{1}{2} (\mathbf{k}_i - \mathbf{k}_j) \cdot \mathbf{r}_{ij}\right]. \tag{15}$$

Similarly the operation of ∇_i^2 on f_{ij} and ϕ_{ik} gives another term

$$U_F = \frac{-1}{\Omega^2} \sum_{k_i, k_j, k_k} \frac{\hbar^2}{2m_i} g_3(\mathbf{k}_i, \mathbf{k}_j, \mathbf{k}_k, \mathbf{r}_{ij}, \mathbf{r}_{ik}) \frac{\nabla_i f_{ij} \cdot \nabla_i \phi_{ik}}{f_{ij} \phi_{ik}} d^3 r_{ij} \, d^3 r_{ik}, \tag{16}$$

and the total energy is

$$E = T + W + W_F + U + U_F. \tag{17}$$

The terms W and W_F can be combined by redefining the effective potential (9) for Fermi systems as

$$V_F(\mathbf{k}_i, \mathbf{k}_j, \mathbf{r}_{ij}) = v(\mathbf{r}_{ij}) - \frac{\hbar^2}{m_{ij}} \left\{ \frac{\nabla^2 f_{ij}}{f_{ij}} + 2 \frac{\nabla f_{ij} \cdot \nabla \phi_{ij}}{f_{ij} \phi_{ij}} \right\}. \tag{18}$$

4. Constrained Variation

Most of the results presented in Paper I are obtained with this simple method. The physical assumption here is that the contribution of farther neighbors of a particle i to the instantaneous potential $\sum_j v_{ij}$ seen by i should mostly be included in the average field of which ϕ_i is an eigenfunction. Hence distant neighbors should not be strongly correlated, and the effect of their correlation on the energy should be small. If this effect is neglected one can work with correlation functions satisfying the conditions

$$\text{if } f_{ij} \neq 1 \quad \text{then all } f_{ik} = 1 \tag{19}$$

when ij are nearest neighbors. It is difficult to handle (19) because it couples various f_{ij}. In practice (19) is approximated as a healing constraint on a single f_{ij} as a function of r_{ij} as follows (Pandharipande, 1971):

$$f_{ij}(|r_{ij}| > d) = 1, \quad \text{and } (\partial f_{ij}/\partial r_{ij})(|r_{ij}| = d) = 0. \tag{20}$$

The healing distance d is chosen such that, on the average, there is only one particle within a distance d of an average particle. With this constraint correlations are at times allowed between second and more distant neighbors, and at times even the first neighbors are treated as uncorrelated. We hope that these effects cancel.

All higher order direct diagrams are zero if (19) is valid, and we will show that higher order exchange diagrams like B.4 of Figure 1, which can contribute even when (19) is valid, are small. Thus the energy can be calculated with only the two-body term in the cluster expansion, which is

$$\frac{E}{N} = \tfrac{1}{2}\varrho \int \left(v - \frac{\hbar^2}{m}\frac{\nabla^2 f}{f}\right) f^2 \, d^3 r \tag{21}$$

for Bose fluids. (We omit subscripts wherever unnecessary.) Before minimizing E with respect to variations in f, the part of v that contributes to the average field and hence does not induce correlations, must be subtracted from (21). The constraint (20) gives for this part denoted by $\lambda(r)$,

$$\lambda(|r| > d) = v(r), \tag{22}$$

and we assume $\lambda(|r| < d)$ to be a constant λ_0 to be determined from (20). The equation

$$\delta \int_0^d \left\{-\frac{\hbar^2}{m} f \nabla^2 f + vf^2 - \lambda_0 f^2\right\} d^3 r = 0 \tag{23}$$

gives directly the two body Schrödinger equation for $r < d$

$$-\frac{\hbar^2}{m} \nabla^2 f + vf = \lambda_0 f \tag{24}$$

and λ_0 can be obtained from the boundary condition (20). The effective potential

V_{ij} is clearly

$$V_{ij}(|r_{ij}| < d) = \lambda_0$$
$$V_{ij}(|r_{ij}| > d) = v_{ij}(r_{ij}). \qquad (25)$$

The f and λ_0 are functions of d, which is determined from

$$\varrho \int_0^d f^2 \, d^3r = 1, \qquad (26)$$

where the left-hand side is simply the average number of particles within d as calculated in lowest order. The two Equations (24) and (26) are solved simultaneously.

In Fermi fluids let us first consider pairs of distinguishable fermions like those of spin up and spin down particles of a baryon type, or of two different baryons of any spin direction. Such pairs are not exchanged in the antisymmetrization of the wave function. If the two particles are in states $\mathbf{k}_i$ and $\mathbf{k}_j$, the lowest order contribution of their interaction to the potential energy $(W + W_F)$ is given by Equations (10) and (18) as

$$(W + W_F)_{ij} = \frac{1}{\Omega} \int \phi_{ij}^2 f^{ij^2} \left\{ v - \frac{\hbar^2}{m} \left[\frac{\nabla^2 f^{ij}}{f^{ij}} + 2 \frac{\nabla f^{ij} \cdot \nabla \phi_{ij}}{f^{ij} \phi_{ij}} \right] \right\} d^3r, \qquad (27)$$

where f^{ij} is the correlation function for particles in model states $\mathbf{k}_i$, $\mathbf{k}_j$, and ϕ_{ij} is given by (15). Since

$$\frac{\nabla^2 \phi_{ij}}{\phi_{ij}} = -[\tfrac{1}{2}(k_i - k_j)]^2 \equiv -k^2, \qquad (28)$$

we can write (27) as

$$(W + W_F)_{ij} = \frac{1}{\Omega} \int \psi^* \left(v - \frac{\hbar^2}{m} k^2 - \frac{\hbar^2}{m} \nabla^2 \right) \psi \, d^3r, \qquad (29)$$

with

$$\psi = f^{ij} \phi_{ij}, \qquad (30)$$

and the constraint (19) as

$$\frac{\nabla \psi}{\psi}(\mathbf{r}, |r| = d) = \frac{\nabla \phi_{ij}}{\phi_{ij}}(\mathbf{r}, |r| = d). \qquad (31)$$

It is convenient to decompose ϕ_{ij} and ψ in partial waves

$$\phi_{ij}(r) = \sum_{l,m} kr j_l(kr) P_l^m(\theta, \phi)$$
$$\psi(r) = \sum_{l,m} kr u_l(r) P_l^m(\theta, \phi). \qquad (32)$$

If v is spherically symmetric the various l states are not coupled, and the contribution of each can be minimized separately following the procedure described for the Bose fluids. The equation for u_l is

$$-\frac{\hbar^2}{m}\frac{\partial u_l^2}{\partial r^2} + \frac{l(l+1)}{r^2} u_l + vu_l = \left(\frac{\hbar^2}{m} k^2 + \lambda_0^l\right) u_l \tag{33}$$

and λ_0^l is determined from the boundary condition

$$\frac{1}{u_l(d)}\frac{\partial u_l(d)}{\partial r} = \frac{1}{j_l(kd)}\frac{\partial j_l(kd)}{\partial r}. \tag{34}$$

Both the correlation function f and the effective interaction V_F depend upon k and l, and may be written as

$$f_{ij} = \sum_l f^l(k, r) P_{ij}^l \tag{35}$$

$$V_{F_{ij}} = (r < d) = \sum_l \lambda_0^l(k) P_{ij}^l \tag{36}$$

$$V_{F_{ij}} = (r > d) = v_{ij}, \tag{37}$$

where the projection operators P_{ij}^l operate only on the model wave function $\phi_{q,i}\phi_{p,j}$ by definition.

The same procedure can be followed in calculating the contribution of interaction between parallel spin fermions of the same type in states $\mathbf{k}_i$ and $\mathbf{k}_j$. Adding the exchange contribution to (29) gives

$$(W + W_F)_{ij\uparrow\uparrow} = \frac{1}{\Omega} \int [\psi^*(\mathbf{r}) - \psi^*(-\mathbf{r})] \left(v - \frac{\hbar^2}{m} k^2 - \frac{\hbar^2}{m}\nabla^2\right) \psi(\mathbf{r}) \, d^3r, \tag{38}$$

and on substituting the partial wave expansion (32) in (38) the even l-state contributions cancel, and those of odd states are doubled. The variational calculation of f_{ij} and $V_{F_{ij}}$ is unaffected because (33) is obtained by individually minimizing the contribution of each partial wave. The d is given by

$$\frac{1}{\Omega^2 N} \sum_{ij} \langle \phi_i\phi_j - \phi_j\phi_i\delta_{\uparrow\uparrow} | \left\{ \sum_l f^{l2}(\mathbf{k}, \mathbf{r}) P^l \right\} \tilde{O}(d) | \phi_i\phi_j \rangle = 1, \tag{39}$$

where $\tilde{O}(d) = 1$ for $r < d$ and zero otherwise. Equations (33), (34) and (39) are solved simultaneously by iteration.

The above method is simple enough to study complex systems like dense hyperonic matter, and still sufficiently general to treat the small differences in baryon-baryon interactions, interactions in different angular momentum states, and baryon masses. It is also possible to generalize it to treat non-central forces, particularly the strong tensor force in the neutron-proton interaction (Pandharipande, 1972; Pandharipande and Garde, 1972).

However, there are approximations in this model justified by purely physical arguments. In particular, the healing distance d is obtained from a lowest order calculation of the number of particles within d. This may not be valid when $d \gg r_0$, where r_0 equals d in uncorrelated systems. In neutron star matter $d \sim 1.2\, r_0$ whereas in liquid helium $d \sim 1.4\, r_0$, because helium atomic cores are much harder than the baryonic cores. The variational property $E \geqslant$ true E_0 is lost due to the lowest order calculation of E. In subsequent sections we describe what we hope are exact calculations of E with sufficiently large values of d so that the effects of the constraint are negligible. A comparison of the two calculations should ascertain the region of validity of the first.

5. Integral Equation for Summing Direct Diagrams

Some of the diagrams that contribute to the radial distribution function in a Bose system are shown in Figure 2. A sum over all particles labeled by numbers 1, 2, ... is implied, and it simply gives a density ϱ with factors to account for double counting. Thus the contribution of diagram E.2 of Figure 2 is

$$f_{mn}^2 \varrho \int F_{1m} F_{1n} \, d^3 r_1 \equiv f_{mn}^2 S_{mn}. \tag{40}$$

The functions F and S are shown in Figure 3 for a typical f with $d = 2\, r_0$, in liquid ^{4}He near equilibrium density. Since the magnitude of S is larger than that of F, and they are of opposite sign, the contribution of diagram E.3

$$= f_{mn}^2 \varrho \int F_{1m} S_{1n} \, d^3 r_1 \tag{41}$$

is larger than that of (40). It can be easily seen that E.4 contributes more than E.3, etc. The diagrams E.2, E.3 ... are called single chains, and in dense fluids their contribution does not decrease with the number of particles in the chain.

The contribution of diagram E.5

$$= f_{mn}^2 \varrho \int F_{m1} F_{n1} S_{n1} \, d^3 r_1, \tag{42}$$

and is of the same order as that of E.2 because $S \gtrsim 1$ in the range of F. Thus diagrams of type E.5–7 in which additional chains are added to connect two particles in a chain are also of the same order as those of single chains. Diagrams in which any two points of a chain may be connected by many chains are called hypernetted chains (HNC), and E.8 is a typical HNC diagram. No two chains, or subchains, are connected by an F.

The simplest diagram in which two chains connecting m and n are connected by an F is E.9. Its contribution $\hat{S}$ is small (Figure 6), particularly in the region where f_{mn}^2 is appreciable, because all four particles have to be within the range of F. One would thus expect contributions of diagrams like E.10 involving E.9 as a subdiagram to be small.

Van Leeuwen *et al.* (1959) have shown that the sum of all diagrams in the Bose case is given by the consistent solution of

$$\ln\left\{\frac{g_{mn}}{f_{mn}^2 \exp(E_{mn})}\right\} = \varrho \int \left\{g_{m1} - 1 - \ln\left[\frac{g_{m1}}{f_{m1}^2 \exp(E_{m1})}\right]\right\}(g_{n1} - 1)\,d^3r_1. \tag{43}$$

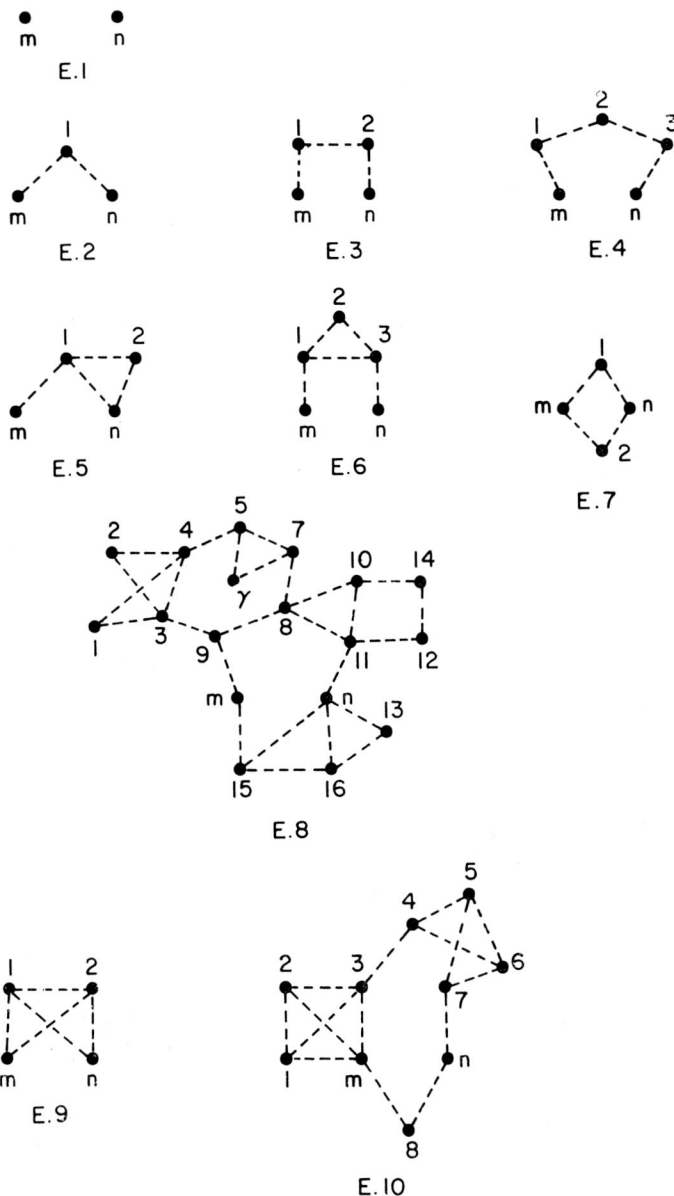

Fig. 2. Irreducible diagrams that contribute to Bose $g(r)$.

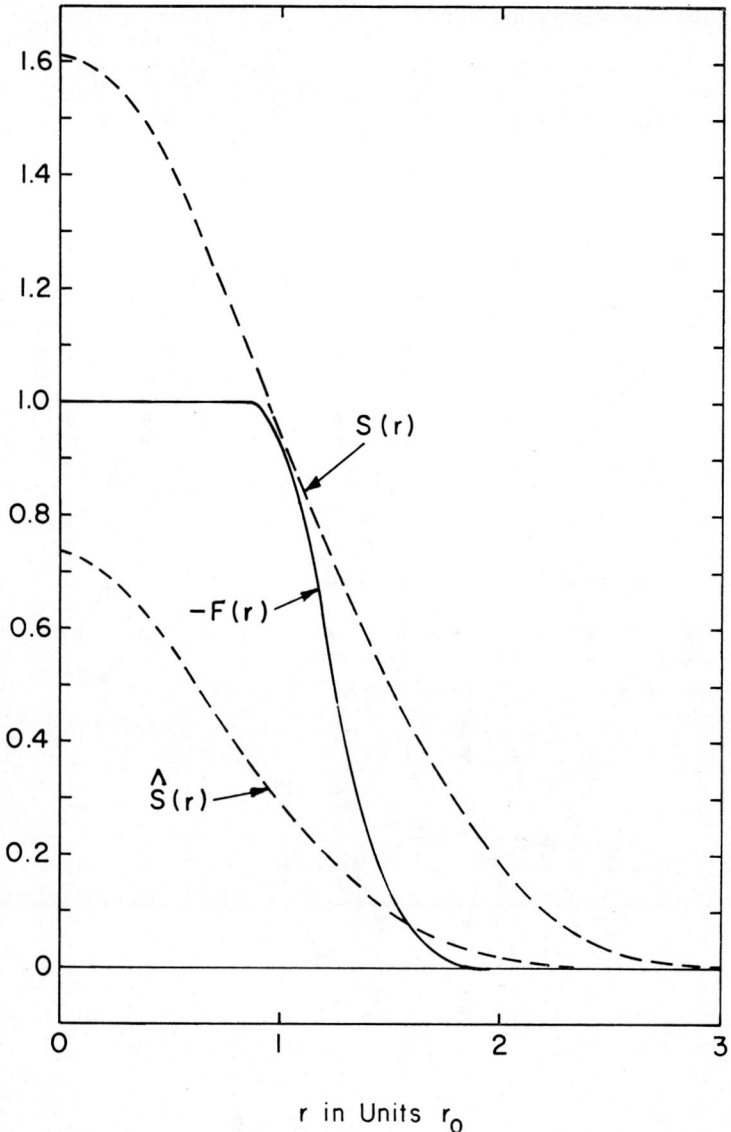

Fig. 3. The functions, F, S, and $\hat{S}$ in liquid ^{4}He at $\varrho = 0.35$ atoms σ^{-3} and $d = 2\,r_0$ with the Lennerd-Jones potential.

If E_{mn} is set equal to zero one obtains the sum of all HNC diagrams. When $E_{mn} = \hat{S}_{mn}$ the equation sums all HNC diagrams plus the diagram E.9 and the HNC's formed with E.9 as an element. We call this approximation HNC/4. The difference between HNC and HNC/4 is rather small (Figure 6) and hence more complicated diagrams in which many particles have to be simultaneously correlated with all others can be neglected.

It is clear that summing all HNC diagrams connecting ij, jk, and ki gives for the three-boson distribution function

$$g_3(\mathbf{r}_{ij}, \mathbf{r}_{ik}) = g_{ij} g_{jk} g_{ki}. \tag{44}$$

We use (44) to obtain g_3 in both HNC and HNC/4 approximations.

6. Exchange Diagrams

The state dependence of the correlation function, that is equations (35) to (37), and that of the potential in the case of neutron matter, makes exact higher order cluster calculations in Fermi fluids very complex. We note that the short range part of the n-n potential, which is mainly responsible for the correlations, is similar in all states (Paper I). The l and k dependence of f is also not too strong for the neutron case, and hence higher order Fermi fluid calculations are simplified considerably by neglecting the differences between $f(\mathbf{k}_m, \mathbf{k}_n)$ and $V_F(\mathbf{k}_m, \mathbf{k}_n)$ in all but the lowest order two-body clusters.

The sum of direct $g(k_m, k_n)$ diagrams can be easily calculated by the integral Equations (43) when the F^{ij} are approximated by an average F,

$$F = \frac{1}{N^2} \sum_{ij} f^{ij2} - 1 \equiv f^2 - 1. \tag{45}$$

Let g_B be the solution of (43) with the average f, and let

$$h = g_B/f^2. \tag{46}$$

Then the approximate contribution of all direct diagrams to $g(k_m, k_n)_D$ is

$$g_B(\mathbf{k}_m, \mathbf{k}_n)_D = (f^{mn})^2 h. \tag{47}$$

Figure 4 shows diagrams in which only particles in states k_m and k_n are exchanged, all f^{ij} other than f^{mn} are approximated by the average f, and the sums over states other than k_m and k_n are performed. As discussed previously, all HNC (or better, HNC/4), diagrams connecting m and n must be summed to any number of particles, and the crossed line of diagram G.4 denotes such a sum. The contribution of G.4 is obviously

$$g_B(\mathbf{k}_m, \mathbf{k}_n)_{ex} = -\exp[i(\mathbf{k}_m - \mathbf{k}_n) \cdot \mathbf{r}] g_B(\mathbf{k}_m, \mathbf{k}_n)_D. \tag{48}$$

Exchange of a particle in state m or n with another particle gives diagrams of type H.1 in Figure 5, and their average contribution is evaluated by summing over $\mathbf{k}_m$ and $\mathbf{k}_n$. The sum of all direct diagrams connecting m and 1 is represented by the double broken line, and equals $(g_B - 1)$. Sum over exchanges k_1 and k_n gives the square of The Slater function $S(k_F r)$ [not to be confused with S of (40)],

$$S(k_F r) = \frac{1}{N} \sum_i e^{i k_i \cdot r}$$
$$= \frac{3}{k_F^3 r^3} [\sin(k_F r) - k_F r \cos(k_F r)]. \tag{49}$$

The sum of diagrams with and without F_{1n} gives f_{1n}^2 which is converted to g_{B1n} by summing all chains connecting 1 and n. Thus the double full line $1n$ denotes

$$F_{ex}(r_{1n}) = \tfrac{1}{2} g_B S(k_F r_{1n})^2, \tag{50}$$

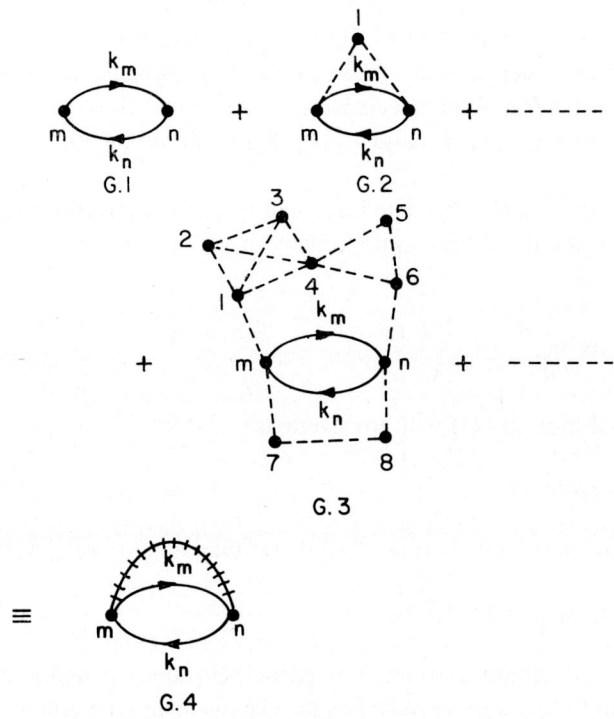

Fig. 4. Fermi $g(k_m, k_n)$ diagrams with one exchange.

where the factor $1/2$ comes from requiring that the spins of particles 1 and n must be parallel. The

$$g_{H.1}(\text{average}) = -2 g_{B\varrho} \int (g_{Bm1} - 1) F_{ex}(r_{1n}) \, d^3 r_1,$$
$$\equiv f^2 h_1 \tag{51}$$

exchanges of m giving the factor 2.

Exchanging m and 1 in H.1 gives the three-particle exchange diagrams $g_{H.1\,ex}$. Full

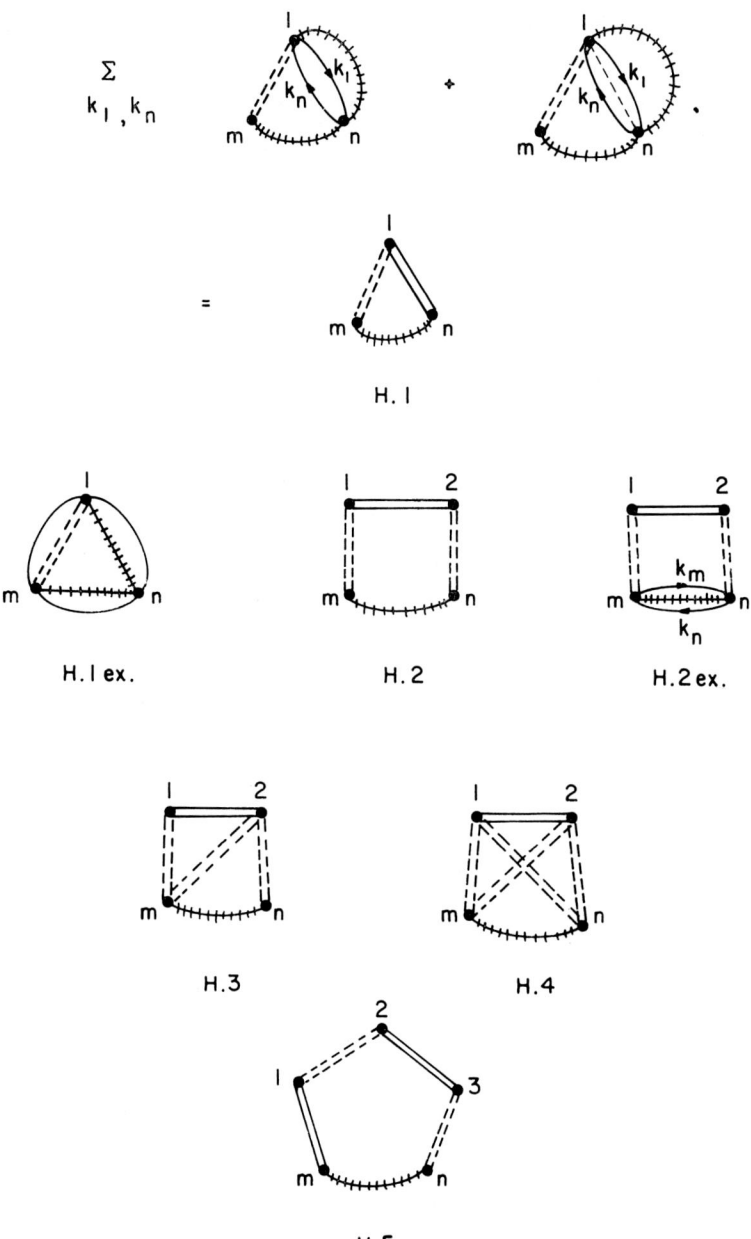

Fig 5. Fermi g diagrams with exchanges in the chains.

lines in this diagram represent the Slater functions, and their contributions

$$g_{H.1\,ex}(\text{average}) = 2S(k_F r_{mn}) g_{B\varrho} \int (g_{Bm1} - 1) S(k_F r_{m1}) g_{Bn1} \times$$
$$\times S(k_F r_{n1}) d^3 r_1 \equiv f^2 h_{1\,ex} \tag{52}$$

cancels $g_{H.1}$ at small r_{mu} when m and n have parallel spins.

In this notation it is quite simple to write the contributions of more complicated exchange diagrams; that of H.2 in Figure 5 is, for example,

$$g_{H.2}(\mathbf{k}_m, \mathbf{k}_n) = - g_B(\mathbf{k}_m, \mathbf{k}_n)_D \varrho^2 \int (g_{Bn1} - 1) F_{ex}(r_{12}) (g_{Bn2} - 1) \times$$
$$\times d^3 r_1 d^2 r_2 \equiv h_2 f^{mn2}. \tag{53}$$

Figure 6 shows the functions $g_{2-\text{body}}(=f^2)$, $\Delta g_{HNC}(=g_{BHNC}-f^2)$, $\Delta g (=\Delta g_{BHNC/4} - \Delta g_{BHNC})$, $g_{H.1}$, and $g_{H.2}$ (average) in liquid ^{3}He at experimental equilibrium density.

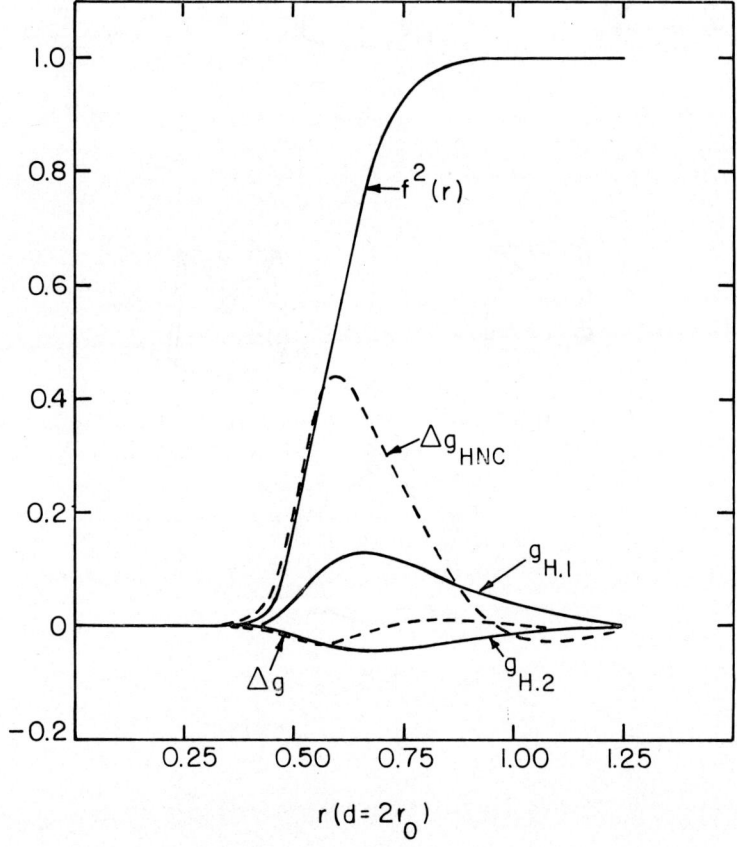

Fig. 6. Contributions to the radial distribution function in liquid ^{3}He at $\varrho = 0.277$ atoms σ^{-3} and $d = 2 r_0$ from various classes of diagrams.

The F_{ex} is a rather small function because g is small where $S(k_F r)$ is unity and vice versa. Hence the diagrams with exchanges in the chains give a small contribution. We neglect diagram H.4 because its contribution should be much smaller than Δg which itself is small. The contributions of H.1, H.2, and H.3 should be treated together because all these diagrams involve only one F. The maximum contribution of H.5 containing two F_{ex} is only 0.016 at $r \sim 0.6\, d$. Since H.5 is expected to be the largest of all diagrams with two F_{ex}, all more complicated exchange diagrams can be safely neglected.

The three-particle distribution function $g_3(\mathbf{k}_i, \mathbf{k}_j, \mathbf{k}_k, \mathbf{r}_{ij}, \mathbf{r}_{ik})$ for various exchanges between i, j and k can be easily expressed in terms of the functions h with the superposition approximation (44).

7. Conclusions

The energy of liquid ^{4}He, as calculated in the HNC/4 approximation, is shown in Figure 7 as a function of (d/r_0). It is very insensitive to d in the neighborhood of $2\, r_0$. Hence most of our calculations are carried out with $d = 2\, r_0$.

Table I compares energies of a pure neutron gas as calculated with the Reid potential (Pandharipande, 1971) in the HNC and HNC/4 approximations. The maximum

TABLE I

The $E(\varrho)$ for neutron gas with the Reid potential

ϱ (n fm^{-3})	(E/n) in MeV	
	HNC	HNC/4
0.2	19.77	19.77
0.6	77.08	77.00
1.0	170.5	169.7
1.4	293.5	292.0
1.8	440.4	438.8
2.2	607.3	606.8
2.6	790.5	793.4
3.0	987.9	997.4
3.4	1197.5	1217.4
3.8	1417.8	1452.4
4.2	1647.3	1701.6
4.6	1885.3	1963.4
5.0	2130.6	2237.6

difference is $\sim 5\%$, and we hope that the effect of more complicated direct diagrams will be $\ll 5\%$. The energy change on the inclusion of the exchange diagram H.5 is only $\sim 1-2\%$. When $d = 2\, r_0$ the higher order clusters contribute a substantial fraction $\sim 50\%$ of the total energy as shown in Table II.

Results of the constrained variational calculation in lowest order, and of (we hope) exact energy calculations with unconstrained d are compared in Figures 8, 9, and 10. The two calculations agree very well in neutron matter, whereas the simple lowest

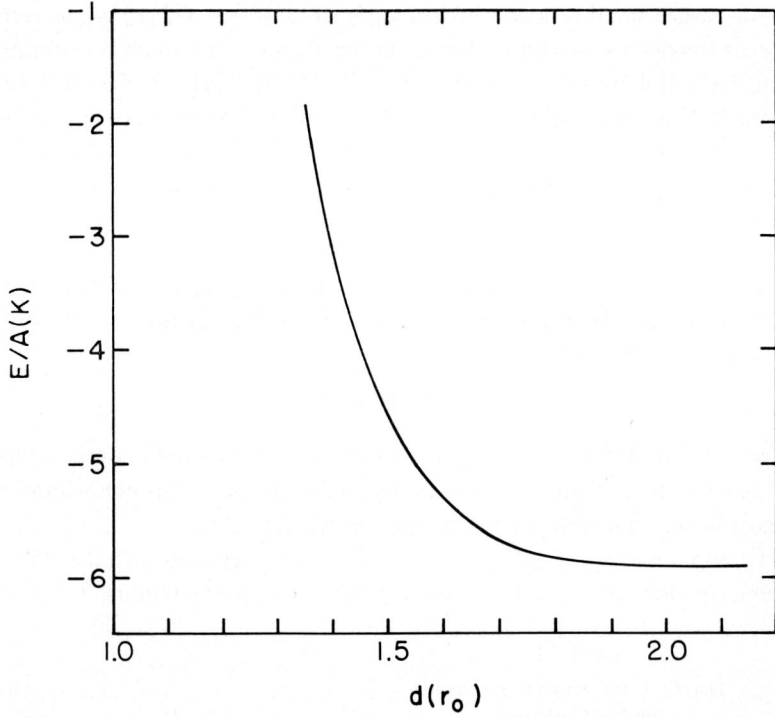

Fig. 7. Variation of energy with d in liquid ^{4}He at $\varrho = 0.35$ atoms σ^{-3} with Lennard-Jones potential.

TABLE II

The contributions to $E(\varrho)$ in neutron gas with Reid potential

ϱ (N fm^{-3})	T	HNC/4 contribution to energy in MeV neutron					
		$(W+W_F)_{L.O.}$	$\Delta(W+W_F)_{H.O.}$	U	U_F	All H.O.	E
0.2	40.7	−21.4	−0.6	0.6	0.5	0.5	19.77
0.6	84.6	−26.4	+0.6	11.3	6.9	18.7	77.00
1.0	119	−11.0	16.3	30.8	14.6	61.8	169.7
1.4	149	21.9	44.5	54.5	22.4	121	292.0
1.8	176	70.6	82.5	79.8	29.9	192	438.8
2.2	201	133	129	106	37.3	272	606.8
2.6	225	209	182	133	44.6	359	793.4
3.0	248	296	241	160	51.8	453	997.4
3.4	269	394	307	189	59.0	554	1217.4
3.8	290	502	378	216	66.2	661	1452.4
4.2	310	618	456	246	73.5	773	1701.6
4.6	329	743	538	273	80.8	892	1963.4
5.0	348	875	625	300	89.9	1015	2237.6

order calculation overestimates the binding energies and equilibrium densities of liquid ^{3}He and ^{4}He by $\sim 20\%$.

Table III shows the total energy, the lowest order, and higher order contributions to it at $d/r_0 = 1.2$, 1.6, and 2.0. The total energy is much less sensitive to d than its decomposition into lowest and higher cluster contributions. The higher order contributions increase by a factor of 1.5–3.0 whereas the total energy decreases only by 10–15% and d is increased from 1.2 to $2 r_0$.

The $d = 1.2 r_0$ in lowest order calculations. Thus correlations with $d = 1.2 r_0$ are mainly two-body correlations because of the average there is only one neutron within

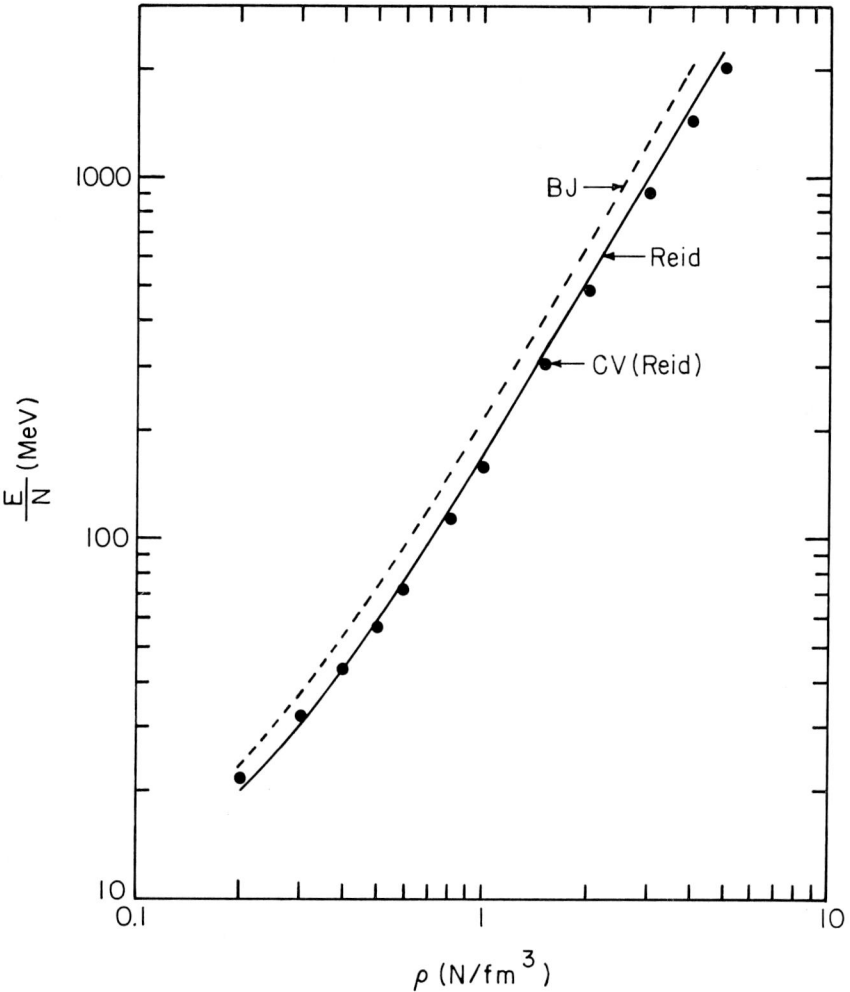

Fig. 8. Neutron gas energy: The full and broken lines give the results of 'exact' energy calculation with Reid and the modified Reid potentials, and the points show results of lowest order constrained variation (from Pandharipande, 1971).

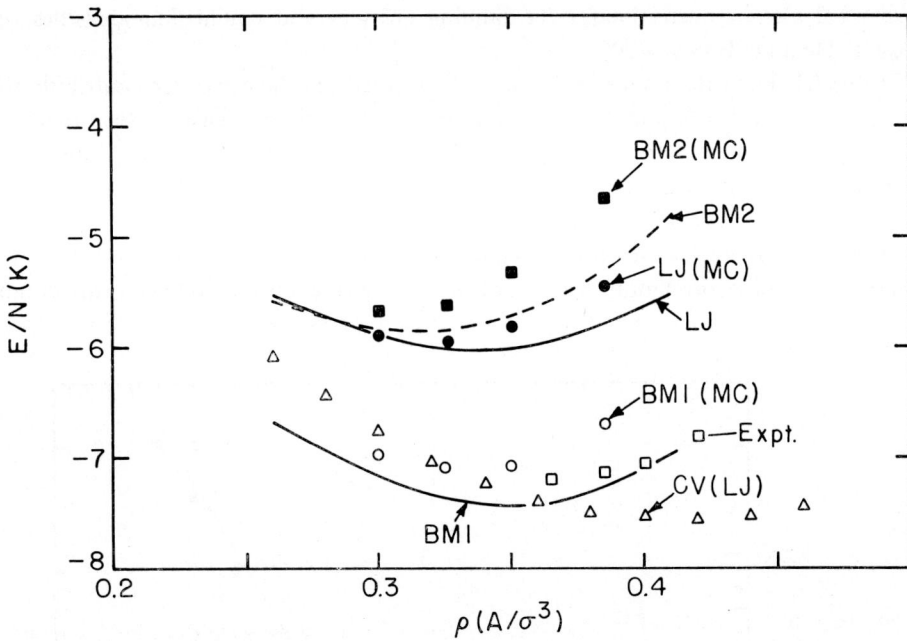

Fig. 9. Energy of liquid ^{4}He: The potentials and the Monte Carlo results are from Murphy and Watts (1970). The curves give results of present calculations, and those of lowest order constrained variation are from Pandharipande (1971).

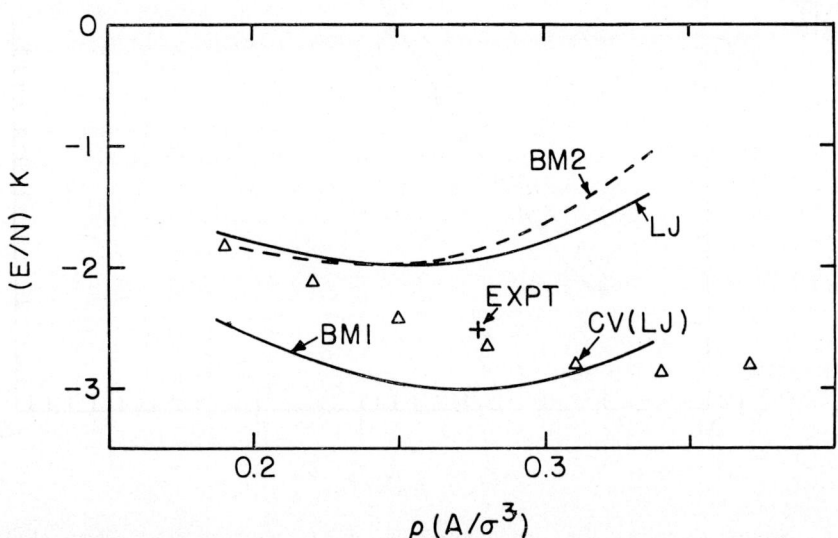

Fig. 10. Energy of liquid ^{3}He with various potentials.

TABLE III

The neutron-gas energy with Reid potential at various values of d

ϱ (N fm^{-3})	(E/N) MeV	1.2	1.6	2.0
1.0	Total	204.4	177.3	169.7
	L.O.	156.3	123.7	108.0
	H.O.	38.1	53.6	61.7
3.0	Total	1174.0	1037.7	997.4
	L.O.	982.5	710.6	544.3
	H.O.	191.5	327.1	453.1
5.0	Total	2467.7	2266.0	2237.6
	L.O.	2117.5	1595.0	1223.0
	H.O.	350.2	671.0	1014.6

correlation volume of any given neutron. The small higher order cluster contributions at $d = 1.2\, r_0$ come from events in which two (or more) neutrons come within the correlation volume of a neutron. In lowest order calculations these are neglected assuming that they cancel events in which the neutron can have a larger correlation volume because there is nothing within $1.2\, r_0$ of it. The actual higher order calculations show that the occurrence of two or more neutrons in the correlation volume gives a somewhat greater contribution than the occurrence of none.

Our main result is the justification, for neutron matter, of the constrained lowest order calculation, to 5% or better, by the more exact calculation of higher order clusters. In Paper I the constrained lowest order method was used.

This work was supported, in part, by the National Science Foundation.

References

Baym, G., Pethick, C., and Sutherland, P.: 1971, *Astrophys. J.* **170**, 299.
van Leeuwen, J. M. J., Geoneveld, J., and de Boer, J.: 1959, *Physica* **25**, 792.
Murphy, R. D. and Watts, R. O.: 1970, *J. Low Temperature Phys.* **2**, 507.
Pandharipande, V. R.: 1971, *Nucl. Phys.* **A178**, 123.
Pandharipande, V. R.: 1972, *Nucl. Phys.* **A181**, 33.
Pandharipande, V. R. and Bethe, H. A.: 1972 (to be published).
Pandharipande, V. R. and Garde, V. K.: 1972, *Phys. Letters* **39B**, 608.
Schiff, D. and Verlet, L.: 1967, *Phys. Rev.* **160**, 208.

PROPERTIES AND SYNTHESIS OF HEAVY NUCLEI AND PROPERTIES OF NEUTRON STAR MATTER

J. ROBERT BUCHLER

Belfer Graduate School of Science, Yeshiva University, New York, N.Y., U.S.A.

Abstract. The nuclear Thomas-Fermi model which is based on nuclear matter calculations has been successfully applied to the study of the bulk properties of nuclei. It is ideally suited for extrapolation into the region of very neutron-rich and of superheavy nuclei. It is therefore a valuable approach for r-process calculations as well as for the study of neutron star matter at subnuclear densities.

Many physical situations require the knowledge of the properties of nuclei located far away from the region of stability. The standard semi-empirical mass formulae give a good fit to the known nuclei, but their extrapolation into the neutron-rich and superheavy region is subject to considerable uncertainty. This is due to the fact that only the coefficients of the four leading terms can be reliably fitted by a least squares procedure. In the neutron-rich regions, mass formula terms which are negligible in the vicinity of the beta stability valley, and therefore ill determined, can significantly influence estimates of nuclear properties. It is therefore of interest to have an alternate simple way to get a grasp on these nuclei without having to do, for instance, a full Brueckner-Hartree-Fock calculation.

The idea of reproducing the bulk properties of nuclei, i.e., their binding energies, radii and surface thicknesses, by means of a statistical or Thomas-Fermi approach, goes back to the earlier days of nuclear physics. The earlier approaches were semi-phenomenological in the sense that a whole set of parameters had to be fitted to give agreement with experiment. The present idea is to make maximum use of the results of realistic nuclear matter calculations, i.e. the binding energy of a homogeneous mixture of neutrons and protons as a function of neutron excess and of density. This allows us to reduce the number of free parameters to just one.

The Thomas-Fermi approach is based on a theorem by Hohenberg and Kohn (1964) which states that for a many-body system the energy is a *functional* of the density alone and that the ground state density distribution is the one which minimizes this functional at constant number of particles. When the density is slowly varying, the functional may be expanded in terms of the gradients of the density. The exact form of the functional is of course unknown. According to Brueckner and collaborators (see, e.g., Brueckner *et al.* (1971) and references therein; Barkat *et al.* (1972)), we therefore approximate the energy of a nucleus by the statistical expression

$$E[\varrho_n, \varrho_p] = \int (dr)^3 \left\{ \varepsilon[\varrho_n(r), \varrho_p(r)] + \frac{e}{2} \varrho_p(r) \Phi(r) + - \frac{3}{4} \left(\frac{3}{\pi}\right)^{1/3} e^2 [\varrho_p(r)]^{4/3} + \eta [\nabla \varrho(r)]^2 \right\}. \quad (1)$$

The first term ε is a *function* of the densities and represents the energy per unit volume of *homogeneous* nuclear matter of neutron density ϱ_n and proton density ϱ_p. We have used for ε the nuclear matter results of Brueckner *et al.* (1968b), together with a recent neutron gas calculation by Buchler and Ingber (1971). The second and third terms are the direct and exchange Coulomb energies, where the Coulomb potential $\Phi(r)$ is of course a functional of ϱ_p. Finally, the last term represents the gradient correction which takes into account the change in correlation energy due to the density inhomogeneity. Here $\varrho = \varrho_n + \varrho_p$ represents the total nucleon density. As shown by Bethe (1968), the quantity η is simply related to the 'long range' part of the nuclear force and should be only weakly dependent on density. In view of the uncertainty involved in calculating η, we have chosen to treat it as a constant determined by a fit to the nuclear binding energies over the range of known nuclei. The constant η is the only phenomenological parameter of the Thomas-Fermi model and its fitted value is in good agreement with Bethe's theoretical estimate.

The problem now consists in minimizing the quantity

$$E[\varrho_n, \varrho_p] - \lambda_n \int \varrho_n(r)(\mathrm{d}r)^3 - \lambda_p \int \varrho_p(r)(\mathrm{d}r)^3, \qquad (2)$$

where the Lagrange parameters λ_n and λ_p are the neutron and proton chemical potentials, respectively, and are chosen so that

$$\int \varrho_n(r)(\mathrm{d}r)^3 = N, \quad \text{and,} \quad \int \varrho_p(r)(\mathrm{d}r)^3 = Z. \qquad (3)$$

This minimization leads to a system of non-linear equations of the form

$$2\eta \nabla^2 \varrho = \frac{\partial \varepsilon}{\partial \varrho_n} + \frac{\partial \varepsilon}{\partial \varrho_p} + e\Phi - \left(\frac{3}{\pi}\right)^{1/3} e^2 \varrho_p^{1/3} + \lambda_n + \lambda_p, \qquad (4a)$$

$$0 = \frac{\partial \varepsilon}{\partial \varrho_n} - \frac{\partial \varepsilon}{\partial \varrho_p} - e\Phi + \left(\frac{3}{\pi}\right)^{1/3} e^2 \varrho_p^{1/3} + \lambda_n - \lambda_p, \qquad (4b)$$

$$\nabla^2 \Phi = -4\pi e\varrho_p \qquad (4c)$$

with the condition that the densities $\varrho(r)$ have a continuous derivative everywhere. As a boundary condition, we require that $\mathrm{d}\varrho/\mathrm{d}r|_{r=R} = 0$, where R represents the point where the density vanishes.

The density profiles for a series of nuclei are shown in Figure 1. The ^{16}O nucleus is seen to consist mostly of surface; its central density never reaches the saturation density of nuclear matter (0.2 fm^{-3}). The heavier nuclei exhibit a dip at the center and a bulge in the outer regions due to the Coulomb repulsion between the protons. The protons also provoke a somewhat smaller bulge in the neutron density because of the symmetry energy.

The vanishing of the density rather than an exponential drop is typical of the Thomas-Fermi model and is a drawback as far as the study of the tails of the distribution are concerned, and in particular the study of the neutron halo. The proton density with our functional (1) vanishes with non-zero slope (for proton-rich nuclei it is the neutron density which vanishes with non-zero slope). This could be remedied by including in the energy functional a density difference gradient term of the form $\theta\{\nabla[\varrho_n(r)-\varrho_p(r)]\}^2$. Because of numerical difficulties, such a term has been treated as a correction to the energy rather than having been included in the minimization of the functional. Higher order terms in the gradient expansion have been omitted. The *a postiori* justification is that their contribution to the energy is negligible for the density profiles which we have obtained.

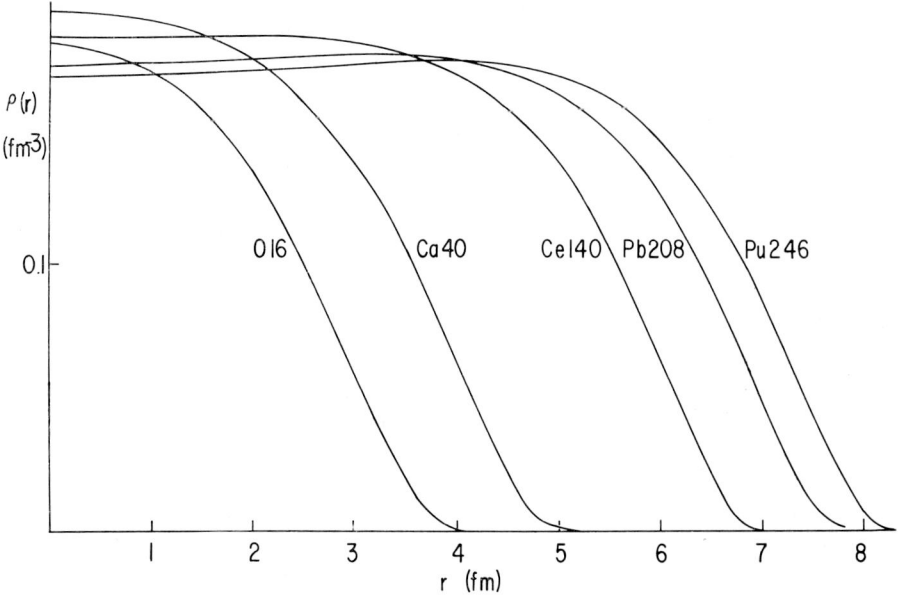

Fig. 1. Density profiles for various nuclei.

The computed binding energies per particle for the *known* spherical nuclei are in good agreement, to within 0.1 MeV (except for ^{16}O for which the discrepancy is ~0.5 MeV) with the liquid drop mass formula of Myers and Swiatecki. (Since our Thomas-Fermi model is a semi-classical approach which cannot reproduce shell effects, our results should not be compared to the *actual* binding energies of nuclei but to their 'liquid drop' part.) The proton radii turn out to be somewhat (a few percent) smaller and the surface diffuseness somewhat (5–10%) larger than the electron scattering experiments seem to indicate (Brueckner *et al.*, 1968a). The overall agreement, however, is remarkable considering that we have only *one* parameter in our theory.

Binding energy contours are shown in Figure 4 of Brueckner *et al.* (1971). In the region of known isotopes ($Z \leqslant 90$, $N \leqslant 160$) including the transuraniums, the agreement is quite good. Deviations between the liquid drop model predictions and the Thomas-Fermi results are quite significant for neutron-rich and superheavy nuclei.

Deformed nuclei have also been studied by Brueckner *et al.* (1970) with the restriction to ellipsoidal shapes. The density profiles exhibit a uniform surface diffuseness, as expected. For small deformations the binding energies are quite close to those of the liquid droplet models.

With this success of the Thomas-Fermi model at hand, we feel confident in extrapolating to the region of unknown nuclei. I will go on to describe two different applications: first, the possibility of producing superheavy nuclei in an r-process; and second, the nature of clusters of nucleons in neutron star matter.

In order to investigate the possible synthesis of superheavy elements during astrophysical processes, Brueckner *et al.* (1972) have recently incorporated the nuclear Thomas-Fermi binding energies in an r-process code by Seeger and Schramm. Their preliminary results show that, with the introduction of these Thomas-Fermi masses, significant changes occur in the results of r-process calculations. Most important, all indicate a strongly decreased probability for superheavy abundances relative to calculations based on the 'Myers-Swiatecki' mass formula. In the medium-heavy to superheavy element region, the r-process path with the energy density input is about 3 nucleon numbers closer to beta stability than the path corresponding to conventional mass formulas. This causes a pronounced shift in the isotopes of maximum abundance (the 'center line' in the r-process path). With the exception of the $N=184$ closed shell, the isotope of maximum abundance for a given Z is about 5–10 neutrons poorer than those calculated with the Myers-Swiatecki mass formula. When the r-process is terminated due to fission, there are fewer neutron-rich nuclei available to beta decay into the superheavy region, so that the overall yield of superheavies is decreased.

At present, there is a controversy about the realistic value of the parameter κ, which is the ratio of the surface symmetry coefficient to the regular surface energy coefficient. Calculations were therefore performed for both the value of Myers and Swiatecki, $\kappa=1.79$, and a larger value of $\kappa=2.84$. The differences in the results for the two values of κ are as follows: whereas with $\kappa=1.79$ one could hope for superheavy element production in the r-process, one cannot imagine this with $\kappa=2.84$. For this large value of κ the barriers are just too small to allow any high Z, neutron-rich nuclei to survive. Even for the small κ value, the abundances were smaller than those computed by Schramm and Fowler (1971). At the same temperature and density, the r-process cycle time is more than double, and the abundances of most of the isotopes of the element 112 are reduced by more than four orders of magnitude. However, it is possible to almost equal the superheavy abundances of Schramm and Fowler by increasing the neutron flux four orders of magnitude and decreasing the temperature by 10%.

The other application of the Thomas-Fermi model which I will describe is the study of neutron star matter. By neutron star matter we mean cold, neutral, and catalyzed matter; cold, because the typical thermal energy of neutron stars is several orders smaller than the chemical potentials of the constituent particles; catalyzed, because we restrict ourselves to densities for which the relaxation time to complete equilibrium with respect to strong, electromagnetic and weak interactions is small. In other words, the question we ask is the following: at a given average baryon density, what particles are contained in the system and how do they spatially arrange themselves in their ground state configuration. It is well known that at low densities ($\varrho \lesssim 10^6$ g cm^{-3}) neutron star matter consists of ^{56}Fe (or ^{62}Ni) nuclei arranged in a crystalline structure surrounded by a sea of electrons. At a density of about 10^7 g cm^{-3}, the chemical potential of the electrons reaches 1 MeV and it becomes energetically favorable for protons to beta-capture; there results a gradual neutron enrichment of the nuclei as the density of matter increases. Since too great an enrichment would excessively raise the symmetry energy, there is at first a concomitant increase in the proton number of the nuclei. The chemical potential of the neutrons becomes positive at a density of about 3.7×10^{11} g cm^{-3}, so that some of the neutrons become unbound and form an embedding sea. At a density of about 10^{14} g cm^{-3} the clusters disappear and the system becomes a homogeneous mixture of neutrons, protons, and electrons.

In previous studies of neutron star matter in the clustered phase (cf. review articles by Cameron (1970), and Canuto (1973), it has been assumed that the clusters resemble ordinary nuclei sufficiently well that they can be described by a semi-empirical mass formula, or more recently by a compressible liquid droplet model with a variable surface coefficient to take into account the reduction of surface energy due to the surrounding neutron gas (Baym et al., 1971). Such an approach, however, suffers from the disadvantage that the results one obtains are very sensitive to the specific functional dependence of the mass formula on A and Z. It also implies a dichotomy between a sharply defined nucleus and a surrounding gas. The Thomas-Fermi model, on the other hand, allows a consistent treatment of the nuclear surface by permitting the density profiles of the nuclei to change and to adjust smoothly to the embedding neutron fluid.

We will first concern ourselves with the clustered regime. Clustering occurs because the protons want to make maximum use of nuclear symmetry energy. In the spirit of the Wigner-Seitz model, Buchler and Barkat (1971a, b) therefore assume that the system is composed of identical non-interacting spherical cells which are centered on the bulges of the proton distribution and which are electrically neutral. The energy per cell is then expressed in terms of the same integral Thomas-Fermi expression as previously described [Equation (1)], but supplemented by the electrons' energy. The radius of the cells is determined by the requirement of charge neutrality, which is the new constraint, instead of (3). The Coulomb lattice energy is thus automatically taken into account to good accuracy. The minimization of the energy functional has now to be performed with respect to the neutron, proton, and electron densities, and proceeds as previously. Some of the resulting densities as a function of distance from the center of the cell are shown in Figure 2. The numbers near the center denote the chemical potential of the neutrons (in MeV), which is a monotonically increasing function of the overall matter density. The numbers in parentheses denote the radius of the cell (in fm) which shrinks with increasing density. Note a decrease of the central densities and an increase in the surface diffuseness of the clusters with increasing outside neutron density, indicating a gradual dissolution of the clusters. In Figure 3 we show

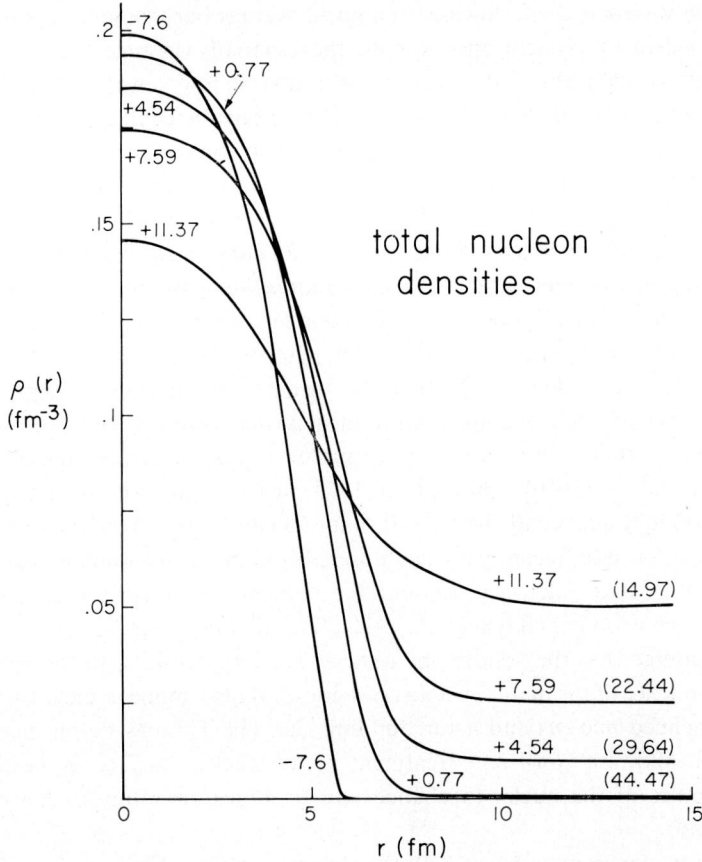

Fig. 2a. Total nucleon density distributions $\varrho(r)$ as a function of radial distance, labelled according to the corresponding neutron chemical potential λ_n. Shown in parentheses are the cell radii in fm.

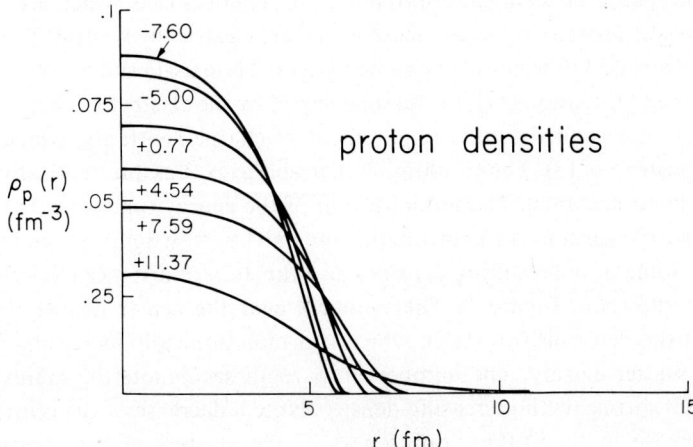

Fig. 2b. Proton density distribution $\varrho_p(r)$.

the dependence of Z, A, and of the density ϱ on the chemical potential λ_n of the neutrons. The proton number Z of the clusters first increases from $Z \simeq 29$ to a maximum of $Z \simeq 35$ and then decreases until the clusters gradually evanesce. Our calculations have not been carried beyond our last computed point because our approximation of spherical symmetry of the cells breaks down. Concurrently with us, Bethe et al. (1970) and later Baym et al. (1971) have been finding much larger nuclei (Z increasing to 151). This controversy has very recently been resolved; Ravenhall et al. (1972), using a one-dimensional (planar) model for the nuclear surface together with the sim-

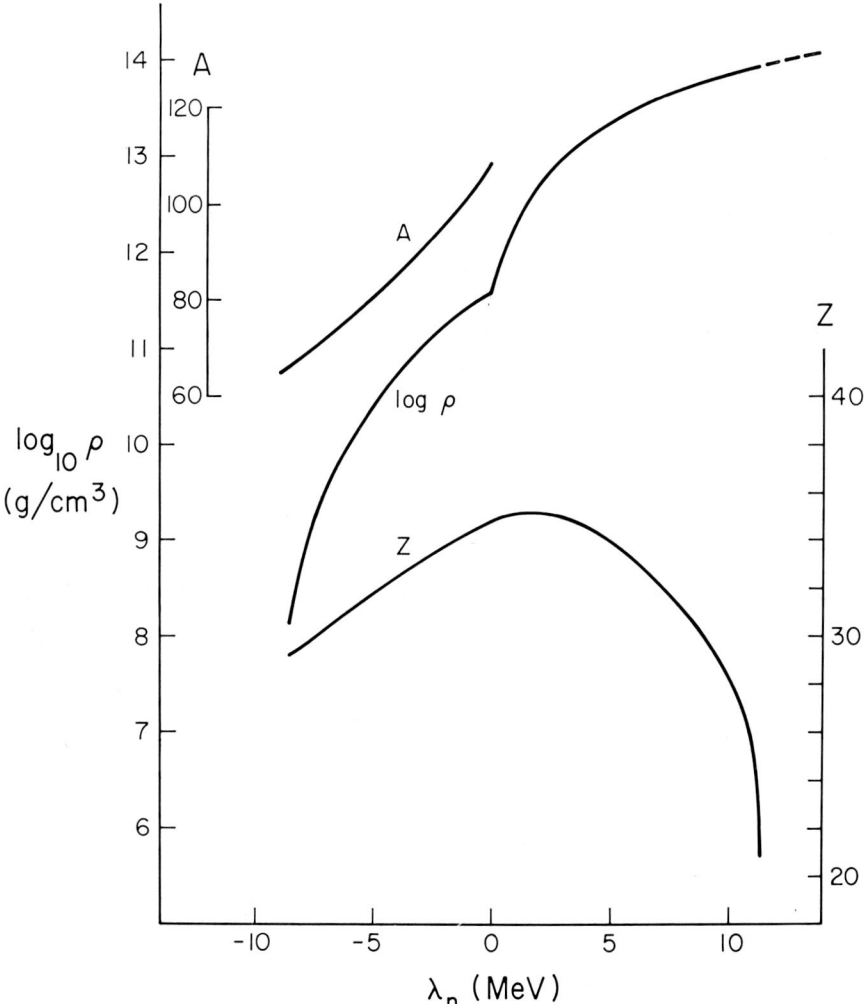

Fig. 3. Proton number Z, atomic number A of the clusters, and average $\log_{10}$ (mass density in g cm^{-3}) as a function of the neutron chemical potential λ_n. The dashed line corresponds to the homogeneous system, showing that the transition is smooth, which contrasts the 'neutron drip' phase transition ($\lambda_n = 0$) at 3.7×10^{11} g cm^{-3}.

ple Skyrme nucleon-nucleon interaction, also find very small nuclei ($Z \leqslant 38$). It is gratifying that a Hartree-Fock calculation (cf. paper by Negele in this volume), which is much more involved and time-consuming, when averaged over shell effects, gives results in good qualitative agreement with our Thomas-Fermi calculation. The small quantitative discrepancy can probably be accounted for by the use of a different nucleon-nucleon force and different nuclear matter results on which both approaches rely.

At higher densities, the clusters have dissolved and we have a homogeneous system of neutrons, protons, and electrons. The various physical quantities, like the energy per particle, the chemical potential of the protons, and the average neutron and proton densities join smoothly with the corresponding quantities of the homogeneous phase, as the calculations by Buchler and Ingber (1971) indicate. Although the exact nature of the phase transition is still uncertain, numerically it seems to be very smooth in contrast to the 'neutron drip' transition which occurred at 3.7×10^{11} g cm^{-3} ($\lambda_n = 0$). In addition to the neutrons, protons, and electrons, muons also appear in the system at a density of 2.3×10^{14} g cm^{-3}, i.e., slightly before nuclear matter density (3.4×10^{14} g cm^{-3}). The percentage of protons increased from about 2.25% at 10^{14} g cm^{-3} to 5.5% at 4×10^{14} g cm^{-3}. It should be pointed out that these numbers are very sensitive to the symmetry energy of the nuclear matter calculations and the nucleon-nucleon potential used in its computation.

Shell effects have necessarily been ignored in our statistical approach, but it is likely that they may produce shifts and jumps in Z. Since a direct confrontation with observation is not possible, the final test for the Thomas-Fermi calculation will be a good quantum mechanical treatment. However, even a reliable Brueckner-Hartree-Fock calculation, because of its far greater complexity, is barely feasible at present.

In summary, the Thomas-Fermi model is capable of reproducing the bulk properties of a wide range of nuclei. It has the advantage of being based on realistic nuclear matter calculations and, therefore, can easily be kept *à jour* with improvements in the nucleon-nucleon potential. We feel that the reliability of its predictions in *r*-process calculations and in the properties of neutron star matter by far exceeds that of conventional semi-phenomenological liquid-drop formulas.

This concludes our brief review of the nuclear Thomas-Fermi model and of some of its applications of astrophysical interest.

We wish to thank the authors of Brueckner *et al.* (1972) for letting us report their results prior to publication. The support of NSF, NASA and the Luxembourg Ministère des Affaires Culturelles is gratefully acknowledged.

References

Barkat, Z., Buchler, J. R., and Ingber, L.: 1972, *Astrophys. J.* **176**, 723.
Baym, G., Bethe, H. A., and Pethick, C. J.: 1971, *Nucl. Phys.* **A175**, 225.
Bethe, H. A.: 1968, *Phys. Rev.* **167**, 879.
Bethe, H. A., Börner, G., and Sato, K.: 1970, *Astron. Astrophys.* **7**, 279.
Brueckner, K. A., Buchler, J. R., Clark, R. C., and Lombard, R. J.: 1968a, *Phys. Rev.* **181**, 1543.

Brueckner, K. A., Coon, S. A., and Dabrowski, J.: 1968b, *Phys. Rev.* **168**, 1184.
Brueckner, K. A., Clark, R. C., Lin, W. F., and Lombard, R. J.: 1970, *Phys. Rev.* **C1**, 249.
Brueckner, K. A., Chirico, J. H., and Meldner, H. W.: 1971, *Phys. Rev.* **C4**, 732.
Brueckner, K. A., Chirico, J. H., Jorna, S., Meldner, H. W., Schramm, D. N., and Seeger, P. A.: 1972 (in preparation).
Buchler, J. R. and Barkat, Z.: 1971a, *Astrophys. Letters* **7**, 169.
Buchler, J. R. and Barkat, Z.: 1971b, *Phys. Rev. Letters* **27**, 48.
Buchler, J. R. and Ingber, L.: 1971, *Nucl. Phys.* **A170**, 1.
Cameron, A. G. W.: 1970, *Ann. Rev. Astron. Astrophys.* **8**, 179.
Canuto, V.: 1973, *Ann. Rev. Astron. Astrophys.* (in press)
Hohenberg, P. C., and Kohn, W.: 1964, *Phys. Rev.* **136**, B864.
Ravenhall, D. G., Bennett, C. D., and Pethick, C. J.: 1972, *Phys. Rev. Letters* **28**, 978.
Schramm, D. N. and Fowler, W. A.: 1971, *Nature* **231**, 103.

DISCUSSION

Bethe: The decision between our theories, in my opinion, *has* been made by the work reported by Dr Negele in the first paper of this conference. At low densities, of order $10^{12}-10^{13}$, Z is 40 to 50, closer to the original value of Baym, Bethe and Pethick than to yours. At *high* density, near 10^{14}, the value of Z decreases, as you have predicted, but not as strongly.

Buchler: I would like to refer back to the graph in Negele's talk which compared the Z values for the different approaches. Negele's results, to me at least, seem to be in good qualitative agreement with ours. The discrepancy, I believe, could arise from the use of different nuclear matter calculations by him and by us; in particular, the arbitrary multiplication factor used by Negele to bring the nuclear binding energy in agreement with the currently accepted experimental value.

Bethe: The work of Negele and Vautherin, published in the *Physical Review* (C5), has shown that it is not sufficient to describe nuclear matter by ϱ and $\nabla^2 \varrho$ as you (and earlier Thomas-Fermi theories) have done, but that one has to add a third quantity, the kinetic-energy density

$$t = \sum_k (\nabla \phi_k)^2$$

which is independent of the first two. They have shown how this follows from a careful consideration of the mixed density $\varrho(\mathbf{r}_1, \mathbf{r}_2)$ which is very important for nuclear matter energy.

Buchler: As far as the second question is concerned, the Thomas-Fermi inhomogeneity correction to the kinetic energy, which arises purely from antisymmetrization of the wavefunction, is the well-known Weizsäcker correction

$$\frac{\hbar^2}{72\,M} \left[\frac{(\nabla \varrho_\mathrm{n})^2}{\varrho_\mathrm{n}} + \frac{(\nabla \varrho_\mathrm{p})^2}{\varrho_\mathrm{p}} \right] \equiv \frac{\hbar^2}{72\,M} \frac{(\nabla \varrho)^2}{\varrho}.$$

It plays essentially the same role as the term you mentioned. Its numerical effect has been found to be small and we feel that a 'renormalization' of our parameter η should adequately take care of it. We realize that the Thomas-Fermi model is a very simple model and has its limitations, but its very simplicity and success make it very attractive.

'STATISTICAL BOOTSTRAP' EQUATION OF STATE FOR COLD ULTRA-DENSE MATTER

J. CRAIG WHEELER

Harvard College Observatory, Cambridge, Mass., U.S.A.

Abstract. The statistical bootstrap theory of hadrons predicts a particle level density which increases with mass like $\varrho(m) = cm^a \exp bm$. The motivation for this level density is explored and then it is used to derive an equation of state for zero-temperature ultra-dense ($> 10^{17}$ g cm^{-3}) matter. The nature, uses, and limitations of the equation of state are discussed.

1. Introduction

Developments in the theory of particle physics over the last several years have had surprising implications concerning the multiplicity of particle states at energies far beyond the reach of present day experiments. A natural impulse is to apply this new found knowledge to physically interesting situations involving high temperatures and/or high densities. In particular, knowledge of the variety of particle species and their interactions under extreme conditions can be used to construct an equation of state applicable to the early stages of the universe in the context of a big bang cosmology or to the structure and stability of hadron stars.

This paper is oriented toward conditions of high density and negligible temperature and so is most directly concerned with the subject of hadron stars. While the equation of state itself follows straightforwardly from two key assumptions regarding the density of particle states and the manner in which their interactions are treated, the interesting physics is involved in the motivation of these assumptions and the degree of their validity. Thus these assumptions will be explored here even though they are of a much more general nature than the specific topic of the equation of state for cold ultra-dense matter. The derivation of the equation of state will then be sketched and its nature, uses, and limitations will be discussed.

2. The Density of Particle States

On aesthetic grounds one might like to think that there were some finite number of fundamental building blocks out of which all strongly interacting particles (hadrons) could be constructed. In fact some recent work indicates a quite different situation. Rather than there being any truly fundamental particles the hadrons may be their own constituents. In this picture not only will hadrons not be broken down into a few prime constituents at high energies but, in addition, the number of different particles will increase exponentially with mass-energy.

That the density of particle states would behave like

$$\varrho(m) \equiv \frac{dn}{dm} \underset{m \to \infty}{\longrightarrow} cm^a e^{bm}, \tag{1}$$

was first predicted by Hagedorn (1965). Hagedorn used a bootstrap type argument of the sort illustrated below in which it is supposed that each hadron is composed of other hadrons in some statistical sense. An independent suggestion that something like (1) is true came with the establishment of the concept of duality – that the Regge-pole contribution is in itself an average description of the full scattering amplitude. Dolen et al. (1968) argued that Regge-pole contributions could not be added coherently to resonance contributions to the scattering amplitude, as in the interference model, without resulting in multiple counting. Duality says that resonances and Regge-poles are related by finite-energy sum rules so that one formalism is equivalent to the other.

Krzywicki (1969) and Fubini and Veneziano (1969) used duality models to derive the degeneracy of particle states as a function of energy. They found the same result as did Hagedorn, the number of particle states increases drastically with mass, agreeing asymptotically with (1). For a recent review of duality and the Veneziano model see Kaĭdalov (1972).

The derivation of the level density based on statistical bootstrap arguments is conceptually much simpler than that based on duality models and so will be used here to illustrate the origin of (1). The following discussion closely follows that of Frautschi (1971).

The current crude bootstrap model assumes that an effective potential serves to confine the constituents of a hadron in a volume with radius of order 10^{-13} cm. Inside this box the constituents are presumed to circulate freely. The density of levels in the box is then identified with the density of hadron states. This treatment assumes that all attractive interactions are accounted for in a statistical sense by counting all the resonance states in the box. In reality we know that a pair of particles in the box will interact *via* some potential. For an attractive potential, as the particles get closer their wave function oscillates more rapidly and for every extra oscillation corresponding to a phase shift of 180° the effect is to create another state in the box. Thus while an exact treatment would count states of motion of all the particles in their mutual potential, an approximate treatment is to count states of motion of the original particles plus the states of motion of all resonances, all considered to be non-interacting. Just such a treatment is shown to be rigorously true in certain idealized cases by Dashen et al. (1969) and Veneziano (1971). They show that narrow resonances in the scattering amplitude are manifested as free particles in the corresponding statistical ensemble.

What of repulsive interactions? Leung and Wang (1971) outline a possible theoretical argument that indicates that any repulsive baryon-baryon interaction would be roughly cancelled by a corresponding attractive interaction. The idea is that while interactions mediated by mesons of the ω family would be repulsive, interactions would also occur with coupling constants of the same order mediated by members of the f family which would provide an attractive interaction. Thus Leung and Wang argue that on the average not only repulsive but all baryon-baryon interactions might be ignored in certain cases.

In a more general way Hamer and Frautschi (1971) argue that repulsive interactions just serve to eliminate possible resonances from the system leaving the attractive interactions to generate the multiplicity of resonances which dominate the physics of the situation. More specifically, they point out that there are no limits to the rate of increase with energy of the number of states which attraction can generate whereas, while repulsion can serve to decrease the number of states up to a point, causality puts a finite limit on the rate of decrease. Therefore as long as channels exist which provide attractive interactions there will be an exponential increase in the density of states which will swamp any effect of repulsion in other channels.

Thus the assumption that the hadron constituents can be treated as moving freely in a box as long as all possible states are counted can be seen to be a fair first approximation. This model should be particularly satisfactory under proper conditions, for instance a very dilute collection of hadrons. On the other hand, in the context we have in mind, dense zero temperature hadron matter, we will require the hadrons to be closely packed and baryons to be degenerate. The possible effects of interactions will then have to be re-examined.

The problem, therefore, is to compute all the states in the box which composes the hadron volume. The number of states of one particle in a box of volume V with momentum between p and $p+dp$ is

$$\frac{V \, d^3 p}{h^3}. \qquad (2)$$

By extension, the density of states for n independent particles with total energy m is

$$\varrho_n(m) = \left(\frac{V}{h^3}\right)^{n-1} \prod_{i=1}^{n} \int d^3 p_i \delta\left(\sum_{i=1}^{n} E_i - m\right) \delta^3\left(\sum_{i=1}^{n} \mathbf{p}_i\right). \qquad (3)$$

The momentum delta function and the omitted factor of V/h^3 correspond to taking the center of mass at rest.

For the bootstrap model of hadrons, the hadrons are assumed to be compounds of other hadrons, for instance, schematically,

$$\pi = (\pi\pi + \pi K + KK + \cdots) + (\pi\pi\pi + \pi\pi K + \cdots) + \cdots.$$

To compute the level density corresponding to this case two conditions must be met. The statistical condition is that the level density of hadrons $\varrho_{out}(m)$ is given by the phase space volume of an arbitrary number of noninteracting internal constituents confined in the volume V which have level density $\varrho_{in}(m_i)$. This level density is

$$\varrho_{out}(m) = \sum_{n=2}^{\infty} \frac{1}{n!} \left(\frac{V}{h^3}\right)^{n-1} \prod_{i=1}^{n} \int d^3 p_i \int dm_i \varrho_{in}(m_i) \times$$

$$\times \delta\left(\sum_{i=1}^{n} E_i - m\right) \delta^3\left(\sum_{i=1}^{n} \mathbf{p}_i\right). \qquad (4)$$

Here the integral over mass sums the contribution to the phase space volume of each mass interval. The factor $1/n!$ eliminates the effect of double counting (cf.

Frautschi, 1971). Equation (4) is incorrect in that it does not take into account the Pauli principle and hence two or more fermions are allowed in the same state. Such states are statistically insignificant and do not contribute to the asymptotic level density in the non-degenerate case. The effects of the Pauli principle on the level density have not yet been included in the computation of the present equation of state. The possible effects of this correction will be discussed.

The second condition necessary to compute the hadron level density is the bootstrap condition, that the internal constituents are just the hadrons themselves, i.e.

$$\varrho_{\text{out}}(m) = \varrho_{\text{in}}(m). \tag{5}$$

Frautschi originally applied this relation only in the asymptotic region of large m. Hamer and Frautschi (1971) call that relation the 'strong asymptotic bootstrap condition.' They discovered that (5) could be applied at *all* masses above a certain threshold which they take to be the two pion threshold in the simplest case. Hamer and Frautschi call (5) the 'strong bootstrap condition.' The 'weak asymptotic bootstrap condition' is that originally used by Hagedorn (1965) where ϱ_{out} is consistent with ϱ_{in} only to within a power of the mass.

Frautschi (1971) then shows that the solution to (4) and (5) is of the form

$$\varrho = cm^a e^{bm}. \tag{6}$$

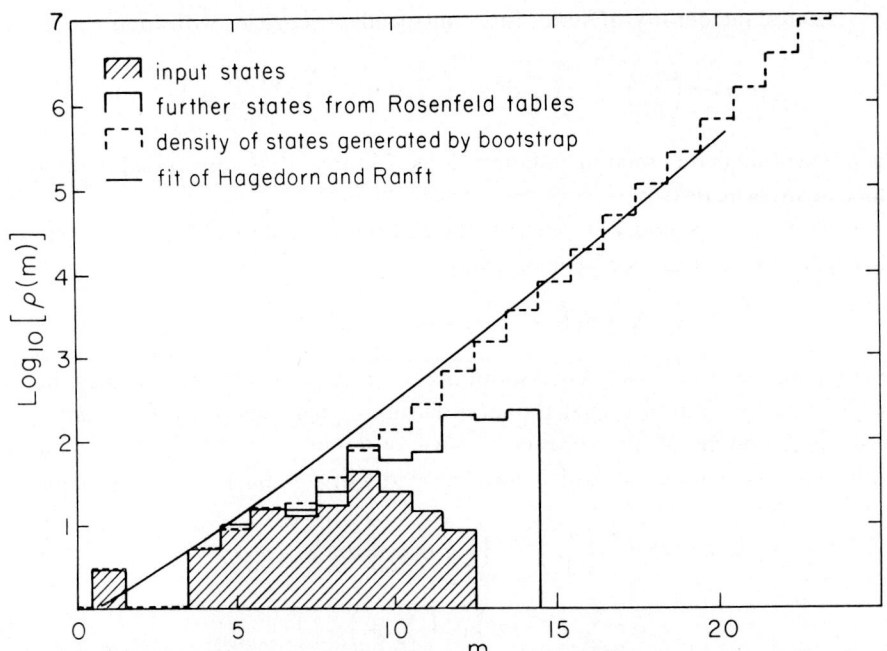

Fig. 1. The bootstrap density of all hadron states compared with experiment. Also shown is the fit of Hagedorn and Ranft (1968). This figure is from Hamer and Frautschi (1971). (Reprinted with the kind permission of the authors and *Physical Review*.)

He also shows that if one considers only hadrons of a particular baryon number B, change Q, strangeness S, etc., the parameters a and b are unique, i.e.,

$$\varrho_{BQS...} = c_{BQS...} m^a e^{bm}. \tag{7}$$

Figure 1 shows the results of the statistical bootstrap model as computed by Hamer and Frautschi (1971) and the theoretical solution as given by Hagedorn and Ranft (1968). The experimental data can not yet distinguish an exponential mass spectrum from a power law. With a level density of the form of (6) Hamer and Frautschi can fit their results down to $m \sim 10\, m_\pi$.

The various theories give estimates of the values of the parameters as indicated in in Table I. Frautschi (1971) finds that the requirement of a consistent solution to (4) and (5) imposes an upper limit to a,

$$a < -5/2. \tag{8}$$

Frautschi's analysis shows that it is not necessary to discuss thermodynamics in order to derive the level density; however, Hagedorn's (1965) thermodynamic analysis indicates that for local thermodynamical equilibrium to hold in high energy collisions a lower limit is placed on a,

$$a > -7/2. \tag{9}$$

TABLE I

Parameter values for the statistical bootstrap level density,
$\varrho(m) = cm^a \exp bm$

$a > -7/2$	Hagedorn
$a < -5/2$	Frautschi
$a = -5/2, -3, -7/2 ...$	Veneziano Model
$a \simeq -3$	Hamer and Frautschi
$a = -3$	Nahm
$b^{-1} \sim 160$ MeV	Hagedorn
$b^{-1} \sim 174$ MeV	Veneziano Model ($a = -3$)
$b^{-1} \sim m_\pi \sim 140$ MeV	Hamer and Frautschi
$c \sim (0.9\text{ BeV})^{3/2}$	Hagedorn ($a = -5/2$)
$c \sim (1\text{--}4)\, m_\pi^2 \sim (0.2\text{ BeV})^2$	Hamer and Frautschi ($a = -3$)

The Veneziano model (Fubini and Veneziano, 1969; Bardakci and Mandelstam, 1969; Huang and Weinberg, 1970) allows a series of discrete values

$$a = -5/2, -3, -7/2.... \tag{10}$$

Consistent with all these results Hamer and Frautschi (1971) compute that the value of a does not differ significantly from $a = -3$ and report that W. Nahm has established this unique value analytically by extending the work of Frautschi.

Hamer and Frautschi have shown that if *all* exotic resonances ($|B| > 1$) are included in the bootstrap spectrum the value of a decreases by one half for each extra internal quantum number included, i.e., Q, S etc. In the present work all exotics will be excluded.

The value of b was first estimated by Hagedorn (1968) and Hagedorn and Ranft

(1968) to be

$$b^{-1} \sim 160 \text{ MeV}. \tag{11}$$

They obtained this number by noticing that the transverse momentum distribution of secondaries in very high energy collisions goes as $\exp(-|p_\perp|/160 \text{ MeV})$. Hagedorn's statistical model interprets the constant in the denominator as b^{-1}. The Veneziano model gives $b^{-1} = 174$ MeV for $a = -3$ (cf. Huang and Weinberg, 1970) and Hamer and Frautschi (1971) find, as expected on dimensional grounds, $b^{-1} \sim m_\pi \simeq 140$ MeV with the value roughly linearly proportional to the radius of the hadron volume assumed.

Hagedorn (1967) estimates a value of the parameter c for the complete hadron spectrum of $(0.9 \text{ GeV})^{3/2}$ for a value of $a = -5/2$. He does this by fitting the derived level density to the known hadron distribution, as far as it goes. Hamer and Frautschi find a value of $c \sim m_\pi^2$ for $a = -3$ from their numerical computations.

The statistical bootstrap model outlined here has some advantages over the Veneziano model. The latter strictly treats only multi-particle interactions among mesons while the bootstrap model incorporates baryons as well. The bootstrap model also gives an interesting constraint on the parameters a, b, c, V namely

$$1 = \sum_{n=2}^{\infty} \frac{x^{n-1}}{(n-1)!}, \tag{12}$$

where

$$x = \frac{(2\pi)^{3/2} cVm_\pi^{a+5/2}}{-(a+5/2) h^3 b^{3/2}}. \tag{13}$$

The solution to (12) is just $x = \ln 2$ so that

$$V = \frac{\ln 2 h^3 b^{3/2} m_\pi^{1/2}}{2(2\pi)^{3/2} c}, \tag{14}$$

for $a = -3$. Equation (14) implies that some of the parameters of the model will change if the volume of the hadron is altered as it could conceivably be under conditions of high density. In fact Hamer and Frautschi find that, in general, $a = $ constant but that

$$b \propto V^{1/3}, \tag{15a}$$

$$c \propto V^{(a+1)/3}, \tag{15b}$$

for given input physics. The nth term in (12) gives the probability that a given break-up channel will have n components. Thus 69% of the resonances will couple to only two components, 24% to three components and 7% to four or more components, with the average number being 2.4.

Chiu and Heimann (1971) have incorporated the effects of spin on the hadron spectrum. For the most part this extra sophistication does not alter the basic results of the present work.

3. The Equation of State

With the level density of hadron states at hand, it is straightforward to calculate the equation of state. To be consistent we must assume that hadron interactions in the macroscopic physical system of interest can be treated in the same manner as were the internal constituents of each hadron. That is, we must assume that the level density implicitly accounts for all hadron interactions and hence that the hadrons can be formally treated as free particles. The validity of the level density of (6) or (7) in the context in which it is used here will be examined later.

In general the partition function for a non-interacting ensemble of particles whose mass distribution is given by the level density $\rho(m)$ is (taking $c=k=1$),

$$\ln Z = \frac{4\pi V}{h^3} \left[\int \varrho_{\text{bosons}}(m) \, dm \int \ln\left(\frac{1}{1-\Lambda}\right) p^2 \, dp + \int \varrho_{\text{fermions}}(m) \, dm \int \ln(1+\Lambda) p^2 \, dp \right], \quad (16)$$

where

$$\Lambda = \exp\{[\mu - (p^2 + m^2)^{1/2}]/T\}. \quad (17)$$

The parameters of the equation of state, pressure, number density and mass-energy density are obtained from the partition function,

$$P = T \frac{\partial \ln Z}{\partial V}\bigg|_{T,\mu}, \quad (18)$$

$$n_i = T \frac{\partial \ln Z}{\partial \mu_i}\bigg|_{T,V}, \quad (19)$$

$$\varepsilon = T^2 \frac{\partial \ln Z}{\partial T}\bigg|_{V,\mu} + \mu T \frac{\partial \ln Z}{\partial \mu}\bigg|_{T,V} + \int \varrho_{\text{bosons}}(m) \, n(m) \, m \, dm. \quad (20)$$

Note that for temperatures much greater than the Fermi energy all particles will obey Maxwell-Boltzmann statistics so that using the level density of (6),

$$\ln Z \propto \int_0^\infty m^{3/2} \varrho(m) \, e^{-m/T} \, dm \propto \int_m^\infty m^{3/2+a} e^{(b-1/T)m} \, dm. \quad (21)$$

Thus the partition function does not converge and thermodynamics is not valid if $T > b^{-1}$. This is the source of Hagedorn's (1965) by now famous statement that the exponential level density implies a finite limiting temperature to any thermodynamical system of order 160 MeV. As energy is put into the system it goes into the rest mass of new particles rather than into the kinetic energy of existing particles.

For application to hadron star matter we must make the opposite assumption, namely that the temperature is much less than the Fermi energy, which we idealize by specifically assuming $T=0$. In this case we find

$$P = \frac{\pi}{6h^3} \int_0^\infty \varrho_{\text{fermion}}(m) \, m^4 f(x) \, dm, \quad (22)$$

$$\varepsilon = \int_0^\infty \varrho_{\text{bosons}}(m)\, n(m)\, m\, dm + \frac{\pi}{6h^3}\int_0^\infty \varrho_{\text{fermions}}(m)\, m^4 g'(x)\, dm, \qquad (23)$$

where

$$x(m) = \frac{[\mu^2(m) - m^2]^{1/2}}{m}, \qquad (24)$$

and f and g' are the standard Chandrasekhar functions (the g' contains the rest mass contribution to the energy density),

$$f = x(1+x^2)^{1/2}(2x^2 - 3) + 3\ln[x + (1+x^2)^{1/2}], \qquad (25)$$
$$g'(x) = g(x) + 8x^3 = 3x(1+x^2)^{1/2}(2x^2 + 1) - 3\ln[x + (1+x^2)^{1/2}]. \qquad (26)$$

The formalism of Ambartsumyan and Saakyan (1960) then gives the chemical potentials as

$$\mu_{\text{lepton}} = \mu_{e^-} = \mu_{\mu^-} = m_\pi \qquad (27)$$
$$\mu_{\text{meson}} = -m_\pi Q_{\text{meson}} \qquad (28)$$
$$\mu_{\text{baryon}} = \mu_{BQ} = \mu_n B_{\text{baryon}} - m_\pi Q_{\text{baryon}} \qquad (29)$$

in terms of the charge (Q) and baryon number (B) where the chemical potential of the neutron is the same as for all $B=1$, $Q=0$ particles. Since baryons with mass greater than their chemical potential μ_{BQ} are prohibited and since there is a lowest mass particle for given B and Q, the integrals for the equation of state variables split into a sum over B and Q of integrals with finite limits. The lepton contribution is fixed at a negligible level and dropping it allows algebraic elimination of the meson contribution giving the following expressions for pressure, energy density and baryon number density;

$$P = \frac{\pi}{6h^3}\Sigma_{B,Q}\int_{m^0_{BQ}}^{\mu_{BQ}} \varrho_{BQ}(m)\, m^4 f(x)\, dm, \qquad (30)$$

$$\varepsilon = \frac{4}{3}\frac{\pi m_\pi}{h^3}\Sigma_{B,Q}Q \int_{m^0_{BQ}}^{\mu_{BQ}} \varrho_{BQ}(m)\, m^3 x^3\, dm +$$

$$+ \frac{\pi}{6h^3}\Sigma_{B,Q}\int_{m^0_{BQ}}^{\mu_{BQ}} \varrho_{BQ}(m)\, m^4 g'(x)\, dm, \qquad (31)$$

$$n_B = \frac{4\pi}{3h^3}\Sigma_{B,Q}B\int_{m^0_{BQ}}^{\mu_{BQ}} \varrho_{BQ}(m)\, m^3 x^3\, dm. \qquad (32)$$

Using the level density

$$\varrho_{BQ} = c_{BQ} m^a e^{bm}, \qquad (33)$$

gives a 'peak' in the integrands as the combined result of the exponential rise and the cutoff at μ_{BQ}. Thus the main contribution to the integrals always comes from masses within about b^{-1} of the cutoff. The ratio of the (Fermi) kinetic energy to the mass is then of order $1/b\mu_{BQ}$. The smallest μ_{BQ} can be is about m_n in which case $1/b\mu_{BQ} \sim 6$. Assuming that μ_{BQ} is always much greater than b^{-1} implies the particles will always be non-relativistic which means $f(x) \simeq 8/5 \, x^5$ and $g'(x) \simeq 8 \, x^3$.

The 'peak' in the integrands also allows the integrals to be approximated by the asymptotic value of a modified Bessel function of argument $b\mu_{BQ}$ which is again taken to be $b\mu_{BQ} \gg 1$.

Equations (30–32) then become

$$P = \frac{1}{b\mu_n} \Sigma_{B,Q} \frac{\varepsilon_{BQ}}{B}, \tag{34}$$

$$\varepsilon = \Sigma_{B,Q} \varepsilon_{BQ}, \tag{35}$$

$$n_B = \frac{\varepsilon}{\mu_n}, \tag{36}$$

where the partial density $\varepsilon_{B,Q}$ is given by

$$\varepsilon_{B,Q} = \frac{(2\pi)^{3/2}}{h^3} \mu_n c_{BQ} B b^{-5/2} \mu_{BQ}^{3/2+a} \exp b\mu_{BQ}. \tag{37}$$

The exponential

$$e^{b\mu_{BQ}} = e^{b(\mu_n B - m_\pi Q)} \tag{38}$$

dominates ε_{BQ} and so the class of baryons with highest B will dominate the sum in (35) giving

$$\varepsilon \simeq \Sigma_Q \varepsilon_{B_{max},Q}. \tag{39}$$

Using Hagedorn's (1965) interpretation of b^{-1} as the limiting temperature T_0 then gives

$$P = n_B T_0, \tag{40}$$

$$= \frac{\varepsilon}{\mu_n} T_0, \tag{41}$$

taking, in the absence of any evidence to the contrary, $B_{max} = 1$. See Wheeler (1971) for the details of this derivation.

The form of (40a) is not surprising since the assumptions made here imply that noninteracting nonrelativistic particles with Fermi kinetic energy $\sim b^{-1} = T_0$ dominate the equation of state. Note from (37) and (38) that the chief dependence is $\varepsilon \propto e^{\mu_n}$ so that asymptotically (40b) will become

$$P \propto \frac{\varepsilon}{\ln \varepsilon}. \tag{42}$$

This result has been derived independently by Hagedorn according to Rhoades and

Ruffini (1971). A thorough discussion of the physics associated with the exponential hadron level density has been given by Lee et al. (1971) followed by discussions of the zero temperature case by Leung and Wang (1971, 1972). See also the contribution by Wang in this volume.

In particular by taking $a = -3$, neglecting $m_\pi Q$ with respect to μ_n in the polynomial of μ_{BQ} and eliminating μ_n by (40b), the expression for the equation of state becomes

$$P = \frac{(2\pi)^{3/2}}{h^3} \frac{T_0^2}{(\varepsilon/P)^{3/2}} c_{\text{eff}} e^{\varepsilon/P}, \tag{43}$$

where

$$c_{\text{eff}} = \Sigma_Q c_{1,Q} e^{-m_\pi Q}, \tag{44}$$

which has dimension (mass)^2.

Hamer and Frautschi (1971) find a value for the parameter c of between $(1-4) m_\pi^2$. A value for c_{eff} of m_π^2 implies that $\varepsilon \sim 5 \times 10^{14}$ g cm^{-3}, a typical nuclear density, when $\mu_n = m_n$.

Using $c_{\text{eff}} = T_0^2 = m_\pi^2$, the approximate numerical representation of (43) is

$$P \simeq 5 \times 10^{33} (\varepsilon/P)^{-3/2} \exp(\varepsilon/P) \text{ dyne cm}^{-2}, \tag{45}$$

where for convenience of expression the units of ε are taken to be erg cm^{-3}. Figure 2

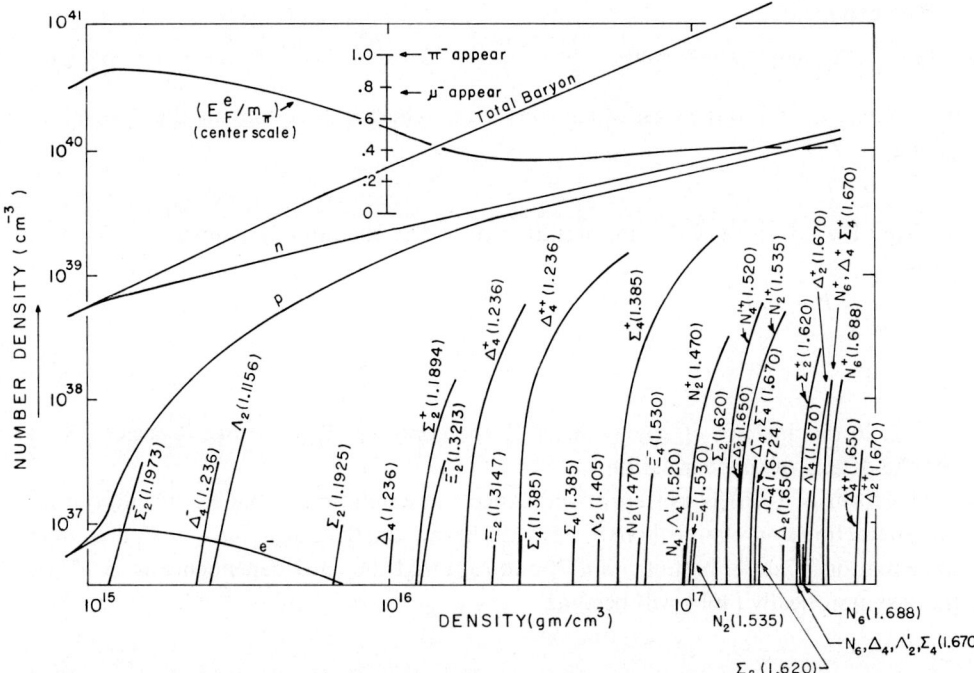

Fig. 2. The number densities of all known baryons with mass less than 1.7 BeV as a function of density when treated as independent particles, as given by Leung and Wang (1971). (Courtesy C. G. Wang.)

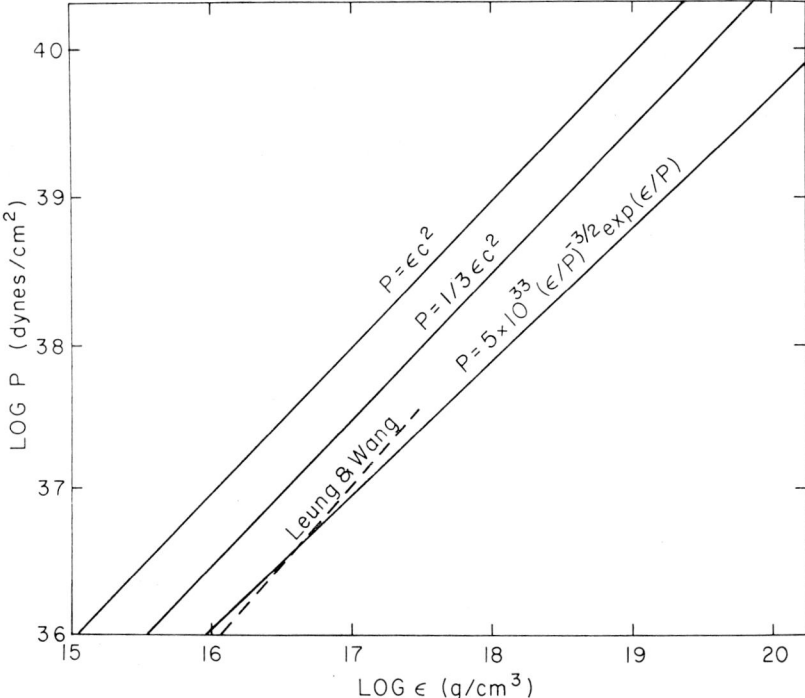

Fig. 3. The statistical bootstrap equation of state is shown in comparison with two common limiting cases, that for free relativistic particles ($P = 1/3\ \varepsilon$), and the causality limit ($P = \varepsilon$). Also shown are the numerical results of Leung and Wang (1971) based on the known hadron spectrum.

shows the number densities of the known resonances as a function of density as given by Leung and Wang (1971) who incorporated all known particles in a noninteracting gas in the spirit of the statistical bootstrap treatment. Equation (45) is plotted in Figure 3 along with the results of the calculations of Leung and Wang (1971).

As mentioned previously Hamer and Frautschi (1971) find that the form of the level density in (6) gives a good fit at masses as low as $10\ m_\pi$, corresponding to a density of about 10^{17} g cm^{-3}. Leung and Wang (1972) have come to the same conclusion in fitting the known hadron spectrum with a power law. Thus, while there may be criticisms of the applicability of the level density of (6) at any density, there is no question that the equation of state (45) becomes inapplicable at densities less than about 10^{17} g cm^{-3}. Fortunately, this is close to the regime where the known resonances can be employed as indicated in Figure 2 and hence the present analytic result can be used to extrapolate the numerical study of Leung and Wang (1971). Some adjustment, for instance of the parameter c_{eff}, might smooth the extrapolation. No such attempt has been made at present.

4. Discussion

The direct use of the equation of state derived here is limited by its region of applicability. Although there are still uncertainties in the details, it is generally agreed that the maximum central density expected in stable hadron stars is about 10^{15} g cm^{-3}. Thus a central density greater than 10^{17} g cm^{-3} would imply a mass far greater than the stability limit.

The value of the statistical bootstrap concept to hadron stars is that it provides a high density boundary condition on the equation of state. The equation of state at sub-nuclear densities is by now fairly well established (cf. the contributions by Negele and Buchler in this volume). Unfortunately the region around the critical central density where the maximum allowable mass of a hadron star occurs is still plagued by insufficient knowledge of the nuclear physics. Thus until these problems are solved the best one can do is to extrapolate to possible equations of state at much higher densities. The statistical bootstrap equation of state gives a serious alternative high density limit to be considered along with those used in the past, mainly on the basis of simplicity, i.e. $p = 1/3\ \varepsilon$ for a strictly non-interacting gas of relativistic particles or $p = \varepsilon$, the causality limit. Both the latter limits tend to imply a relatively 'stiffer' equation of state than the statistical bootstrap case in the region around the critical density.

Leung and Wang (1971) have used their bootstrap equation of state to extrapolate in a straightforward way from sub-nuclear densities and have computed a maximum neutron star mass of around one half solar mass. Their result is all the more intriguing in light of the fact that a recent estimate of a neutron star mass, that of Wilson (1972) for the X-ray pulsar Centaurus X-3, is very low, one tenth of a solar mass. On the other hand Börner, in this volume, argues that the mass of the Crab pulsar is $\sim 1.2\ M_\odot$. In any case the extrapolation of the equation of state is not unique since a 'kink' in the equation of state at just the right density due to some quirk of the nuclear potential might raise the limiting neutron star mass considerably from the result of Leung and Wang and still be consistent with the statistical bootstrap results at higher densities. The limiting mass of a neutron star is ultimately a problem for nuclear physics, not hadron physics.

Some cosmologists, noting the difficulty in forming galaxies in the context of a hot big bang model have considered a cold big bang which expands from the singularity at zero temperature (cf. Layzer, 1969). The equation of state discussed here would seem to be relevant in such a situation.

The work outlined here can only be regarded as a first step. Many potential difficulties can be discovered. A few of them and their possible effects will be mentioned.

As noted previously the expression for the level density (4) does not incorporate the Pauli principle and is thus actually inappropriate to problems dealing with a degenerate Fermi gas. Frautschi (1971) has checked that inclusion of the Pauli principle does not significantly alter the asymptotic expression for the level density for cases where degeneracy is not important. Thus there is certainly hope that the exponential level density which is the mainstay of the present work will be retained when the degenerate case is handled more rigorously.

Sawyer (1972) has discussed one possible effect of the Pauli principle. This is that due to the high Fermi level of its decay products the mass of a given resonance may be shifted upward from the mass determined in the laboratory where the resonance is free to decay with no phase space restrictions. Leung and Wang (1972) consider this point and argue that the exponential level density could still be expected to result although perhaps based on an 'effective' baryon mass spectrum. The basic idea is that, although the mass of each resonance might be shifted upward by some fraction, this effect would not have any ultimate bearing on the basic form of the level density since an exponential increase will always eventually dominate a geometric increase. The question of whether the level spacing between baryons, i.e., T_0, could change with density is examined in a somewhat different context below.

Another serious point is that in practice the hadron spatial distribution is anything but that of a dilute gas. Frautschi et al. (1972) have worked out the effective volume of the hadrons in the limit as $T \to T_0$ and find that it is somewhat larger than the actual volume assumed for the hadrons. Thus, although the situation is marginal the results are reasonably consistent with the assumptions of the statistical bootstrap model.

The situation is more severe in the zero temperature case. Writing out (36) again neglecting $m_\pi Q$ in the power of μ_{BQ} gives

$$n_B = \frac{(2\pi)^{3/2}}{h^3} \frac{T_0^{5/2}}{\mu_n^{3/2}} \Sigma_{B,Q} c_{BQ} B^{-1/2} \exp(\mu_n B - m_\pi Q)/T_0. \tag{46}$$

Eliminating a weighting factor B gives the number density of hadrons (with $B_{max} = 1$)

$$n_h = \frac{(2\pi)^{3/2}}{h^3} \frac{T_0^{5/2}}{\mu_n^{3/2}} e^{\mu_n/T_0} c_{eff}, \tag{47}$$

with c_{eff} as in (43). Frautschi's constraint (14) gives an expression for the assumed hadron volume

$$V = \frac{\ln 2 \, h^3 m_\pi^{1/2}}{2(2\pi)^{3/2} T_0^{3/2} c}. \tag{48}$$

The ratio of the actual assumed volume to the effective volume $V_{eff} \equiv 1/n_h$, taking $m_\pi \sim T_0$ and $c \sim c_{eff}$, is then

$$\frac{V}{V_{eff}} = \frac{\ln 2}{2} \left(\frac{\mu_n}{T_0}\right)^{-3/2} e^{\mu_n/T_0}. \tag{49}$$

This number should be less than unity for self-consistency. However, even at $\mu_n = m_n$ (49) gives $V/V_{eff} \simeq 9$. In actuality there are more than 9 hadrons in the volume assumed for one.

At this point the pessimist will argue that application of the exponential level density simply makes no sense. A more optimistic outlook would be to say that the basic results are reasonable even though there is some scrambling of construct with constituent. The constituent states were, after all, assumed to be completely overlapping within the hadron volume. A third alternative would be to take advantage of the result of Hamer and Frautschi (1971) that the parameter T_0 actually scales with the hadron volume V. A consistent scheme can be constructed in which the basic results given here hold true but where one regards $T_0 = T_0(V)$.

Suppose, for instance, that in the degenerate case the hadron volume continually adjusts itself to the condition which holds naturally in the high temperature case, namely $V \sim V_{\text{eff}}$. By requiring

$$V_{\text{eff}} = \alpha V, \tag{50}$$

where $\alpha \geqslant 1$ and employing the empirical scaling law of Hamer and Frautschi (1971)

$$T_0 = T_{0,0} \left(\frac{V}{V_0} \right)^{-1/3}, \tag{51}$$

where $T_{0,0}$ and V_0 are the standard values, one finds the following relation

$$\frac{\ln 2}{2} \alpha \left(\frac{T_{0,0}}{\mu_n} \right)^{3/2} = \left(\frac{V}{V_0} \right)^{1/2} \exp \left[-\frac{\mu_n}{T_{0,0}} \left(\frac{V}{V_0} \right)^{1/3} \right]. \tag{52}$$

Solving (52) gives

$$T_0 \simeq \left(1 - 2 \ln \frac{\alpha}{1.03} \right) \mu_n. \tag{53}$$

Thus (40b) becomes

$$P = \left(1 - 2 \ln \frac{\alpha}{1.03} \right) \varepsilon. \tag{54}$$

That is, the equation of state again becomes $P \propto \varepsilon$, with causality intact if $\alpha = V_{\text{eff}}/V > 1.03$, and positive pressure if $\alpha < 1.03\, e^{1/2} \simeq 1.7$.

Perhaps the real value of the study of the statistical bootstrap model in the context of cold ultradense matter is not that it will teach us more about the nature of hadron stars but that it will lead us to new insights into the nature of hadron physics itself.

References

Ambartsumyan, V. A. and Saakyan, G. S.: 1960, *Soviet Astron. – A.J.* **4**, 197.
Bardakci, K. and Mandelstam, S.: 1969, *Phys. Rev.* **184**, 1640.
Chiu, C. B. and Heimann, R. L.: 1971, *Phys. Rev.* **D4**, 3184.
Dashen, R., Ma, S. K., and Bernstein, H. J.: 1969, *Phys. Rev.* **187**, 345.
Dolen, R., Horn, D., and Schmid, C.: 1968, *Phys. Rev.* **166**, 1768.
Frautschi, S.: 1971, *Phys. Rev.* **D3**, 2821.
Frautschi, S., Steigman, G., and Bahcall, J.: 1972, *Astrophys. J.* **175**, 307.
Fubini, S. and Veneziano, G.: 1969, *Nuovo Cimento* **64A**, 811.
Hagedorn, R.: 1965, *Nuovo Cimento Suppl.* **3**, 147.
Hagedorn, R.: 1967, *Nuovo Cimento* **52A**, 1336.
Hagedorn, R.: 1968, *Nuovo Cimento Suppl.* **6**, 311.
Hagedorn, R. and Ranft, J.: 1968, *Nuovo Cimento Suppl.* **6**, 169.
Hamer, C. and Frautschi, S. C.: 1971, *Phys. Rev.* **D4**, 2125.
Huang, K. and Weinberg, S.: 1970, *Phys. Rev. Letters* **25**, 895.
Kaĭdalov, A. B.: 1972, *Soviet Phys. Usp.* **14**, 600.

Krzywicki, A.: 1969, *Phys. Rev.* **187**, 1964.
Layzer, D.: 1969, *Astrophysics and General Relativity*, vol. II, Gordon and Breach, New York.
Lee, H., Leung, Y. C., and Wang, C. G.: 1971, *Astrophys. J.* **166**, 387.
Leung, Y. C. and Wang, C. G.: 1971, *Astrophys. J.* **170**, 499.
Leung, Y. C. and Wang, C. G.: 1972, *Astrophys. J.* **181**, 895.
Rhoades, C. E. and Ruffini, R.: 1971, *Astrophys. J. Letters* **163**, L83.
Sawyer, R.: 1972, *Astrophys. J.* **176**, 205.
Veneziano, G.: 1971, *Proceedings International Conference on Duality and Symmetry in Hadron Physics* (ed. by E. Gotsman), Weizmann Science Press of Israel, Jerusalem.
Wheeler, J. C.: 1971, *Astrophys. J.* **169**, 105.
Wilson, R. E.: 1972, *Astrophys. J. Letters* **174**, L27.

A SIMPLE EQUATION OF STATE OF MATTER AT SUPER-NUCLEAR DENSITIES*

Y. C. LEUNG

Southeastern Massachusetts University, North Dartmouth, Mass. 02747, U.S.A.

and

C. G. WANG

Massachusetts Institute of Technology, Cambridge, Mass. 02139, U.S.A.

Abstract. The qualitative features of the equation of state of matter at supernuclear densities are deduced through a careful examination of the nature of particle interactions at short distances and by the introduction of an 'effective mass spectrum.' it is found that the equation of state begins to take on a particularly simple form (the 'asymptotic form') at a relatively low matter density of 10^{17} g cm^{-3}. A brief review of various approaches to calculate the equation of state for a neutron star is also given.

1. Introduction

In a system of particles with strong interaction dynamics, the two models very often employed are the potential interaction and the dual resonance approach. The former has been applied very extensively, and most successfully, to systems of many nucleons; the latter has been applied to systems of a few (most successfully zero) baryons. In the construction of an equation of state for a neutron star, the matter at nuclear densities can be described very accurately by the potential model of two-neutron correlations, while at much higher densities, the proliferation of excited particle states suggests the dominance of single particle dynamics. Unfortunately, the region of density which is most sensitive to the mass of a neutron star is at a matter density of 10^{15} to 10^{17} g cm^{-3}, and it is not covered by the usual potential interactions in nuclear physics, or by the dual resonance model in high energy physics. A major effort is clearly required to bring together techniques from different fields of physics to explain this wonderful object.

2. Difficulties in the Theoretical Models

Figure 1 shows some results of neutron matter calculations (Leung and Wang, 1971). The equations of state derived from various nuclear potentials are fairly consistent with each other for matter near nuclear densities, but they begin to diverge with each other at higher densities as the effect of the nuclear potential core becomes important. There are various forms of potential core; some are infinitely stiff at a finite core radius, some have an inverse radius dependence on the repulsion, some have finite and constant repulsion, and some have no explicit repulsion. It is not clear that one form of potential core is less arbitrary than the other. There are many corrections that should

---

* Presented by C. G. Wang.

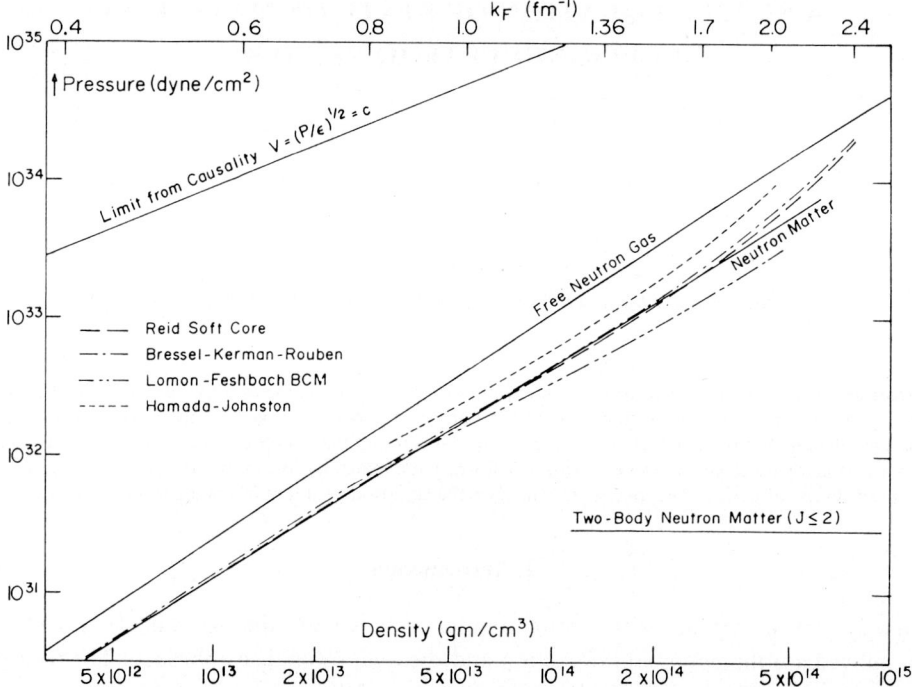

Fig. 1. Equations of state for neutron matter calculated from four sets of nuclear potentials.

be included in the consideration; the condensation energy of the neutrons, the clustering effect of the protons and the small contribution from the electrons, the many-body forces, the correlations from higher than the two-body terms, the relativistic corrections, the potential contribution due to the high angular momentum partial waves, the non-local consideration of the momentum transfer at short distances. Besides the first two items, all corrections become increasingly important as the nucleons get closer. There are also some basic questions that may be asked about the potential approach. Consider, for example, the Yang scaling model with 0.7 fm for the proton dimension and 0.6 fm for the pions. Surely when the internucleon distance approaches this dimension, some serious corrections should be made. The most important difficulty perhaps arises from the lack of a well developed potential to operate among hadrons in general. One may certainly consider all baryons as if they were nucleons, but there is no compelling reason to do so.

Figure 2 is an overall equation of state for degenerate matter. Starting with a very repulsive potential core and extending the many-body calculations from the nuclear physics, one obtains the maximum mass of a neutron star of about 1.5 $M_\odot$, while without an explicit repulsion in the nuclear core, the limiting mass may be as low as 0.3 $M_\odot$. Note that the equation of state is limited by the so-called 'causality relation' $P = \varepsilon c^2$; a stiff equation of state at matter densities higher than 10^{15} g cm^{-3} would not support a neutron star heavier than about 1.5 $M_\odot$ (unless one let the equation of state

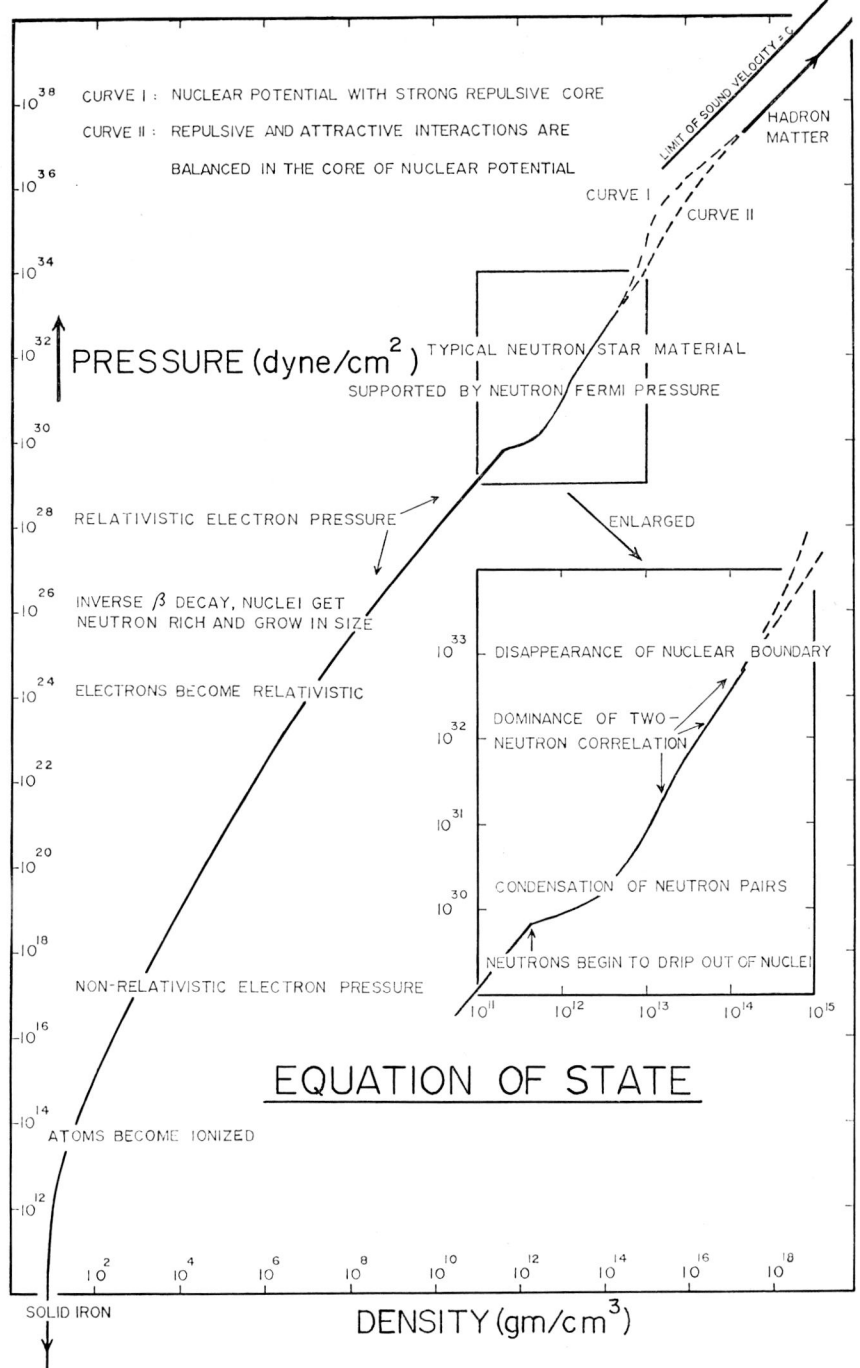

Fig. 2. An overall description of the equation of state for the cold degenerate matter maintained at neutral charge. Curve (1) can support a neutron star with maximum mass of about 1.5 $M_\odot$, while curve (2) can support only 0.3 $M_\odot$.

go beyond the causality limit) – only a stiff equation of state at matter density around 10^{14} g cm^{-3} would do, but this is where the usual nuclear physics operates and a judgment can be easily made.

Raymond Sawyer in a recent series of articles (Sawyer, 1972a, 1972b), cast two serious doubts upon the recent calculations made on the equation of state of matter at supernuclear densities. His first objection is that the states of the decay channel of the excited baryons may be occupied and their self-energy should therefore be shifted upward. The second objection is that the pions in a very dense matter may form a degenerate sea so as to considerably reduce the Fermi energy of the baryons. These objections apply to the potential approach as well as to the dual resonance or the bootstrap considerations. We concur with Sawyer's observation, and suspect that some future considerations uncommon to the usual nuclear physics or high energy physics may yet to be raised. Faced with such extraordinary uncertainties, we come back to a re-examination of the baryons spectrum hoping to learn some general properties of the equation of state of matter at supernuclear densities.

3. The Baryon Mass Spectrum

A list of known baryons with masses below 2 GeV can be found in an article by Söding et al. (1972). This list is fairly complete and is transcribed in Appendix A giving the mass and multiplicity of each. If we compare the cumulative number of baryons, N, below a certain mass with the square of that mass, m^2, normalized to that of the neutron mass, m_N^2, we can see from Figure 3 that it can be approximated by a simple power law:

$$N = c_1 (m^2/m_N^2) a. \tag{1}$$

In Figure 3, $\log N$ is plotted against $\log(m^2/m_N^2)$ and the data show remarkable linearity with $c_1 = 4.0$ and $a = 2.9$. From (1) we can deduce the baryon spectrum:

$$\Delta N(m) = c_0 (m^2/m_N^2)^\alpha \Delta(m^2/m_N^2), \tag{2}$$

where ΔN denotes the number of baryons within the mass interval $\Delta(m^2 = m_N^2)$, and $c_0 = ac_1$, $\alpha = a - 1$.

Just as the free neutron equation of state served to illustrate the qualitative features of the equation of state of nuclear matter, the equation of state derived from free baryons by including all known baryons would provide a qualitative guide to that of supernuclear matter. Such an attempt was first considered by Ambartsumyan and Saakyan (1960) who made use of all hyperons and resonances known at that time with and without some form of interparticle interaction. Based on repulsive vector meson exchange, Zel'dovich (1961) has given

$$P = (v - 1) \mathscr{E} c^2. \tag{3}$$

where P is the pressure, $\mathscr{E}$ the energy density and $v = 2$ in his asymptotic limit for repulsive interactions. However, Harrison (1965) claimed that due to multiple-meson exchange, the parameter v would be brought back to the more conventional range, $1 \leqslant v \leqslant 4/3$.

The equation of state derived on the basis of free particles cannot of course be realistic and there are serious objections to it since baryons interact strongly at such interparticle distances. The neglect of strong interactions certainly cannot be justified. Two issues have been raised. First, we know that the mass of a particle resonance such as one of those listed in Appendix A is determined experimentally when the particle is free to decay strongly. However (Sawyer, 1972a), if the resonance is inhibited from decay due to the lack of phase space, as is the case when it is immersed in the Fermi sea of dense matter, would its mass remain the same? Second, strong interactions among baryons and among baryons and mesons (Sawyer, 1972b, Scalapino, 1972) can contribute substantial interaction energy to the system. How much is this interaction energy?

Since there is no accurate way of computing the mass shifts of resonances due to

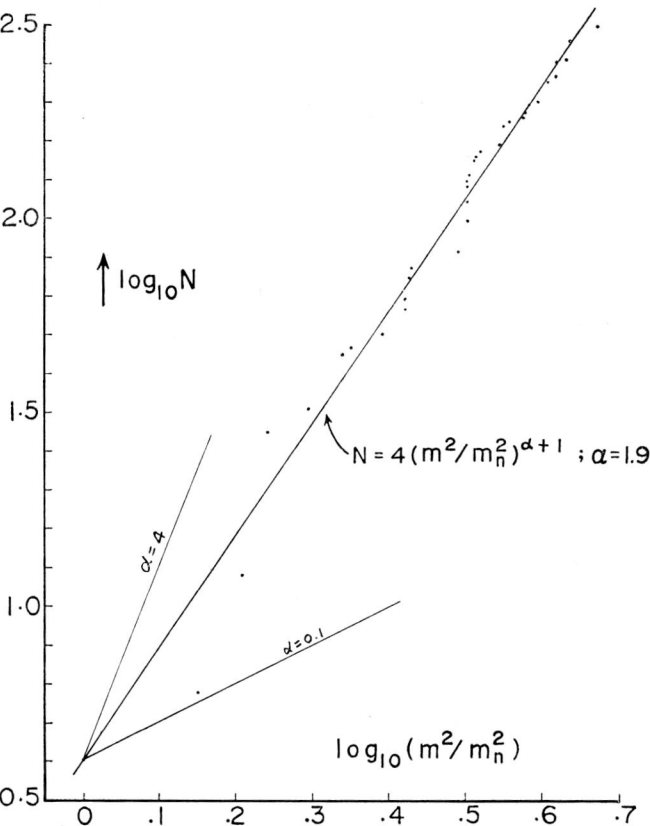

Fig. 3. The cumulative number of baryons N below a certain mass m is plotted against the square of the mass normalized with respect to the square of the nucleon mass m_N on a log-log plot. The spectrum is very sensitive to the index α.

alterations in their lifetimes, and the knowledge of particle dynamics is at present inadequate to determine the interaction energy, the problem has reached an impasse. We shall nevertheless give some plausible arguments below which would then allow us to understand the qualitative nature of the equation of state at supernuclear densities. We shall make use of the notion of mass shifts and claim that the equation of state can be derived in terms of a free-particle formalism if the mass of each particle is shifted sufficiently to adjust for interaction. It is, of course, always possible to describe the statistical mechanics of an interacting baryon matter in terms of a free baryon gas provided the masses of particles are adjusted continuously to yield the correct result. This, however, is a very difficult thing to do in general if there are only a few types of baryons involved since the adjusted mass of each would have to be known exactly. Now, the main point of our argument is that since the baryon mass spectrum obeys a power law, it is most probable that the 'effective baryon mass spectrum' (which one would use in conjunction with a free-particle formalism) due to mass shifts as a result of interactions would still obey a power law with possibly a different power α. The power law expresses merely a certain distribution of baryon states and should not be completely obliterated by mass shifts. The plausibility of this conjecture is argued on the basis that the baryon-baryon interaction is not the dominant interaction even at very high densities, in contradistinction to the reasoning behind Zel'dovich's derivation of (3); the dominant interaction belongs indeed to the meson-baryon interaction which to a large extent is revealed by the baryon spectrum. (More explicit considerations of these arguments are given in Section 5). Hence, in the next section, we shall give a simple analytic study of the equation of state of dense baryon matter with an effective baryon mass spectrum given by (2) except that α is treated as a parameter, bearing in mind that even though the analysis resembles that of a free baryon gas, it has in fact a much broader basis of validity.

4. The Equation of State

Let us discuss now the statistical mechanics of a free baryon gas consisting of particles obeying a mass spectrum given by (2). The pressure P and energy density $\mathscr{E}$ of such a gas are given by:

$$P = (c_2/3) \int dm^2 m^{2\alpha} \int_m^\infty \frac{(\varepsilon^2 - m^2)^{3/2} \, d\varepsilon}{e^{(\varepsilon - \mu)\beta} + 1}, \tag{4}$$

$$\mathscr{E} = c_2 \int dm^2 m^{2\alpha} \int_m^\infty \frac{\varepsilon^2 (\varepsilon^2 - m^2)^{1/2} \, d\varepsilon}{e^{(\varepsilon - \mu)\beta} + 1}, \tag{5}$$

where μ is the chemical potential, $\beta = 1/kT$ and $c_2 = c_0 (2\pi^2 \hbar m_N^{2\alpha+2})^{-1}$. (Note that when there are a large variety of baryons present and occurring in different charge states, the role of leptons in the equation of state can be ignored.)

Consider first the degenerate case where $\beta \to \infty$. Then,

$$P = (c_2/3) \int_{m_0^2}^{\mu^2} dm^2 m^{2\alpha} \int_m^{\mu} (\varepsilon^2 - m^2)^{3/2} d\varepsilon, \qquad (6)$$

$$\mathscr{E} = c_2 \int_{m_0^2}^{\mu^2} dm^2 m^{2\alpha} \int_m^{\mu} \varepsilon^2 (\varepsilon^2 - m^2)^{1/2} d\varepsilon$$

$$= 3P + c_2 \int_{m_0^2}^{\mu^2} dm^2 m^{2(\alpha+1)} \int_m^{\mu} (\varepsilon^2 - m^2)^{1/2} d\varepsilon, \qquad (7)$$

where the lower limit of the mass spectrum is denoted by m_0^2, which we may take to be $m_0^2 \simeq 0.8 \, m_N^2$. An integration by parts of the second term in the right-hand side of (7) gives:

$$\mathscr{E} = 3P + 2(\alpha + 1) P + \Delta, \qquad (8)$$

where

$$\Delta = (2c_2/3) \, m_0^{2(\alpha+1)} \int_{m_0}^{\mu} (\varepsilon^2 - m_0^2)^{3/2} d\varepsilon. \qquad (9)$$

In this manner we have reduced the expression for $\mathscr{E}$ to one in terms of P and Δ. It is clear that the expression cannot be further reduced since Δ is only of the order of P, and (Δ/P) diminishes as μ increases at the rate $(m_0/\mu)^{2(\alpha+1)}$. The equation of state can now be written as

$$(\mathscr{E} - \Delta) = (5 + 2\alpha) P, \qquad (10)$$

which simplifies immediately to the form of (3) whenever (Δ/P) becomes negligible, viz.,

$$\mathscr{E} \simeq (5 + 2\alpha) P. \qquad (11)$$

Thus, the equation of state for cold dense matter is determined once α is known. In order for (11) to be useful, it is of course necessary that α should not vary to any extent as μ is varied. This, however, is expected to be the case since α appears as a power in the effective baryon mass spectrum and therefore should be comparatively stable.

To get some qualitative feeling for this analysis, we can adopt $\alpha = 2$, which is the power derived for the known baryon spectrum. Assuming that the power law (2) holds up to a mass of 2.0 GeV we shall consider the chemical potential μ also up to 2.0 GeV. Within the interval $1.5 \text{ GeV} < \mu < 2.0 \text{ GeV}$, corresponding to matter densities $10^{17} \text{ g cm}^{-3} < \varrho < 2 \times 10^{18} \text{ g cm}^{-3}$, the factor (Δ/P) is realized. The equation of state also compares well with the one using the actual mass of the baryons, as shown in Figure 4, testifying to the adequacy of the power law approximation to the baryon spectrum.

The analysis described above is restricted to the case of zero temperature in order to be precise. When the temperature is non-zero some of the mass states beyond the

2 GeV limit will be excited and the mass spectrum may deviate from a simple power law there. The analysis is however not any more difficult when π is finite. From inspection one can easily see that (10) is still correct provided that Δ and P are modified by including the factor $[\exp(\varepsilon-\mu)\beta+1]^{-1}$ within the energy integral and that the upper limit of the mass integral no longer ends at μ^2. Again, when (Δ/P) becomes negligible the equation of state reduces back to (11). There is however little need to consider the case of finite β since the smallest energy scale in this problem is m_0 which corresponds to a temperature of 10^{13} K.

Indeed, one can also consider a more general effective mass spectrum, such as:

$$N = c_0 (m^2)^\alpha (1 + a_1 m + a_2 m^2 + \cdots). \tag{12}$$

There is however very little to be gained by having a more complicated result at this time since we are only interested in estimating roughly at what densities the equation of state will take on the form given by (3).

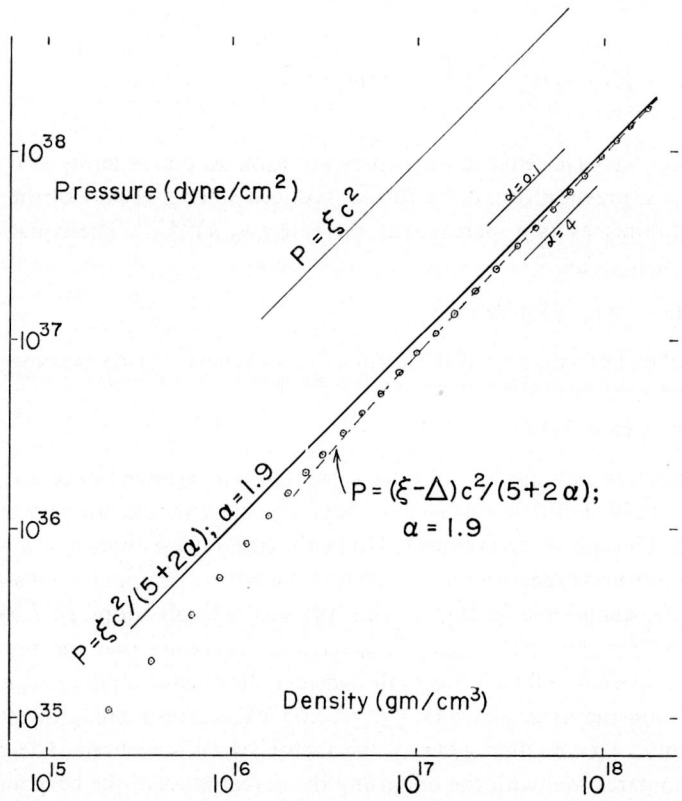

Fig. 4. The free baryon equation of state computed from Equation (10) using a simple power law approximation to the baryon spectrum is drawn in broken line, and the approximate expression given by Equation (11) is drawn in solid line. These are compared with that computed using the exact baryon spectrum. The latter is represented by circles. The equations of state are not sensitive to the index α.

5. Discussion

We try to point out in this article that the equation of state of matter at high densities takes on a particularly simple form given by (11) when supernuclear densities around 10^{17} g cm^{-3} are reached. We arrive at this result from dynamical considerations completely different from those employed before. Whereas it has been customary to assume that the interaction energy coming from inter-baryon repulsion would become dominant at high matter densities when baryons are squeezed very tightly together, we have avoided relying on this assumption completely since there are indications to the contrary (see below). Instead, we lay heavy emphasis on the meson-baryon interaction of which a great deal is revealed to us through the resonance spectrum of the baryons.

In nuclear physics, nuclear potentials are provided with a repulsive core. Such a strong short-range repulsive baryon-baryon interaction is deemed necessary to prevent nuclear collapse and finds its origin theoretically through the exchange of a neutral vector meson (like the ω-meson). With such a repulsive core a baryon-baryon interaction will indeed play the dominant role for matter at high density. We can however infer its absence by extending the same theoretical basis for its presence, i.e., the mechanism of one-meson-exchange. The fact that spin-1 meson exchange produces repulsion and spin-2 meson exchange produces attraction between baryons can be easily seen through analogy with the Coulomb repulsion and gravitational attraction, which can be derived by the exchange of photon (spin-1) and graviton (spin-2) respectively, and in identifying the baryonic charge (i.e., the baryon number) with the electric charge and the gravitational mass. Recent studies in high energy particle physics indicate that the ω-meson (spin-1) and the f-meson (spin-2), which are responsible for strong baryon-baryon forces, lie on 'exchange degenerate linear Regge trajectories,' which means the possible existence of (infinitely) many more even and odd spin mesons of increasingly high mass contributing alternatively to attractive and repulsive forces of comparable strengths between baryons. If this is the case, the net baryon-baryon interaction energy is not expected to be large at matter densities being considered. (At such densities the inter-baryon separation is of course considerably shorter than the ordinary nuclear separation and is well within the repulsive 'core' employed in nuclear physics.)

The dynamical situation is a great deal clearer with regard to meson-baryon interactions since the resonance spectrum provides us with useful information on the structure of the scattering amplitudes. The so-called 'narrow resonance models' in particle physics rely precisely on just such knowledge (see Sivers and Yellin, 1971 for a review). The implementation of the narrow resonance approximation for meson-baryon interaction in the framework of statistical mechanics is just the statement of (4) and (5) as made clear by the work of Dashen *et al.* (1969). Therefore, meson-baryon interactions can be handled reasonably adequately.

The remaining consideration centers upon mass shifts of the Sawyer type, which affect directly the treatment of the meson-baryon interaction discussed above. Since there exists at present no reliable means of estimating mass shifts for all baryon resonances, an exact determination of the equation of state of matter at supernuclear densities is impossible. We can nevertheless aim for a qualitative understanding of the equation of state at these densities. In line with the above discussion we postulate therefore an 'effective baryon mass spectrum' which includes the effects of mass shifts as well as baryon-baryon interactions. These effects are meant to cause minor distortions of the original baryon mass spectrum so that the power law approximation is retained. This then leads to our result given by (10) and (11).

Thus, we see that the equation of state of matter at supernuclear densities departs qualitatively from that of a free Fermi gas due to meson-baryon interaction which greatly undercuts the Fermi pressure in the system, and this is believed to occur at densities as low as 10^{17} g cm^{-3} using the known baryon mass spectrum as a guide.

We are grateful to Dr Philip Morrison who suggested to us the possibility of interpreting particle interactions in terms of a shifted mass within the free-particle formalism, a concept which we find useful here.

Appendix A

We list here the better known baryons which are being made use of in determining the baryon mass spectrum (see Figure 3). The data come from Söding *et al.* (1972) and their notations are being followed here. We use m to denote the mass, m_N the nucleon mass and N the cumulative number of baryons below a certain mass. The statistics for m below 2 GeV is fairly accurate, but incomplete for higher masses.

Name	Mass (MeV)	m^2 (GeV^2)	Spin Multiplets	Isospin Multiplets	Total Components	N
N	939	0.88	2	2	4	4
Λ	1115	1.24	2	1	2	6
Σ	1190	1.42	2	3	6	12
Δ	1236	1.53	4	4	16	28
Ξ	1315	1.73	2	2	4	32
Σ	1385	1.92	4	3	12	44
Λ	1405	1.97	2	1	2	46
N'	1470	2.16	2	2	4	50
N'	1520	2.31	4	2	8	58
Λ	1520	2.31	4	1	4	62
Ξ	1530	2.34	4	2	8	70
N'	1535	2.36	2	2	4	74
Δ	1650	2.72	2	4	8	82
Δ	1670	2.79	4	4	16	98
N	1670	2.79	6	2	12	110
Λ'	1670	2.79	2	1	2	112
Σ	1670	2.79	4	3	12	124
Ω	1672	2.80	4	1	4	128
N	1688	2.85	6	2	12	140
Λ''	1690	2.86	4	1	4	144

Appendix A (continued)

Name	Mass (MeV)	m^2 (GeV^2)	Spin Multiplets	Isospin Multiplets	Total Components	N
N''	1700	2.89	2	2	4	148
Σ	1750	3.06	2	3	6	154
Σ	1765	3.12	6	3	18	172
N''	1780	3.17	2	2	4	176
Λ	1815	3.30	6	1	6	182
Ξ	1820	3.31	2	2	4	186
Λ	1830	3.35	6	1	6	192
N	1860	3.46	4	2	8	200
Δ	1890	3.57	6	4	24	224
Δ	1910	3.64	2	4	8	232
Σ	1915	3.66	6	3	18	250
Ξ	1940	3.76	2	2	4	254
Δ	1950	3.80	8	4	32	286
Σ	2030	4.12	8	3	24	310
Λ	2100	4.41	8	1	8	318
N	2190	4.8	8	2	16	334

References

Ambartsumyan, V. A. and Saakyan, G. S.: 1960, *Astron. Zh.* **37**, 193. (English translation: 1960, *Soviet Astron. – AJ* **4**, 187.)
Dashen, R., Ma, S. K., and Bernstein, H. J.: 1969, *Phys. Rev.* **187**, 345.
Harrison, E. R.: 1965, *Astrophys. J.* **142**, 1643.
Leung, Y. C. and Wang, C. G.: 1971, *Astrophys. J.* **170**, 499.
Sawyer, R.: 1972a, *Astrophys. J.* **176**, 205.
Sawyer, R.: 1972b, *Phys. Rev. Letters* **29**, 382.
Scalapino, D. J.: 1972, *Phys. Rev. Letters* **29**, 386.
Sivers, D. and Yellin, J.: 1971, *Rev. Mod. Phys.* **43**, 125.
Söding, P., Bartels, J., Barbaro-Galtieri, A., Enstrom, J. E., Lasinski, T. A., Rittenberg, A., Rosenfeld, A. H., Trippe, T. G., Barash-Schmidt, N., Bricman, C., Chalorepka, V., and Roos, M.: 1972, *Phys. Letters* **39B**, 1.
Zel'dovich, Ya. B.: 1961, *Zh. Eksperim. i Teor. Fiz.* **41**, 1609 (English translation: 1962, *Soviet Phys. – JETP* **14**, 1143.)

DISCUSSION

Bethe: You have done something similar to Zel'dovich's work, but without strong repulsive interactions. Mind you that without strong repulsion, nuclei would collapse.

Wang: Most of the recent new nuclear potentials have more and more softer cores. I would be very surprised if anyone of them would give rise to a collapsed nuclei.

Bethe: In your calculation of the equation of state, instead of plotting P and E, you should plot energy per baryon, which is more sensitive.

Wang: But P and E is what one uses to calculate the models of neutron stars. The main point of my talk is trying to show that at densities beyond 10^{17} g cm^{-3}, the equation of state becomes insensitive to the detailed models of calculations.

Pandharipande: I differ with your statement that a soft repulsive core would give a much softer equation of state. My calculations with the Reid potential and with the BKR potential give very close results. You have that too in your graph. The problem is whether one should use a static potential.

Wang: My potential calculation has not been extended beyond the nuclear densities. I suppose that the relativistic effects would force all calculations to merge at the causality limit. But surely, at higher densities, the recoil momentum should be very important, therefore the non-static (non-local) contribution should be important. Mind you that Sawyer's criticism applies to the potential approach as well.

Pandharipande: Sawyer's self-energy is the three-body force term. Things like the relativistic corrections, the non-local recoil, the three-body forces, the three-body clustering etc. can be worked out in time. There is no reason to give up the potential approach.

Wang: What do we know about potentials among hadrons in general? Even the potentials among nucleons are difficult enough, and Sawyer's self-energy is not exactly the usual three-body force term. You have to worry about the self-consistency part of the Fermi energy, and you would get just as many coupled equations to solve. I can see some light to work out the nuclear physics from high energy physics, but not the other way around.

Pandharipandi: I would be most happy if your number and mine agree so we all do the right physics. I do not really care what approach one takes.

Wang: Come to think of it, we differ less than an order of magnitude, so there is really hope of convergence.

PIONS IN NEUTRON STAR MATTER

RAYMOND F. SAWYER

Dept. of Physics, University of California, Santa Barbara, Calif. 93106, U.S.A.

Abstract. Arguments are given for pions becoming an important constituent of neutron star matter at densities greater than a critical density lying between 5×10^{14} and 10^{15} g cm^{-3}. It is anticipated that this will lead to a substantial softening of the equation of state.

Most of this work was done in collaboration with D. Scalapino. The final conclusions are not yet firm. What we think we have established is that there will be some very interesting phenomena involving real pions in the ground state of neutron star matter. However we have looked carefully at only a few of the possible ways pions can enter, and there are many other possibilities to be considered. The game is to look for the way in which the energy is minimized and it is likely that someone will soon find a way of using pions to give even a lower energy than that of our 'ground state.'

The possibility of real pions in the ground state of superdense matter is of interest both for the possible effects on the zero temperature equation of state of the matter, and for the effects on the transport properties of the matter. If pions are present they will surely occur in a condensed mode with many pions in the same state. The strong interactions of pions will determine whether such a condensation occurs; free π^-'s would not begin replacing e^-'s in matter of density less than 10^{16} g cm^{-3} under any set of assumptions. However a π^- meson still is, *a priori*, the most likely pion, since it needs to acquire negative potential energy equal only to the pion rest energy minus the electron chemical potential.

Therefore we begin by asking what is known about π^- interactions in nuclear matter. A π^- at rest in a sea of protons and neutrons acquires a potential energy from *s*-wave pion-nucleon interactions equal to (Bethe, 1971),

$$\delta E = 219(\varrho_n - \varrho_p) \text{ MeV}, \tag{1}$$

where ϱ is in fm^{-3}.

Thus a single π^- in dense neutron rich matter has an enormous positive potential energy. This has been quoted as an argument against the development of a π^- phase in neutron star matter. However if we have, for example, an equal number of protons and neutrons, the proton charge neutralized by negative pions, the potential energy from *s*-wave π^- nucleon interactions vanishes and the residual interactions will determine whether the state is energetically preferred over a pure neutron state.

A moving pion will undergo strong *p*-wave interactions in nuclear matter. The optical potential of Auerbach *et al.* (1967), the real part of which is based on low energy *p*-wave phase shifts, gives a good fit to the scattering of a π^- particle from $A=2Z$ nuclei. We can express the effect of the real part of this potential as a modification of the dispersion relation for the pion moving in a uniform medium,

$$\omega^2 = k^2(1 - 6\varrho) + m_\pi^2. \tag{2}$$

At densities slightly greater than nuclear, $\varrho > 0.16 \text{ fm}^{-3}$, this dispersion relation exhibits a striking property: as k is increased from zero, ω decreases. Of course the scattering length approximation underlying (2) is only valid for small k and should be cut off for high k, so when ϱ is only slightly above 0.16 fm^{-3} we would expect a shallow dip in energy followed by a rise as k is increased. But for larger ϱ the pion energy will dip far enough for some moderate value of k to make it economical to trade an electron for a π^-; or even dip to $\omega = 0$. In the latter case the optical potential approach is inapplicable for a number of reasons, but the result is strongly suggestive of an instability toward formation of a large number of π^-'s.

Instead of pursuing the optical potential approach further we turn to an approach based directly on the Hamiltonian for emission and absorption of single pions, which is thought to underlie the p-wave πN scattering amplitudes. Solving exactly for the behavior of a single pion moving in a medium of protons and neutrons is out of the question in this approach, but the problem of many pions condensed in the same mode is much simpler and admits a solution.

For non-relativistic nucleons the Hamiltonian for the interaction of one plane wave π^- mode with momentum $k\hat{z}$ with nucleons is

$$H = H_0 + H_\pi = H_0 + \frac{ifk\sqrt{2}}{m_\pi \sqrt{2\omega_k V}} \times$$

$$\times \int d^3x \, [\tilde{n}(x) \sigma_3 p(x)] \, a(k\hat{z}) \, e^{ikz} + \text{H.C.}, \qquad (3)$$

where $f = 1.1$; $\tilde{n}$ and p are the creation and annihilation operators for neutron and proton; $a(k\hat{z})$ the annihilation operator for a π^- of momentum $k\hat{z}$. H_0 is the Hamiltonian for free pions and nucleons.

If we have a pion condensation in the mode $k\hat{z}$ we can make the c-number replacement for the operators a and a^+,

$$a, a^+ \to \sqrt{N_\pi}.$$

We begin by comparing the lowest energies of the following two systems, with a fixed number of baryons in a volume V, (a) pure neutrons, and (b) neutrons, protons and a condensed pion mode, where the number of π^-'s is constrained to be equal to the number of protons.

From (3) we can see that if we can find a state for the nucleons in which

$$\langle \tilde{n}(x) \sigma_3 p(x) \rangle = \text{const.} \, i\varrho_{nucl} \, e^{-ikz}$$

without undue sacrifice in kinetic energy, then for sufficiently large density the interaction terms in (3) will overcome the free meson energy terms and give a lower energy to the state with condensed pions. In what follows the kinetic energy sacrifice per proton in the condensed state will be seen to be of order $k^2/2 M_{nucleon}$ and to be negligible in the domains of interest.

To solve the problem we minimize $\langle H \rangle$ subject to constraint of neutrality,

$N_p = N_\pi \equiv XN$, where N is the total number of baryons. Writing $\mathscr{H} = H_0 + H_\pi + \mu N_p$, we see that $\mathscr{H}$ can be almost diagonalized by a canonical transformation on the nucleon fields,

$$u(x) = \sqrt{1-\theta^2}\, n(x) + i\theta \sigma_3 p(x)\, e^{ikz},$$
$$v(x) = i\theta n(x) + \sqrt{1-\theta^2}\, \sigma_3 p(x)\, e^{ikz}, \qquad (4)$$

where θ is a known function of the proton chemical potential, μ. Discarding a term (which is small and which we pick up later in perturbation theory) coming from the non-diagonalization of H_0 under the above transformation we obtain

$$\mathscr{H} \cong H_0 + \lambda_-(\mu) \int \bar{u}(x) u(x)\, d^3x,$$
$$+ \lambda_+(\mu) \int \bar{v}(x) v(x)\, d^3x, \qquad (5)$$

where $\lambda_- < \lambda_+$.

The ground state is thus a sea of 'u particles'

$$|\psi_0\rangle = \prod_k^{K_F} \bar{u}_k |\text{vac}\rangle \qquad (6)$$

instead of a sea of neutrons. The parameter μ (and therefore θ) is determined by the requirement that the average charge of the baryons be given by the parameter, X, which fixes the value of the pion field,

$$\langle \psi_0 | N_p | \psi_0 \rangle = NX = \theta^2 X$$
$$N_\pi = NX. \qquad (7)$$

Although the state $|\Psi_0\rangle$ does not have a definite electric charge, the charge fluctuations around the value given by (7) die off in the $V \to \infty$ limit sufficiently fast to make the Coulomb energy per nucleon vanish.

The energy per nucleon can now be calculated from (5) and turns out to be

$$\frac{\text{Energy}}{\text{Nucleon}} = \frac{3}{5}\frac{k_F^2}{2M_N} + X\omega_k + \frac{k^2}{2M_N} - 2\frac{\sqrt{\varrho f}}{\sqrt{\omega_k m_\pi}} X \sqrt{1-X}. \qquad (8)$$

The first term on the right-hand side is the Fermi energy of a non-interacting neutron gas. We shall refer to the sum of the next three terms as the condensation energy. When it is negative the ground state will contain condensed pions. As the density, ϱ, is increased from nuclear densities the condensation energy becomes negative first for a particular value of k, $k = 1.2\, m_\pi$. In this case the onset of the condensation would be at $\varrho = 0.25$ baryons fm^{-3}. The onset in this model is gradual (second class phase transition) beginning at $\varrho = 0.25$ from $X = 0$ and increasing at $\varrho = 1\, \text{fm}^{-3}$ to about $X = 1/2$ (one half protons).

As an example of the potential mischief which could be caused by the pion condensation we can calculate the pressure loss (relative to the pressure of the free Fermi gas) implied by (8). We find a 30% pressure loss at $\varrho = 0.3$ fm^{-3} and a 73% loss at $\varrho = 0.5$ fm^{-3}. However, we think that it is premature to try to estimate effects on neutron star masses and radii, for reasons which should become clear in what follows.

Having given a detailed treatment of the simplest model which shows the condensation, we now give a qualitative discussion of several ways in which the model can be made more realistic, and of several outstanding problems which must be solved before a believable equation of state can be found.

The most essential extension of the model is the inclusion of a realistic nuclear force. It is not immediately obvious, for example, that the demands made on the nucleonic wave function in order to sustain the pion condensation are not completely in conflict with the demands made by the short range repulsions. However, we are spared the indignity of having to do a new nuclear matter calculation in the presence of a condensed pion field by the following extraordinarily lucky circumstance; if the two-body nucleon-nucleon force is spin and isospin independent, as the hard core repulsion and a good deal of the medium range attraction are supposed to be, then the *difference* in energies between the ground state of a pure neutron gas and that of our condensed pion state is exactly the same as we calculated for the case of no nuclear force. This is because the two-body force is invariant under the canonical transformation, (4), in the sense that it has the same form in the quasi-particle fields u and v as in the nucleon fields n and p. The additional terms in the Hamiltonian for the condensed pion case measure only the total number of u and v particles (as in [5]). So the ground state in the condensed state consists of exactly the same configuration of 'u particles' as the pure neutron case does of neutrons, the energies differing by exactly the same amount as in the case of no two-body interactions.

Thus in calculating the condensation energy we need to concern ourselves with the spin and isospin dependent part of the two-nucleon force only. Single pion exchange is the most obvious source of spin and isospin dependent terms. Accordingly we calculate the second order effects of pion emission and absorption (non-condensed pions; that is, all modes).

The wave function of the nucleons does matter for this calculation. We have considered only the case of the free Fermi sea of 'u particles.' The effect of correlations, if the short range repulsions had been taken into account, would probably have been to reduce the second order pion effects. In any event our calculation gives a positive term in the condensation energy which, if added to our earlier result, would delay onset of the condensation until a density between 0.3 baryons fm^{-3} and 0.5 baryons fm^{-3}, depending on a cutoff.

It is of interest to note that only about 60% of this effect is attributable to the one pion exchange 'potential.' There is an additional term of second order in the coupling constant of the non-condensed pions which can be described as a medium dependent self-energy effect. This kind of effect is significant in our calculation and we suspect that similar terms should be included in any high density nuclear matter calculation. They are clearly omitted in variational approaches based on phenomenological two-body forces.

Next we briefly consider the effects of pion-nucleon and pion-pion s-wave interactions. According to (1), s-wave interactions in a pure neutron medium are very unfriendly to π^-'s, so that one would expect the small X onset of condensation to be considerably delayed or perhaps eliminated altogether in favor of a first class phase transition to an $X \approx 1/2$ state at higher density. However, with this concentration of π's we can expect $\pi^-\pi^-$ s-wave forces to be significant also. Information on this subject is purely theoretical. We have taken the formula analogous to (1) for the energy due to $I=2$ s-wave $\pi^-\pi^-$ interactions, using the scattering length of Weinberg (1966) and combined it with the πN term and the free condensation energy formula. This combination of terms turns out to lead to a second order phase transition with onset at a density of $\varrho \approx 0.36$ fm^{-3}.

Thus the pion condensation survives the inclusion of all two-body interactions which are thought to act between the different constituents of the matter. Unfortunately there are many corrections which could become important at high densities and which might favor either a condensed pion ground state or a pure neutron ground state. Three-body forces are one possible source of such effects, but a more important source of differences between the condensed state and the normal state is probably a different kind of term: the s-wave π^--nucleon interaction which underlies (1) is equivalent to a direct interaction of fields $\Psi \psi \phi^* \phi$. The energy of interaction was calculated by putting in the condensed pion field for ϕ and ϕ^*. If only one pion field is replaced by the condensed field the direct interaction term becomes an effective vertex for the emission or absorption of a non-condensed pion, and there will be a new set of terms in the energy coming from the emission and absorption of pions by this mechanism. The second order terms in this case clearly favor the condensation, but we have made no numerical evaluation as yet. Furthermore, if we started from a typical chiral dynamic non-linear Lagrangian for multi-pion emission from nucleons, a nightmarish collection of new terms would result.

Another complication, at densities greater than 1 fm^{-3}, or so (depending on the equation of state), is the intrusion of species of baryon other than neutrons and protons. The Λ, Σ^- hyperons are the first to show up, and since they can communicate via π^- emission and absorption $\Sigma^- \to \Lambda + \pi^- \to \Sigma^-$, they should be as effective as neutrons and protons in sustaining the condensation.

Another class of questions which has not yet been answered completely is whether a single plane wave mode of π^- gives the lowest energy under all circumstances. Perhaps a standing wave mode will be preferred, or perhaps a condensate consisting of a mixture of π^+, π^- and π^0 will take over in some region in density. Let us leave out magnetic energies for the moment (later we discuss how a filamentary structure can solve the magnetic energy problems for the plane wave case) and ask which sort of solution minimizes the strong interaction energy. The following conclusions are only preliminary:

(a) A standing wave mode would have a potential onset at exactly the same baryon density as a plane wave mode, but would be slightly less favored energetically for densities somewhat above the onset density. At sufficiently high densities a standing wave solution apparently has a lower energy than the plane wave solution. The problem in this case cannot be solved analytically and a numerical calculation is in progress.

(b) Even near onset a state with a certain small fraction of π^+ mesons may be preferred, with the fractions of π^+'s growing as the density is increased.

(c) π^0's may develop at higher densities.

(d) There never arises a catastrophic situation in which it would lower the energy to introduce indefinitely many pions, even in the absence of any repulsive pion-pion interactions.

Pending resolution of these questions we shall say only a little about including electromagnetic interactions and building a macrostructure out of condensed pion matter. If it turns out that a plane wave structure is greatly preferred by the strong interactions, then according to our estimates filaments could be formed of diameter from 10 to 20 fm, with currents moving in opposite directions on neighboring filaments. In this configuration the lowering of energy in the condensed state due to strong interactions wins out over the positive magnetic energy and filament surface energy terms. It is not clear what the length of the filaments should be, nor how the problem of connecting the ends should be resolved. If the matter breaks up into fairly small domains, randomly oriented in space, then we could probably use our equation of state for homogeneous matter in the absence of electromagnetic interactions in order to discuss the properties of a neutron star core in the large. If on the other hand the directionality of the matter (determined by the direction of the wave vector for the condensed mode) were preserved over the entire core one would anticipate very exotic mechanical properties, as well as transport properties, for the core of the star.

If the standing wave solutions turn out to be dynamically preferred then we presumably can have a homogeneous phase, instead of the filamentary structure. The matter would still have the directionality in this case.

To summarize, we believe that we have very strong arguments for pions becoming an important constituent of superdense matter at some baryon density between 0.4 fm^{-3} and 1 fm^{-3}. We do not yet know, however, what the best configuration will be to minimize the energy, or what the exact effects will be on the equation of state. However we can anticipate a substantial softening of the equation of state, which should lead to a lower value of the maximum mass of a neutron star.

References

Auerbach, E. H., Fleming, D. M., and Sternheim, M. M.: 1967, *Phys. Rev.* **162**, 1683.
Bethe, H. A.: 1971, *Ann. Rev. Nuclear Sci.* **21**, 93.
Weinberg, S.: 1966, *Phys. Rev. Letters* **17**, 168.

HYPERCOLLAPSED NUCLEAR MATTER

Y. NE'EMAN

University of Tel-Aviv, Tel-Aviv, Israel, and University of Texas, Austin, Tex., U.S.A.

Abstract. We evoke the possible existence of hypercollapsed tightly bound nuclear isomers, due to the existence of an attractive 'heart' deep inside the repulsive core.

1. Introduction

Physical theory has yielded in the past a series of predictions relating to new states of matter, or to novel types of material constituents. Successful predictions have included new chemical elements, previously unknown elementary particles, the world of anti-particles, etc. and recently macroscopic nuclear matter in the form of neutron stars. As yet unfound are magnetic monopoles, quarks, W mesons; superheavy elements and collapsed material in 'black holes' (or coming out of 'white holes'). To this list we are now adding the hypothesis of the existence of *hypercollapsed nuclear matter*.

2. Repulsive Core and Attractive Heart

We were led to the first suggestions of the existence of such tightly bound nuclear isomers by considering the new evidence about a strongly attractive internucleon force at very short ranges, beyond the barrier of the repulsive core (Ne'eman, 1968a, b). It is this region that we shall name the 'attractive heart' inside the repulsive core.

The evidence for the attractive heart comes from high energy physics. Internucleon forces – and not just in two-particle (four-prong) processes – can be represented by the exchange of Regge trajectories (Collins, 1971). For forward and near-forward scattering these are meson trajectories; backward scattering is produced by the exchange of two nucleons. The meson trajectories appear in a Chew-Frautschi plot as straight lines: the spin is linear in M^2 for the various meson states lying at positive momentum-transfer squared t' at integer J, either even or odd.

The quark model suggests that there should be trajectories passing through states with (J = the spin; P = the space parity; C = the charge parity of the neutron component)

$$J^{PC} = 0^{-+}; \quad 1^{--}$$
$$\Downarrow \qquad \swarrow \downarrow \searrow$$
$$1^{+-} \quad 0^{++} \; 1^{++} \; 2^{++}$$

where the upper row corresponds to the two s-wave quark-antiquark combinations. The recent results of duality theory confirm the observation of exchange-degeneracy, which connects 0^{-+} to 1^{+-} and 1^{--} to 2^{++} (our double arrows).

Experimentally, all of these mesons are now known (Söding *et al.*, 1972) except for the 0^{++} set which is as yet uncertain. For reasons of covariance, the 1^{++} set does not play an important role in the near-forward region. For zero-strangeness cases, we have two isoscalar and one isovector state for each J^{PC}; this would imply dealing with five states per J^{PC}, i.e., 30 states altogether (or 18 isospin $\times J^{PC}$ multiplets). Actually, the isoscalar 0^{-+} is relatively weakly coupled to the nucleon, because of the peculiar F/D ratio in the SU(3) coupling. In the 1^{--}, 0^{++}, 1^{++} and 2^{++} the particular form of SU(3) breaking mixes isoscalars so that only one is coupled to the nucleon. This leaves us with the following (Collins, 1971, with masses in parentheses),

	0^{-+}	1^{+-}	1^{--}	0^{++}	2^{++}
$I=1$	$\pi(140)$	$B(1235)$	$\rho(765)$		$A_2(1310)$
$I=0$	$\eta'(960)$		$\omega(785)$		$f(1260)$

Phenomenologically, the couplings of the 2^{++} are about five times stronger (in $g^2/4\pi$) than the 1^{--} couplings (Michael, 1970). This implies an attractive heart which should have some chance of producing a second potential well, closer to the nucleon's center.

The existence of a tensor force in nuclei has been known for a long time. The new high-energy picture points to the possibility that aside from quadrupole moments, the longitudinal component of a 2^{++} exchange might create an attractive heart – depending of course upon the particular many-body situations arising for various baryon numbers. This is plausible, since the elementary interactions, repulsive and attractive, are of the same order of magnitude, and it is only the variation in A and in configurations which can establish clear cut regions of preponderance.

From the experimental evidence it now looks as if Regge trajectories keep rising indefinitely in the Chew-Frautschi plot. This has been idealized in duality as a system of straight lines (e.g., the Veneziano model), perhaps an overstatement. However, it should at least *provide for a further hard core* reappearing at even shorter range. This would help in creating conditions in which the hypercollapsed state would be stable, rather than tend to collapse further if external pressure is applied.

3. Nuclear Isomers

Bodmer (1971) has studied the possible existence of hypercollapsed nuclei in terms of the nuclear Hamiltonian. He assumes this Hamiltonian to contain in addition to the 'normal' $H_A(N)$, a part $H_A(C)$ for the collapsed version corresponding to the same baryon number A.

Before going into any details, it is instructive to consider the analogy with molecular (nuclear) fusion and fission. Taking a deuterium molecule (the N state), we know that it is fusion-favored, i.e., that given the right conditions it can 'collapse' into a He atom (the C state); the D-D system is originally bound by $B_{2,2}(N) < B_{2,2}(C)$ the binding energy in He. On the other hand, at the other end of the periodic table we have fissioning nuclei (or 'molecules'), i.e., with the fission products originally bound by $B_A(C) < B_A(N)$, the binding in a pseudo molecule made of the fission-product atoms.

The radius of the hypercollapsed state $R_c < R_N = r_0 A^{1/3}$ and is either a constant ($\lesssim 0.5$ fm) or a saturating radius $R_c \sim r_c A^{1/3}$ with $r_c < 0.4$ fm. As in the above molecular examples, the most general possibility would correspond to at least one cross-over region A_{crit}, but the requirement of spherical symmetry at the microscopic level appears to invert the sequence:

$$B_A(C) < B_A(N) \quad \text{for} \quad A < A_{\text{crit}}$$
$$B_A(C) > B_A(N) \quad \text{for} \quad A > A_{\text{crit}}.$$

Bodmer estimates A_{crit} to lie somewhere between $A = 16$–40. To reach this value, he treats the second case; assuming the lifetime of a 'normal' stable nucleus to be $\tau_A(N) > 10^{31}$ s (one collapse per mole per year), he estimates the penetrability of the saturation barrier (the hard core) $W(r)$ between normal and collapsed states. Comparing $\tau_A(N) \geqslant 10^{31}$ s with the 'period' τ corresponding to $R_A(N)$, i.e., about 10^{-22} s, he gets (P is the penetration probability, τ^{-1} the number of penetration 'attempts' per second)

$$P = 10^p \leqslant 10^{-22} \div 10^{31} = 10^{-53}.$$

Using various plausible values for the nuclear parameters does not modify p by more than 20%. P can now be related to the equation of state of nuclear matter – which is of course part of our guess. Thus

$$p = \exp\left\{2(2M/\hbar^2)^{1/2} \int_{R_c}^{R_N} [W(r)]^{1/2} \, dr\right\},$$

where $M \approx A \, m_N$ is the mass of N_A, the normal nucleus. It is by extrapolating smoothly from the region around R_N where we know the equation, to the region of a hypothetical R_c, that Bodmer gets his estimate of A_{crit}.

For the binding energy per nucleon in the hypercollapsed state, we have to assume that it will stay in the hundreds of MeV, so as to avoid zero or negative energy nuclei.

4. Unitary Symmetry

The Fermi energy may rise to 0.5 m_N or above. This will then lead to the creation of hyperons and mesons. In the extreme case, we might have the unitary spin equivalent of an α particle: $A = 16$, with all eight octet baryons appearing with spins up and down. The next 'complete shell' would be at $A = 56$, with all components of *8* and *10* etc. The total strangeness (and electric charge) tends to be very low in general (and 0 in the 'complete' cases). Bodmer gets similar results from a rough quark model calculation.

5. 'Superbaryons' and Cosmology

In recent years, a thermodynamical approach to strong interactions (Hagedorn, 1970) has been used to predict results of high energy multiparticle collisions in accelerators and in cosmic rays, and to study the early stages of the expanding universe. This approach provides, it seems, an efficient way of accounting for the strong interactions through the insertion of an infinite spectrum of metastable ('elementary') states, which implies there is a maximal temperature to hadrons. In this approach, the development of concentrations (leading to the ultimate formation of galaxies) appears to be inescapably tied up with the existence of 'superbaryons.' A superbaryon is a quasi-elementary particle, with

$$10^{67} > A > 1.$$

It is required (Hagedorn, 1972) that the decay of a galactic-mass state go through cascade emission of nucleons with the large-A state, appearing as a hadron. Ordinary nuclei do not fit the above description; hypercollapsed nuclear matter might be just what is required.

Note that as an energy source, if $B_A(C) \gg B_A(N)$, we might have a new mechanism in the gradual collapse of normal to hypercollapsed matter.

6. Observation

The search for hypercollapsed matter resembles the search for quarks, in that we are after unusual e/M ratios (here because of new M values instead of e values). Cosmic rays appear to be one possible place to search; present limits on nuclear isomers (as pointed out by W. Webber in a private communication to H. Kasha) do not get under 5%. Neutral collapsed matter would be more difficult to find. One could also search material on the Earth's crust (though hypercollapsed matter might sink gravitationally).

7. Astrophysics

In the main, the existence of hypercollapsed nuclei would modify present estimates of the size of neutron stars, etc. We would have a new region of stability, following white dwarfs and 'normal' neutron stars. Neutron star material would also include dense 'raisins' of hypercollapsed matter. Further catastrophic collapse into a black hole – a possibility raised by the effects of 2^{++} attraction in the center of the neutron star, i.e. superstrong short-range gravitation – would be inhibited by the existence of a new repulsive barrier as predicted by the Chew-Frautschi plot (the ρ or ω Regge recurrences).

Acknowledgments

The author is indebted to Dr R. Hagedorn for an illuminating discussion, and to Dr H. Kasha for an enlightening correspondence.

References

Bodmer, A. R.: 1971, *Phys. Rev.* **D4**, 1601.
Collins, P. D. B.: 1971, *Phys. Reports* **1C**, 103.
Hagedorn, R.: 1970, *Astron. Astrophys.* **5**, 184.

Hagedorn, R.: 1972, lecture given at the Ettore Majorana School of Astrophysics and Cosmology.
Michael, C.: 1970, *Springer Tracts Mod. Phys.* **55**, 174.
Ne'eman, Y.: 1968a, in *Science Year*, World Book Science Annual, Field Enterprises Educationa. Co., Chicago, p. 157.
Ne'eman, Y.: 1968b, in A. Perlmutter *et al.* (eds.), *Symmetry Principles at High Energy*, W. A Benjamin, New York, p. 149.
Söding, P., Desy, J. B., Barbara-Galtieri, A., Enstrom, J. E., Lasinski, T. A., Rittenberg, A., Rosenfeld, A. H., Trippe, T. G., Barash-Schmidt, N., Bricman, C., Chaloupka, V., and Roos, M.: 1972, *Phys. Letters* **39B**, 1.

DISCUSSION

Bethe: I do not understand why it is not possible for a *few* nucleons to go into the condensed state, if that state exists, rather than the entire nucleus. I can well understand that *two* nucleons would not have a condensed state, for the same reason that a deuteron has such a small binding. But *four* nucleons should be very favorable, just like the alpha-particle, and in this case the penetrability should be very high – perhaps $P = 10^{-10}$. Then with the characteristic nuclear time of 10^{-22} s, the lifetime of an ordinary nucleus against collapse would be 10^{-12} s – clearly unacceptable. If nuclear forces again saturate at these smaller distances, i.e. if $R_c = r_c A^{1/3}$, then one should except the quasi-alpha-particle to have essentially the full binding energy per particle for a collapsed nucleus, because for $A = 5$ the p-shell would have to be started, with much diminished binding. Therefore I believe the estimate of 10^{-12} s for a collapse time is reasonable.

If, on the other hand, there is no saturation, i.e. if the radius of collapsed nuclei is $R_c = r_c$ *independent* of A, then it might be necessary to assemble a large number of nucleons before the collapsed state becomes energetically favorable, and then the long collapse times might be realized.

Ne'eman: I agree with the general spirit of your comment which is one of the reasons I mentioned a lower bound, i.e. a minimal baryon number for the collapsed state to appear. This is indicated by its non-appearance in abundance in the form of partly-collapsed heavier nuclei, in which an internal alpha-particle would have collapsed. I think however that your figure of 10^{-10} for the penetrability of the particle is somewhat too large, being based upon the Coulomb barrier rather than upon the forces we have in this case.

The no-saturation picture for the collapsed state might indeed be more plausible because of the non-appearance of collapse for a small number of baryons.

Bethe: If a collapsed state for nuclei exists, this would have a catastrophic effect on neutron stars. Namely, if neutrons are brought together within a distance of order $r_c = 0.4$ fm by *external* forces, i.e. by gravitation, then the collapsed state would automatically be formed, *without* the need of going through a potential barrier. In other words, the equation of state would be very soft. It would not be possible to form neutron matter of a density corresponding to $r_0 = r_c$ which is four nucleons fm^{-3}. In fact, the pressure would have a maximum at a much *lower* density than this, because at r_c the *energy* is supposed to be negative. This would keep the maximum density of a neutron star way below 6×10^{15} g cm^{-3}, and the maximum pressure correspondingly low. Accordingly, the maximum mass of a neutron star would be very low, and the maximum moment of inertia much below 2×10^{44} g cm^2 which is the most probable (minimum) value for the Crab pulsar.

Ne'eman: If indeed we should take the value of 2×10^{44} g cm^2 for the moment of inertia it seems to put a lower bound on the mass of a neutron star. Counting a binding energy of 50 MeV per nucleon in the collapsed state should not make a large difference in our estimate of the star's mass, but its size will of course be 2–3 times smaller. Considering that we do not have independent knowledge of the star's baryon number, we do not know if a size of 5 km, for instance, corresponds to the number of baryons in one solar mass at normal nuclear density, or to perhaps 10 times as many baryons at collapsed distances and a density which is larger than normal even though it is smaller than at intermediate distances.

Bethe: Concerning observations, I understand that cosmic ray physicists, e.g. Dr Peters at Copenhagen, can observe *both* charge and mass. If only nuclei like ^{16}He (which should exist if *some* of the nucleons are converted into hyperons) were to exist, it would stand out very clearly and should be discoverable in much smaller quantities than 5%.

Ne'eman: I agree that the 'hyperonized' types of collapsed nuclei would be easier to discover. However, the entire question of the abundances of such nuclei makes it difficult to get good upper bounds for cosmic rays.

MATTER IN SUPERSTRONG MAGNETIC FIELDS

M. RUDERMAN*

Institute for Advanced Study, Princeton, N.J. 08540, U.S.A.

Abstract. We consider the structure of atoms and atomic chains in the presence of ultra-strong magnetic fields as might be found in pulsars or neutron stars. Some consequences of these models for neutron star surfaces are mentioned.

1. Electrons in Superstrong Magnetic Fields

The enormous range of known 'strong' magnetic fields which exist in the universe is sampled in Table I. In magnetic fields greatly exceeding 10^9 G the nature of matter is qualitatively different from that of our normal experience: magnetic forces on electrons become stronger than Coulomb forces in atoms. The surface of a neutron star may consist of such a rather unique form of matter and this is the prime motivation for considering the structure of atoms, molecules, and above all compact matter, in superstrong fields. The nature of such matter seems reasonably susceptable to detailed analysis, if worth the effort. Any possible effects on observations or electrodynamics of pulsars is much less clear.

TABLE I
Very strong magnetic fields

Location	Field (Gauss)
Iron magnet	10^4
Superconducting magnet	10^5
Solar surface	$1-10$
Outer solar interior	10^3
Magnetic white dwarf surface	10^6-10^7
Neutron star (pulsar) surface	$10^{12}-10^{13}$
Neutron star interior	$? \gg 10^{13}$
Nuclear surface and interior	$10^{15}-10^{16}$

Classically a sufficiently huge magnetic field (**B**) confines a free electron to motion along the field like a bead on a straight wire. Quantum mechanical zero point motion perpendicular to **B** limits the magnetic confinement. 'Weak' electric fields can accelerate electrons only parallel to **B** or give a slow unaccelerated drift velocity perpendicular to **B**.

The essential features of the quantum mechanical description of the motion of electrons in superstrong **B** and weak atomic electric fields is clear in the Bohr model.

* Permanent address: Department of Physics, Columbia University.

A non-relativistic electron of mass m_e in a uniform **B** has an equation of motion

$$\frac{m_e v^2}{\varrho} = \frac{ev}{c} B \tag{1}$$

with ϱ the cylindrical radius perpendicular to **B** and **v** the perpendicular velocity. (Along **B** the electron, of course, moves as a free particle.) The quantization condition is

$$\mathbf{p} \times \boldsymbol{\varrho} = n\hbar \mathbf{B}/B \tag{2}$$

with

$$\mathbf{p} = m_e \mathbf{v} + \frac{e\mathbf{A}}{c}. \tag{3}$$

A convenient gauge for the vector potential is

$$\mathbf{A} = \frac{B\varrho}{2} \hat{\boldsymbol{\phi}}. \tag{4}$$

The quantized energies are

$$E_n = n\hbar \omega_c; \quad n = 1, 2, \ldots, \infty \tag{5}$$

with the 'cyclotron frequency'

$$\omega_c \equiv \frac{eB}{m_e c}. \tag{6}$$

The circular orbits have quantized radii

$$r_n = (2n)^{1/2} \hat{\varrho}$$

with

$$\hat{\varrho} \equiv \left(\frac{\hbar c}{eB}\right)^{1/2} = \frac{2.6 \times 10^{-10}}{B_{12}^{1/2}} \text{ cm}, \tag{7}$$

where B_{12} is the field in units of 10^{12} G. We shall generally consider fields so large that neither thermal energies nor perturbative electric fields are sufficient to give significant occupation to any except the lowest energy state with $E = \hbar \omega_c$. (For $B = 2 \times 10^{12}$ G, $\hbar\omega_c \sim 25$ keV and $r_1 = 3.6 \times 10^{-10}$ cm.)

By choosing other gauges for **A** the centers of the circular orbits can be moved anywhere. This degeneracy is most simply resolved by considering first the effect of a weak uniform electric field **E** perpendicular to **B**. The classical electron will then have a cycloidal motion with an average drift velocity

$$\mathbf{v}_d = \frac{\mathbf{E} \times \mathbf{B}}{B^2} c. \tag{8}$$

In a weak radial electric field, say from an unshielded nucleus at the origin $\varrho = 0$, the electron's uniform cycloidal motion becomes slightly bent along a large circle so that the classical electron's cycloidal motion takes it slowly around the nucleus as shown in Figure 1. These orbits are just a superposition of the circular orbits with origins $\varrho \neq 0$ but with the orbit centers guided to revolve about the origin of **E**. The strength of the (weak) electric field controls only the rate of revolution about the origin and has only a very slight effect upon the electron confinement or energy. For the lowest energy orbits whose guiding center radius $\gg r_1$ the quantization condition becomes

$$2\pi m \hbar = \oint \mathbf{p} \cdot d\boldsymbol{\varrho} \sim \oint \frac{e\mathbf{A}}{c} \cdot d\boldsymbol{\varrho}. \tag{9}$$

The quantization of (9) thus determines the guiding center radii to be

$$\varrho_m = (2m)^{1/2} \hat{\varrho}; \quad m = 1, 2, ..., \infty \tag{10}$$

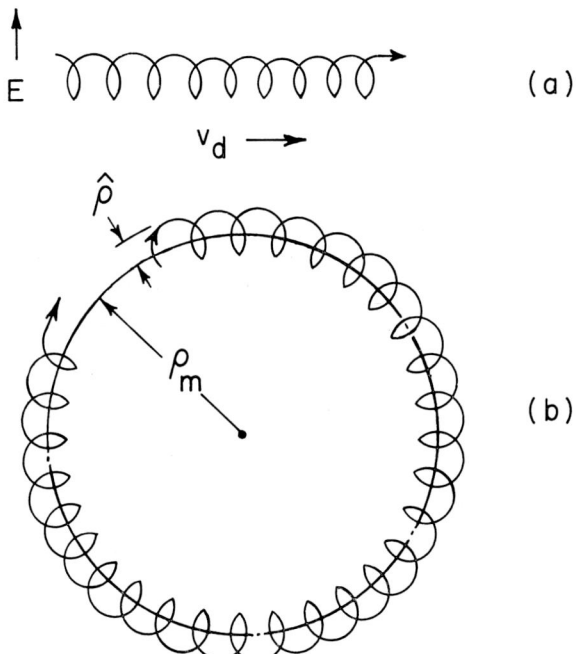

Figs. 1a–b. (a) Motion of an electron in crossed electric and magnetic fields. (b) Motion in a radial electric field with perpendicular **B** out of the plane.

for a set of states all of which in the limit $\mathbf{E} \to 0$ have energy

$$E = \hbar \omega_c \quad \text{(all } m\text{)}. \tag{11}$$

The angular momentum quantum number m mainly determines how far away the electron is from the origin rather than its $m_e v \varrho$. The width of the cycloidal orbit $\sim \hat{\varrho}$ for all m.

The criterion for a 'weak' electric field is that the difference in potential across the orbit satisfy

$$eE\hat{\varrho} \ll \hbar\omega_c. \tag{12}$$

Applied to an atom this is equivalent to

$$\hat{\varrho} \ll a_0, \tag{13}$$

i.e., the magnetic confinement radius is much less than that from the atomic electric fields, the Bohr radius a_0. Equation (13) gives

$$B \gg \left(\frac{e^2}{\hbar c}\right)^2 \frac{m_e^2 c^3}{\hbar e} = 4 \times 10^9 \text{ G}. \tag{14}$$

This is the criterion that the magnetic field be sufficiently large that all electrons in ground state atoms or condensed matter be in the lowest magnetic energy states, given by (11).

The magnetic confinement will force the electrons to be relativistic when $\hat{\varrho} \lesssim \hbar/m_e c$ or $\hbar\omega_c \gtrsim m_e c^2$. These criteria imply that a necessary condition for electrons to be non-relativistic is

$$B \ll \frac{m_e^2 c^3}{\hbar e} = 4.4 \times 10^{13} \text{ G}. \tag{15}$$

The inequalities (14) and (15) are probably well satisfied for neutron star surfaces. (Relativity does not, however, introduce great complications into atomic calculations when fields exceed 4.4×10^{13} G.)

The wave mechanical description of the classical orbit of Figure 1b uses the 'Landau orbital' (Landau and Lifschitz, 1965) solutions of the Schroedinger equation with uniform magnetic field (in the z-direction):

$$\psi_{m,k} = \exp\left(\frac{-\varrho^2}{4\hat{\varrho}^2}\right) \varrho^{|m|} \exp(im\phi) \exp(ikz). \tag{16}$$

This wave function represents a free particle in the z-direction, magnetically confined in the perpendicular direction with a maximum probability at

$$\bar{\varrho} = (2m+1)^{1/2}\hat{\varrho}; \quad m = 0, 1, ..., \infty. \tag{17}$$

The angular momentum $m\hbar$ is mainly $e\varrho \times \mathbf{A}/c$. The energy of the states represented by the wave function of (16) is

$$E = \frac{p_z^2}{2m_e} + \hbar\omega_c + \frac{e\boldsymbol{\sigma} \cdot \mathbf{B}}{m_e c} \tag{18}$$

with $\boldsymbol{\sigma}$ the Pauli spin matrices. All states not represented by (16) are higher in energy by an integral multiple of $\hbar\omega_c$, tens of keV on a pulsar surface. Similarly, parallel and anti-parallel electron spin differ by $\hbar\omega_c$. A complete set of magnetic ground state wave functions is given by (16) with anti-parallel electrons. From (18) the corresponding electron energies are

$$E = \frac{p_z^2}{2m_e} \quad \text{(all } m\text{)}. \tag{19}$$

2. Atoms

The hydrogen atom in a superstrong magnetic field contains an electron whose binding in the z-direction is caused by Coulomb attraction but whose confinement in the perpendicular direction is magnetic. The ground state is composed of an $m=0$ Landau orbital except that $\exp(ikz)$ is replaced by a real function of z chosen to minimize the total energy. The probability distribution is roughly a long cylinder of length l and radius $\hat{\varrho}$ with $l \gg \hat{\varrho}$ (Figure 2). Then, very roughly,

$$E \sim \frac{\hbar^2}{m_e^2 l} - \frac{e^2}{l} \ln\left(\frac{l}{\hat{\varrho}}\right) \tag{20}$$

and minimizing with respect to l yields

$$l \sim \frac{a_0}{\ln\left(\frac{a_0}{\hat{\varrho}}\right)} \tag{21}$$

and

$$E \sim -\frac{\hbar^2}{m_e a_0^2}\left[\ln\left(\frac{a_0}{\hat{\varrho}}\right)\right]^2 \tag{22}$$

(cf. Haines and Roberts, 1969). In the limit of huge fields $a_0 \gg l \gg \hat{\varrho}$ and the binding energy E grows like $\ln^2 B$.

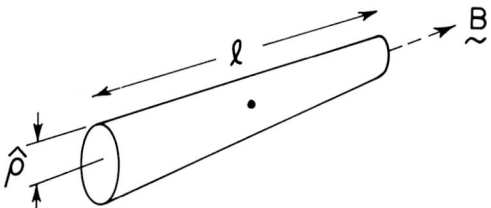

Fig. 2. Schematic hydrogen atom in a huge magnetic field.

(An analogue of the hydrogen atom in superstrong **B** is obtained within some solids with achievable **B** because the dynamic mass of an electron m^* can be enormously smaller than m where the curvature in an energy band is great. Then (14) gives a critical **B** smaller by $(m^*/m)^2$. An additional reduction can come from dielectric reduction of the effective e. Bound electron-hole pairs [excitons] can be qualitatively altered in 10^4–10^5 G fields.) A more quantitative variational estimate of the hydrogen binding energy is given in Figure 3 for the regime $B \gg 10^{10}$ G (Cohen et al., 1970). A free hydrogen atom on a pulsar surface would be expected to have a binding energy of a few hundred eV.

There are two classes of excited states of such simple atoms, neither of which involve excitation out of the ground magnetic state. Instead of the smallest radius cylinder $m=0$, higher values of m corresponding to cylinders of radius $\sim (2m+1)^{1/2}\,\hat{\varrho}$ of (17), can be used for the electron. From (22) the binding energy depends only logarithmically on this radius, so that for superstrong **B** these states are bound almost as tightly as the ground state. The second class of excited state consists in replacing the nodeless

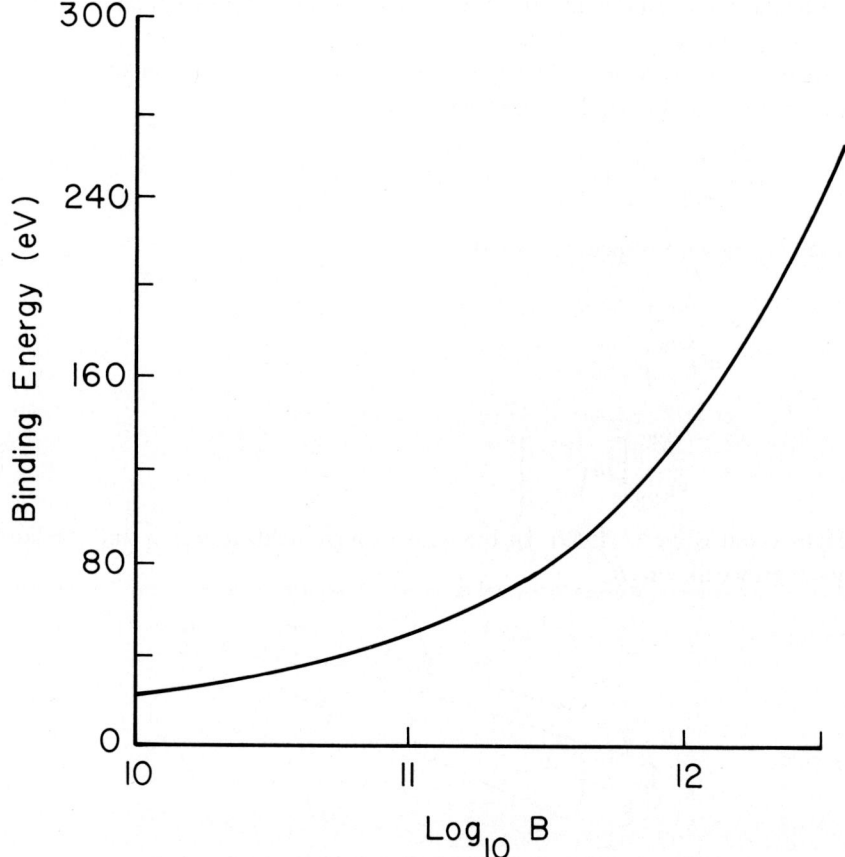

Fig. 3. Binding energy of a hydrogen atom in superstrong magnetic fields.

z-dependent part of the wave function by an orthogonal function with $v=1, 2, \ldots$ etc. nodes. All of these states are relatively very weakly bound (cf. Figure 4) and have a z-extent of order a_0 or greater rather than the l of (21). A simple way to estimate the energies of the second class of excited state has been pointed out by L. Spruch. Normal hydrogenic s-states satisfy the reduced wave equation $-\phi''-(2m_e/\hbar^2 r)\,\phi=E\phi$ with $\psi=\phi/r$. Therefore $\phi(0)=0$. The quasi-one dimensional atom in superstrong **B** satisfies almost the same equation with $r = z$ except that $\psi \leftrightarrow \phi$ and not the ground but the first excited odd state has the node at $z = 0$. Therefore this one node excited state has about the same binding energy as the ground state of the normal atom. Multinode states are less bound as the number of nodes increases.

The simplest multielectron atoms are formed by filling consecutively the $m = 0, 1, 2, \ldots, Z$ Landau orbitals with a single (spin anti-parallel to **B**) electron in each. From Figure 4 it is evident that putting two electrons in the same orbital by introducing excited states in the z-dependent component costs much more energy than putting the electrons slightly further away in ϱ. This continues to remain true only until the distance ϱ becomes comparable to $a_0 Z$; the ground states of heavy multinode wave atoms utilize functions to put many electrons into a Landau orbital.

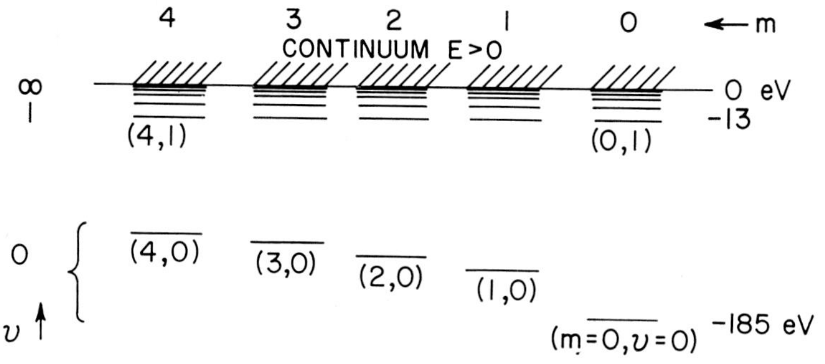

Fig. 4. Energy levels of a hydrogen atom in a 2×10^{12} G field (not to scale).

For the light atoms in the Hartree approximation each electron has its own pure Landau orbital. A variational calculation with such wave functions in a 2×10^{12} G magnetic field gives atoms whose shape and size is modeled in Figure 5 (Cohen et al., 1970). The length of the cylindrical atoms is not sensitive to Z while the radii increase roughly as $(2Z+1)^{1/2}$. These atoms have all electron spins anti-parallel to **B**, huge quadrupole moments, and volumes of order 10^{-3}–10^{-4} that of normal atoms. The ionization energy (E_1) of the last electron of light atoms in $B = 2 \times 10^{12}$ G has been estimated as

$$E_1 \sim 160 + 70 \ln Z \text{ eV}. \tag{23}$$

The $\ln Z$ term is the exchange energy contribution, calculated in perturbation theory. Calculations including exchange *ab initio* are in progress (Kaplan and Glasser, 1972). The ionization energy of (23) is typically two orders of magnitude greater than that for single ionization of normal atoms. It is also a smooth function of Z with none of the usual steep valleys and peaks. The absence of significant 'chemical' differences between atoms with neighboring Z's is a consequence of the one-dimensional nature of the filling of orbitals: there are, for example, no closed shells, which depend upon special degeneracies among three-dimensional angular momentum components.

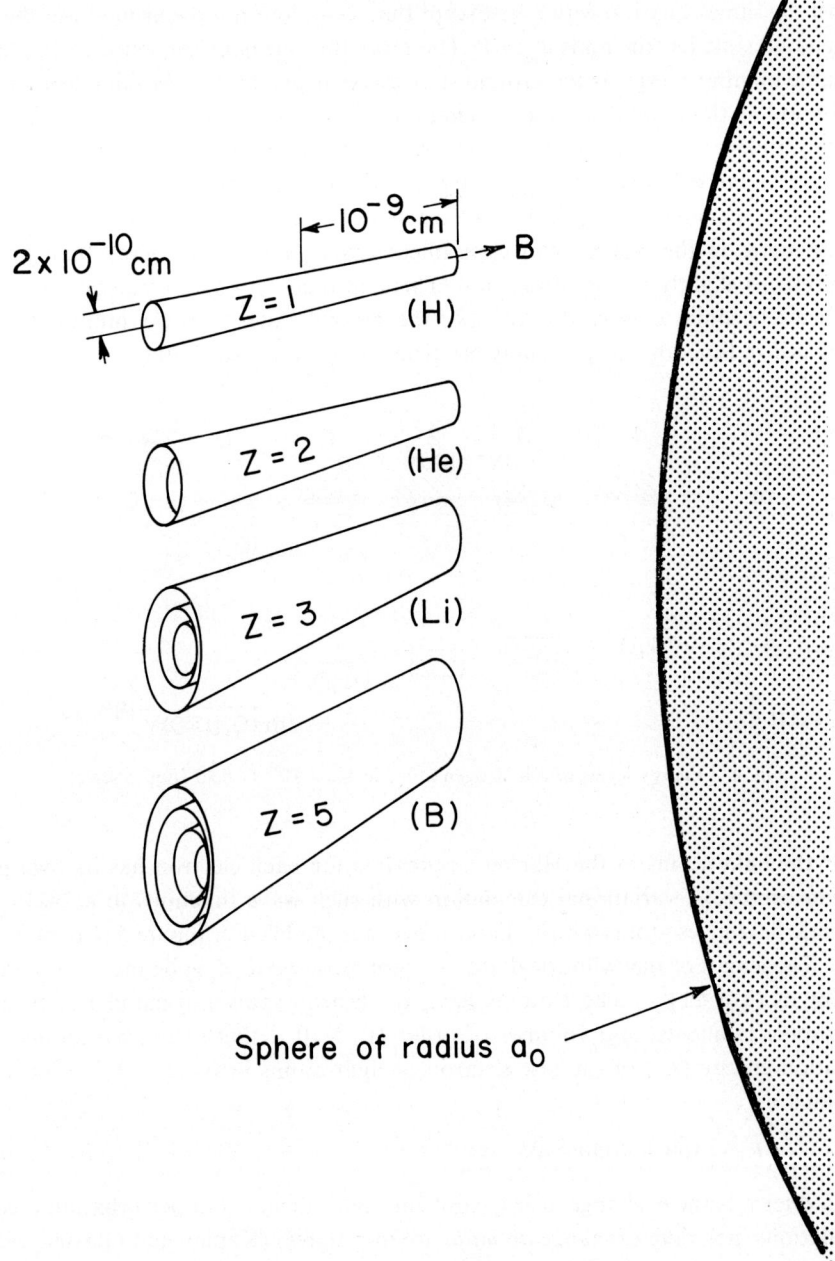

Fig. 5. Shapes and sizes of light atoms in a 2×10^{12} G field.

For heavier atoms the recipe of achieving the lowest energy state of an atom by putting only one electron in each Landau orbital fails. If it did not the last electron would be on a cylinder whose minimal approach to the nucleus is the radius

$$\varrho_z = (2Z + 1)^{1/2} \hat{\varrho}. \tag{24}$$

If, instead, the last electron were put into the closest orbital in its first excited state (i.e., a node at $z=0$ in the z-component part of the wave function), then its typical distance to the nucleus is a_0/Z. The dimensionless ratio

$$\eta \equiv \frac{a_0}{Z\varrho_z} = \left(\frac{B}{4.6 \times 10^9 \, Z^3}\right)^{1/2} \tag{25}$$

characterizes the different kinds of atoms which can exist (at $B \sim 5 \times 10^{12}$ G, $\eta \sim 0.3$ for $Z=26$ and $\eta=33$ for $Z=1$).

(i) The regime $\eta \gg 1$, corresponding to $B \gg 4.6 \times 10^9 \, Z^3$ G, is that described above with one electron per orbital. The total binding energy of such an atom has been estimated (Kadomtsev and Kudryavtsev, 1971a) as

$$E \sim -\frac{9 \, Z^3 e^2}{8 \, a_0} \ln^2 \eta. \tag{26}$$

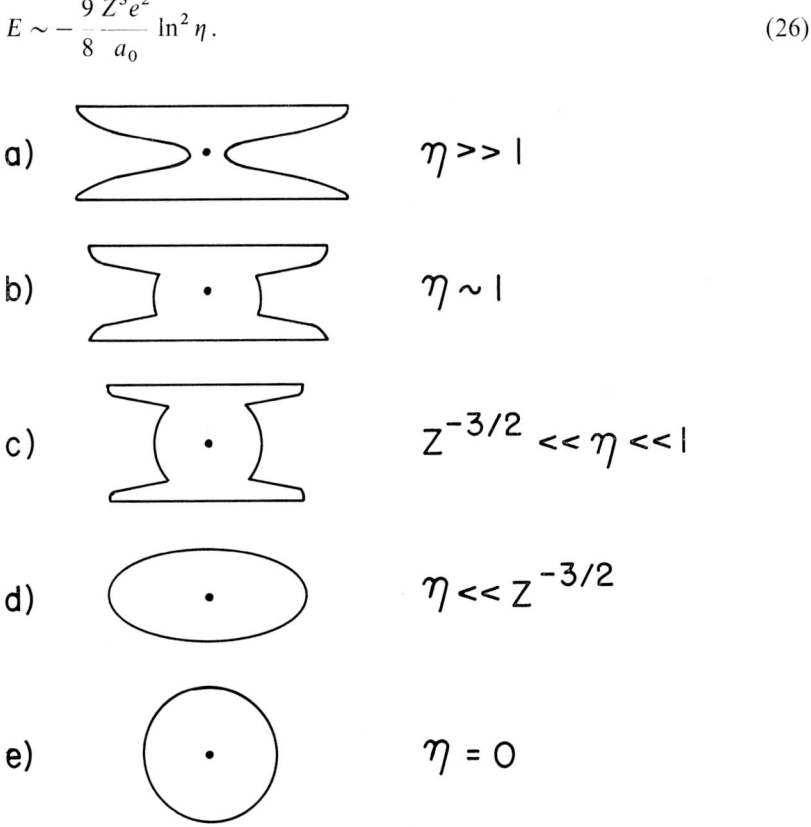

Fig. 6. Shapes of atoms in superstrong magnetic fields as a function of η of Equation (25) (not to scale).

The shape of the atom as revealed by cutting the atom in half with a slice through its symmetry axis parallel to **B** is given in Figure 6a. Dr H.-H. Chen finds this regime appropriate for iron ($Z=26$) in a field of a few times 10^{12} G (Columbia Univ. doctoral thesis 1973).

(ii) The regime $\eta \sim 1$ is characterized by emptying some of the outer orbitals of regime (i) and placing those electrons into excited states in inner orbitals. This intermediate regime has not yet been considered quantitatively. Its presumed shape is that of Figure 6b.

(iii) The regime $1 \gg \eta \gg Z^{-3/2}$ still has sufficiently strong B that no excited magnetic states (n of equation [5]>1) are significantly populated. But all of the inner Landau orbitals contain many electrons, so many that a Fermi-Thomas treatment of the electrons in each level appears reasonable. A Fermi-Thomas treatment of atoms (or ions) in superstrong **B** gives a spherical shape of radius

$$R \sim \frac{a_0}{Z\eta^{4/5}} \tag{27}$$

and total energy

$$E \sim -150 (B_{12})^{2/5} Z^{9/5} \text{ eV} \tag{28}$$

(Kadomtsev, 1970; Mueller et al., 1971). The Landau orbitals furthest from the nucleus have too few electrons to be treated in this approximation. These electrons 'see' a spherical ion core of net charge $\hat{Z} < Z$ which is essentially impenetrable because of the Pauli principle. These extracore electrons fill successive Landau orbitals singly, beginning with that orbital whose cylinder radius $\varrho \sim a_0/\hat{Z}\eta^{4/5}$ (Ruderman, 1971). There are not yet quantitative descriptions for their wave functions. They form a cylindrical sheath surrounding a spherical core to give a shape sketched in Figure 6c. The thickness of the sheath (δ) is of order $\eta^{4/5} R$.

(iv) When $\eta \ll Z^{-3/2}$, $B \ll 4 \times 10^9$ G, and perturbation theory is adequate. The perturbative part of the Hamiltonian is

$$H' = \frac{e\mathbf{B} \cdot \mathbf{L}}{mc} + \frac{e^2 B^2 \varrho^2}{8mc^2}. \tag{29}$$

When spin orbit coupling is negligible its effect on atomic shape is a slight additional confinement perpendicular to **B**.

3. Condensed Matter

Atoms in superstrong fields can bind to each other extraordinarily tightly. Even if all other quantum effects are ignored, the huge quadrupole moments of isolated atoms will result in a very strong electrostatic interatomic attraction for certain orientations. In an orthorhombic configuration near b.c.c. nearest neighbors attract and next nearest repel. But the quadrupole-quadrupole force falls off so rapidly with spacing (r^{-6}) that the attraction dominates. However a quantum mechanical binding analogous to conventional covalent bonding appears to contribute the main binding energy.

In ordinary matter the binding among atoms is essentially determined by the state of the least bound (valence) electrons. When these are shared between neighbors the binding energy is generally very much less than the total binding energy of the atoms. The wavelengths of almost all the atomic electrons, except the valence ones, are very much less than the interactomic spacing so that the state of most of the electrons is insensitive to the environment of the atom which contains them. This situation is very different for atoms in superstrong magnetic fields, especially when $\eta \gg 1$. All of the atomic electrons can be effective in binding, and in the limit $B \to \infty$ the binding between atoms can greatly exceed the total binding energy of an isolated atom (Kadomtsev and Kudryavtsev, 1971b).

The main difference between 'magnetic atoms' and conventional ones is the effect of the Pauli principle when atoms approach each other. Two hydrogen atoms with

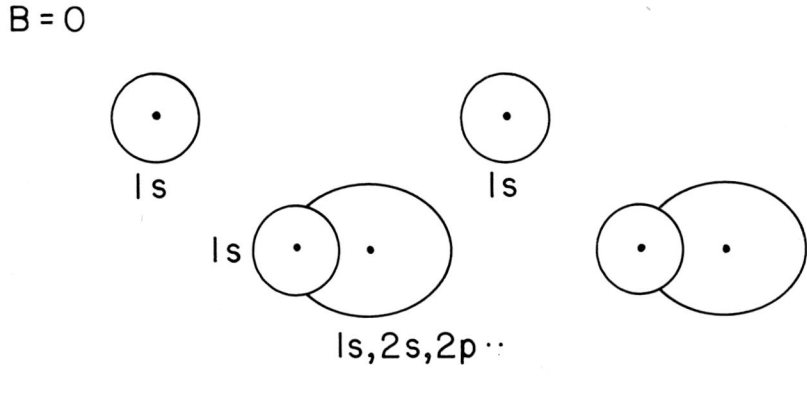

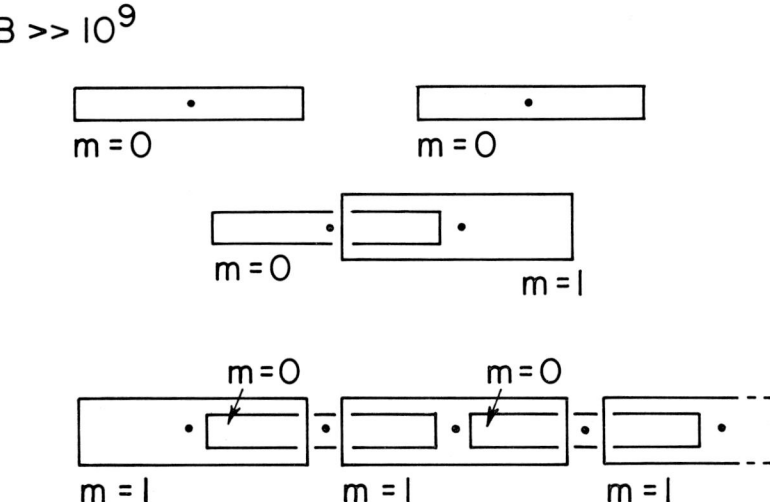

Fig. 7. Binding of magnetic hydrogen atoms into molecules and polymer chains. (All electron spins are antiparallel to **B**.)

parallel electron spins do not bind because in order to have the electrons close to each other one of the electrons has to be promoted out of its normal 1 s state to the 2 s, 2 p, etc. in order to satisfy the exclusion principle. This takes a relatively enormous energy, i.e., one comparable to the binding of the 1 s electron. Magnetic atoms which approach each other along their symmetry axes can satisfy the exclusion principle with a relatively small energy expenditure (Figure 7). An excitation which consists just in changing a Landau orbital from state m to $m+1$ changes the average cylinder radius $\bar{\varrho}$ by only $\delta\varrho \sim (2m)^{-1}\bar{\varrho}$. This changed radius, according to (22), enters only logarithmically into the electron binding energy. In the regime $\eta \gg 1$ the energy needed to excite in this way is negligible next to the electron binding energy. Thus it is inexpensive to satisfy the exclusion principle and both electrons can be shared by two nuclei. A similar argument holds for atoms with $Z>1$: it does not cost an excessive amount of energy to alter Landau orbital quantum numbers m in order to make room for more electrons.

When two atoms join in this way, additional atoms can attach at the ends (cf. Figure 7) to extend the diatom to a chain of arbitrary length (Ruderman, 1971). The magnetic polymer formed in this way can be described as a chain of nuclei of some spacing (l) surrounded by a sheath of electrons which occupy Landau orbitals out to some spatial distance (r). Magnetic confinement holds them near to the symmetry axis along which the nuclei are distributed. The electrons are free to move in the z-direction subject only to the restrictions of Fermi statistics. They behave as if in a one-dimensional metal. For an infinitely long magnetic jelly roll polymer of this sort, the energy per atom (i.e., per cell of Z electrons) is

$$E_a = -\frac{(Ze)^2}{l}\left[\ln\frac{2l}{r} - \left(\gamma - \frac{5}{8}\right)\right] + \frac{2Z^3\pi^2\hbar^2}{3m_e l^2}\left(\frac{\hat{\varrho}}{r}\right)^4 \quad (30)$$

with γ the Euler constant.* A minimization with respect to l and r yields (in the limit $\eta \to \infty$)

$$r \sim \frac{1.3\, a_0}{Z\eta^{4/5}} \quad (31)$$

$$l \sim \frac{2.4\, a_0}{Z\eta^{4/5}} \quad (32)$$

and a total binding per atom

$$E_a \sim -\frac{0.5\, Z^3 e^2}{a_0}\eta^{4/5} \quad (33)$$

This binding energy of a magnetic polymer atom is greater than the estimate of the binding energy of the isolated magnetic atom given in (26) when $\eta \gg 1$. The polymer is,

---

* I am grateful to Mr Hsing-Hen Chen for pointing out that the fraction ¾ in the published result should be as in (30).

therefore, extremely tightly bound. Its density (σ) is

$$\sigma \sim \frac{Z^4 m_n}{6a_0^3} \eta^{12/5} \sim 10^6 \left(\frac{Z}{26}\right)^4 \eta^{12/5} \text{ g cm}^{-3}, \tag{34}$$

enormously greater than the 1–10 g cm^{-3} range for conventional matter.

The nuclei in the surface of a neutron star are probably not hydrogen. Most scenarios for the birth of a neutron star suggest iron peak elements with, perhaps, a very thin covering layer of helium. For helium $\eta \sim 12$ so that the approximations leading to (33) may be valid. (The $|E_a|$ computed there is at least a minimum total binding energy for any η.) Then $\sigma \sim 10^4$ g cm^{-3} and $E_a \sim -800$ eV in a 5×10^{12} G field. But if the nuclear constituent is iron with $Z = 26$, $\eta = 0.3$ and regime (ii) or (iii) obtains. Only the outer

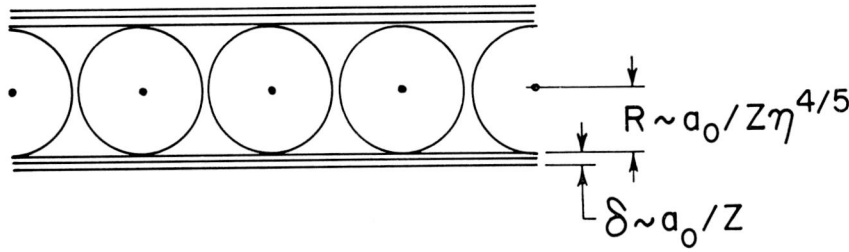

Fig. 8. Magnetic polymer chains in the regime $\eta \lesssim 1$.

sheath electrons contribute to the binding described above with the core electrons remaining almost unaffected (Figure 8). A very rough estimate for an iron magnetic polymer in a 5×10^{12} G field gives $E_a \sim -30$ keV and $\sigma \sim 4 \times 10^4$ g cm^{-3}. (See note added in proof on page 131.)

The strength or Young's modulus of a single polymer chain $\sim E_a/l \sim 10$–10^2 dyne. A bundle of polymers has a modulus $\sim 10^{19}$ dyne cm^{-2}, about 10^7 that of steel.

A polymer chain will attract a neighboring one electrostatically even if van der Waals polarization attraction is ignored. The periodically distributed nuclear charges

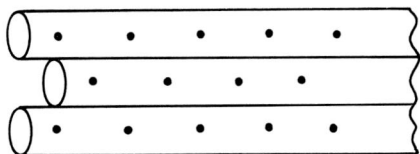

Fig. 9. Arrangement of adjacent magnetic polymer chains in condensed matter.

are imperfectly screened by the surrounding electron sheath. Therefore a magnetic polymer chain will be attracted to one right alongside itself but displaced half a lattice length along the field (Figure 9). (The resulting distribution of nuclei is close to that of a b.c.c. lattice, expected for nuclei in an approximately uniformly distributed electron background.) Matter formed of such chains will have the density σ in the range 10^4–10^5 g cm^{-3}.

4. Neutron Star Surfaces

Even under its almost zero vapor pressure the magnetic matter described in Section 3 remains in its condensed high density form up to very high temperatures; only when kT approaches the energy released when an atom is bound into a magnetic polymer chain does condensed magnetic matter become a gas. For iron this will happen near $T \sim 10^8$ K; even for helium such a T approaches 10^7 K. Both of these temperatures are much higher than estimated surface temperatures of all expect possibly very young pulsars, those whose age is less than or of order 10^3 yr. Therefore if a neutron star surface region were to be undisturbed except for thermal agitation it would have an abrupt edge. Unlike a gaseous surface the pressure can drop to zero even though the matter is dense and hot. The neutron star surface would more closely resemble that of the earth than that of a normal star. The solid (or liquid) magnetic polymer surface of a neutron star has a thickness of only a few meters. Below this the pressure from the weight of this layer compresses the underlying matter to above 10^5 g cm^{-3}, where the electron Fermi energy exceeds $\hbar\omega_c \sim 25$ keV. Then magnetic effects no longer qualitatively alter the matter equation of state. (The nuclei may arrange themselves in a crystalline lattice but this is a solid caused by the extreme pressure.)

The electrical properties of the magnetic polymerized surface are those of a one-dimensional metal: a good conductor parallel to **B**, a poor one perpendicular to it. Like a metal, and unlike an ionized gas, it has a work function which opposes the emission of electrons by appropriately directed electric fields. The electron work function (W) is typically expected to be of order a few hundred eV to a keV. In addition, an electric field whose direction is such that it would pull positive ions from a hot stellar surface plasma may be entirely ineffective in pulling nuclei out of the condensed metallic surface.

In a pulsar older than, say, 10^5 yr the surface temperature is expected to be below 10^5 K so that $kT \ll W$. Then field emission from the surface would be the main source of electrons injected into the pulsar magnetosphere if bombardment of the stellar surface by energetic particles and radiation is ignorable. Only surface electric fields parallel to **B** are effective. Unfortunately, it is just this component that is extremely model dependent in magnetosphere models. A conducting sphere (of radius R with surface dipole field B_s) spinning in a vacuum with angular frequency Ω has a surface electric field component $\mathbf{E}_s^0$ given by

$$\mathbf{E}_s^0 \cdot \mathbf{B}_s \sim \frac{\Omega R}{c} B_s^2. \tag{35}$$

This would give a vacuum electric field component parallel to $\mathbf{B}_s$ of about 10^{12} V cm^{-1} for the Crab pulsar and about 10^{10} V cm^{-1} for an old pulsar whose period is 3 s. The actual electric field component parallel to $\mathbf{B}_s$ may be very much less. Thus, we write

$$\mathbf{E}_s \cdot \mathbf{B}_s \equiv \varepsilon \mathbf{E}_s^0 \cdot \mathbf{B}_s, \tag{36}$$

where $\varepsilon = 1$ for no magnetosphere, $\varepsilon = 0$ for an exactly corotating magnetosphere and

ε is near zero for models with almost corotating magnetospheres. The field emission electron current from the electric field of (35) and (36) is

$$j \sim 10^3 \, F^2 \exp\left(-\frac{30 \, WP}{\varepsilon B_{12}}\right), \qquad (37)$$

where j is the current in A cm^{-2}, F is $\mathbf{E}_s \cdot \hat{\mathbf{B}}_s$ in V cm^{-1}, W is the magnetic metallic surface's work function in keV, and P is the pulsar period in seconds. (For most models of the Crab pulsar $j \sim 10^5$–10^7 A cm^{-2}.) From (37) long period pulsars ($P \gtrsim 3$ s) might have their electric current emission turned off and be unobservable. Faster pulsars could have sufficient field emission of electrons but the outgoing flow of negative charge would be balanced by electron inflow along other channels rather than positive ion outflow as conventionally assumed. Unfortunately uncertainties about the external flow of energetic particles and radiation onto the stellar surface preclude firm conclusions at present.

Note added in proof. Dr Chen in his thesis (*loc. cit.*) finds the additional binding energy for a chain of iron atoms in 2×10^{12} G to be about 10 keV/atom. The lattice spacing is 0.42 Å; 8 electrons per atom remain localized around each iron nucleus while the other 18 move freely in the surrounding cylindrical sheath of radius 0.25 Å.

Acknowledgment

This research was supported in part by the National Science Foundation.

References

Cohen, R., Lodenquai, J., and Ruderman, M.: 1970, *Phys. Rev. Letters* **25**, 467.
Haines, L. and Roberts, D.: 1969, *Am. J. Phys.* **37**, 1145.
Kadomtsev, B.: 1970, *Zh. Eksperim. Teor. Fiz.* **58**, 1765 [*Soviet Phys. JETP* 1970, **31**, 945.]
Kadomtsev, B. and Kudryavtsev, V.: 1971a, *Pis'ma Zh. Eksperim. Teor. Fiz.* **13**, 15 [*Soviet Phys. JETP Lett.* 1971, **13**, 9].
Kadomtsev, B. and Kudryavtsev, V.: 1971b, *Pis'ma Zh. Eksperim. Teor. Fiz.* **13**, 61 [*Soviet Phys. JETP Lett.* 1971, **13**, 42].
Kaplan, J. and Glasser, M.: 1972, *Phys. Rev. Letters* **28**, 1077.
Landau, L. and Lifschitz, E.: 1965, *Quantum Mechanics*, Addison-Wesley, Reading, Mass.
Mueller, R., Rav, A. R., and Sprunch, L.: 1971, *Phys. Rev. Letters* **26**, 1136.
Ruderman, M.: 1971, *Phys. Rev. Letters* **27**, 1306.

QUANTUM CRYSTALS IN NEUTRON STARS

V. CANUTO* and S. M. CHITRE**

Institute for Space Studies, Goddard Space Flight Center, NASA, New York, N.Y., U.S.A.

Abstract. Using the many-body techniques appropriate for quantum crystals it is shown that the deep interior of a neutron star is most likely an orderly arrangement of neutrons, protons and hyperons forming a solid. It is shown that a liquid or gas arrangement would produce higher energy. If so, a neutron star can be viewed as two solids (crust and core) permeated by a layer of ordinary or (perhaps) superfluid liquid. Astronomical evidence is in favor of such a structure: the sudden jumps in the periods of the Crab and Vela pulsars that differ by a factor of $\sim 10^2$ can be easily explained by the star-quake model. If the Crab is less massive than Vela (i.e., if it is not dense enough to have a solid core), the star-quakes take place in the crust whereas for Vela they occur in the core.

1. Introduction

It is not an exaggeration to say that the equation of state of matter at densities much beyond nuclear density is not adequately understood. An approach adopted by Banerjee *et al.* (1970), following a suggestion by Bethe, was to assume that when the nuclear forces become sufficiently repulsive, a possible minimum energy state could be achieved by keeping the nucleons as far away from one another as possible, i.e., by localizing them at lattice sites. This was a preliminary attempt to explore the viability of neutron crystallization and the lattice calculation was done in the harmonic approximation employing the classical Debye model. Shortly afterwards Pandharipande (1971) calculated the binding energy of dense neutron gas up to a density of the order of 7×10^{15} g cm^{-3}. He used the Reid soft-core potential in the framework of the lowest order variational approach by expanding the trial wave function as a product of single particle wave functions and the short-range correlation *à la* Jastrow. His calculations yielded energies which were lower approximately by a factor of 1.5–2 compared to the energy of Banerjee *et al.* (1970) over the density range $7.5 \times 10^{14} \leq \varrho \leq 6 \times 10^{15}$ g cm^{-3}. This clearly demonstrated a need to undertake a quantum mechanical treatment for the neutron lattice. Such a treatment is expected to lower the energy compared to the classical calculation by spreading the wave function around each lattice site. Moreover, the classical harmonic oscillator treatment is not adequate for a satisfactory description of a lattice in which the zero-point kinetic energy of oscillating particles becomes comparable with their potential energies. We must, therefore, work in the framework of a quantum mechanical formulation by including the effects of short-range correlation.

Recently Anderson and Palmer (1971) and Clark and Chao (1972) adopted an empirical approach based on de Boer's quantum mechanical law of corresponding states in order to examine the possibility of crystallization of neutron matter. In

---

* Also with the Dept. of Physics, City College of New York, N.Y.C.
** NAS-NRC Senior Research Associate. On leave of absence from the Tata Institute of Fundamental Research, Bombay.

this approach the solidification pressure of neutron matter is estimated by scaling the known low temperature properties of quantum solids like ^{3}He to those characteristic of nucleons. Both these computations yielded a solidification pressure of the order of 5×10^{27} atmospheres and the corresponding density for the onset of solidification of neutron matter in the vicinity of 5×10^{14} g cm^{-3}. Quite apart from the fact that the nucleon-nucleon forces are spin, state, and isospin dependent, which can scarcely be embodied in a single effective potential, it is not altogether justified simply to scale the potential. Such a scaling, even though it may give an over-all agreement between the well-depth and the core-radius of various potentials, almost certainly cannot produce a good fit to the tail of the potential – features which, in fact, are critical from the point of view of the stability of the lattice. We have displayed in Figure 1 the singlet $^1S_0$ nucleon-nucleon potential along with the appropriately scaled Lennard-Jones potential from which it can be readily seen that even though the core radius and the depth can be fitted to a desired accuracy, the quantitative difference in the relative displace-

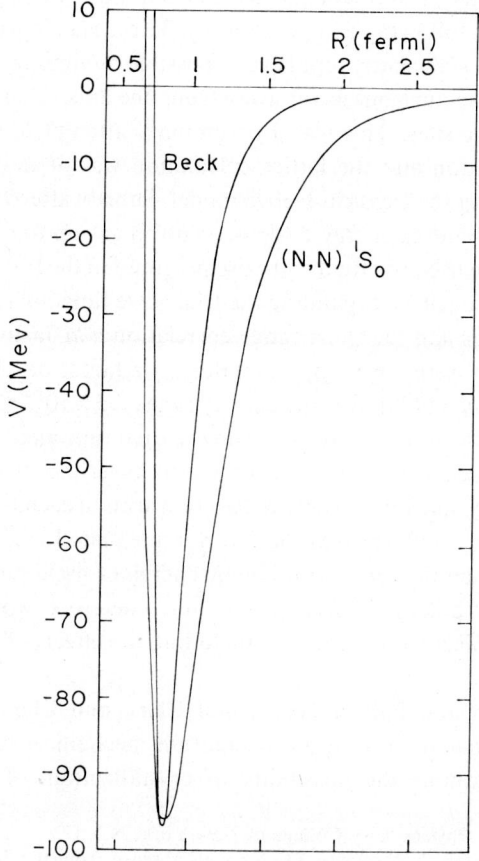

Fig. 1. Singlet $^1S_0$ Reid potential and scaled L-J potential.

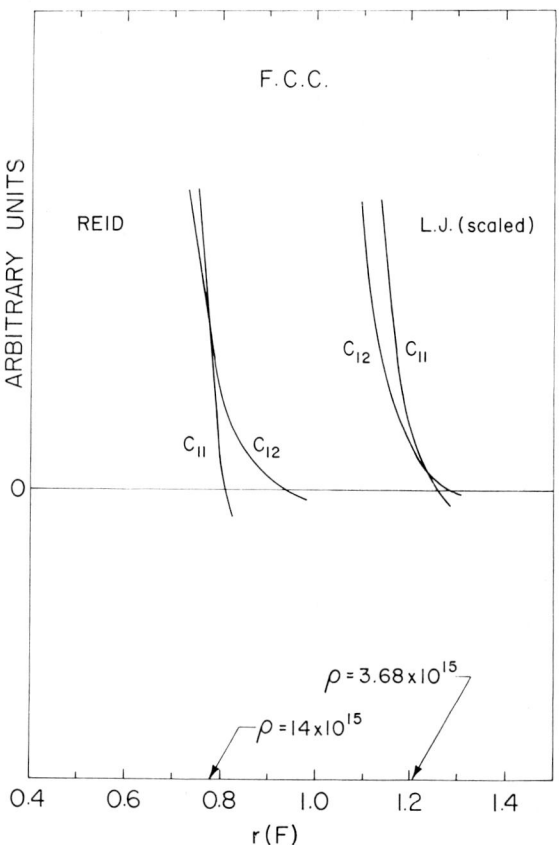

Fig. 2. Elastic constants C_{11}, C_{12} computed for an FCC lattice using Reid and L-J (scaled).

ment of the points of inflection and different slopes of the two potentials are quite noteworthy. The elastic constants computed by using the nuclear and scaled potentials are shown in Figure 2 where the interesting feature is the point at which the elastic constants go soft. On the scaled LJ potential the elastic constants indicate an onset of the solid phase at a density $\sim 3.7 \times 10^{15}$ g cm^{-3}, while the elastic constants computed using the Reid soft-core potential do not permit the solid to exist until a density approximately four times higher is reached. We feel that simply estimating the solidification pressure does not by any means provide the complete answer, rather we must test a given lattice structure for stability to determine whether the solid state is permissible on mechanical grounds.

2. Many-Body Treatment

The many-body treatment which we have used is an extension of the T-matrix approach to quantum crystals recently discussed by Brandow (1972), and an excellent review can be found in a paper by Guyer (1969). The Slater determinant for a system of N

particles is built up by single-particle wave functions which we take to be of Gaussian form:

$$\phi(r) = \alpha^{3/2}\pi^{-3/4}e^{-\alpha^2/2|r-R|^2}, \tag{1}$$

where R is the coordinate of the lattice site around which the particle is supposed to perform an oscillatory motion. Clearly (1) is the eigenfunction of an harmonic oscillator potential $U(r)$, centered around the lattice site R, i.e., $U(r) = 1/2\, m\omega^2 |r-R|^2$, where the frequency ω enters the wave function through the parameter $\alpha^{-1} = (m\omega/\hbar)^{-1/2}$ which represents the spread of ϕ around the lattice site. For determining α we do a Hartree calculation by taking the single particle potential to be given by

$$U(r_1) = \sum_j \phi_j^*(r_2)\, V(r_1 - r_2)\, \phi_j(r_2)\, d^3r_2, \tag{2}$$

where the two-body nucleon-nucleon potential is taken to be Reid's phenomenological soft-core potential. The index j runs over successive neighbors (shells) from a given particle located at R_1. The values of α obtained by solving (2) are then used to start the full HF equations which finally give the form of ϕ.

Among the various degrees of sophistication that one can use to write the equation for the correlated or perturbed wave function ψ we shall employ the one first employed by Guyer and Zane (1969) where ψ is assumed to satisfy an equation analogous to the Bethe-Goldstone equation of motion, i.e.,

$$[T(1) + T(2) + U(1) + U(2) + V(12)]\psi = \varepsilon\psi. \tag{4}$$

Here $T(1)$ and $T(2)$ are respectively the kinetic energies, $U(1)$ and $U(2)$ are the harmonoscillator potentials of particles 1 and 2, and $V(12)$ is the two-body interaction potential. Equation (4) is basically the equation of motion of two particles, each moving in a harmonic oscillator potential centered around two different lattice sites and in addition, interacting through a two-body potential. If one uses for $U(1)$ and $U(2)$ the form given by (2), then (4) becomes

$$[T_R + T_r + \tfrac{1}{4}m\omega^2(r-\Delta)^2 + m\omega^2(R-\delta)^2 + V(r)]\psi = \varepsilon\psi. \tag{5}$$

Here

$$T_R = -\frac{\hbar^2}{4m}\nabla_R^2, \quad T_r = -\frac{\hbar^2}{m}\nabla_r^2, \quad r_1 - r_2 = r, \quad r_1 + r_2 = 2R,$$

$$R_1 - R_2 = \Delta, \quad R_1 + R_2 = 2\delta.$$

Since the nuclear interaction potential is highly angular momentum dependent, we write the full angular momentum decomposition of ψ in the form:

$$\psi(\mathbf{r}) = \sum_{l=0}^{\infty}\sum_J (2l+1)\, i^l \sqrt{\frac{4\pi}{2l+1}}\, (l0SM_s|JM_s) \sum_{l'} \psi_{ll'}^{JS}(r)\, Y_{l'JS}^{M_s}(\Omega) \tag{6}$$

in the usual notation. The second summation over l' is introduced to take into account the presence of tensor forces. Equation (5) contains a $\cos\theta$ term which, much like the Stark effect, couples l with $l\pm 1$, but of course does not couple spins and therefore it is to be expected that the singlet states $^1S_0, ^1P_1, ^1D_2$ will be coupled by such a term and analogously the triplet states $^3S_1, ^3\tilde{S}_1, ^3P_0, ^3P_1, ^3P_2, ^3D_1, ^3\tilde{D}_1, ^3D_2$ will be linked together. After inserting (6) into (5) and with a certain amount of algebraic manipulation, we get ($x=r/r_0, d=\Delta/r_0, a=\alpha r_0$):

SINGLET

$$\begin{cases} h_0'' + (E - U_0) h_0 + a(x, d) h_1 & = 0 \\ h_1'' + (E - U_1) h_1 + \tfrac{1}{3} a(x, d) [h_0 - 2h_2] & = 0 \\ h_2'' + (E - U_2) h_2 - \tfrac{2}{3} a(x, d) h_1 & = 0. \end{cases} \qquad (7)$$

where the following notation has been used

$$^1S_0: \quad r\psi_{00}^{00} = h_0, \quad a(x, d) = \tfrac{1}{2} a^4 x d$$
$$^1P_1: \quad r\psi_{11}^{10} = h_1$$
$$^1D_2: \quad r\psi_{22}^{20} = h_2$$

TRIPLET

$$h_3'' + (E - U_3) h_3 - \sqrt{8}\, V_T \frac{2\mu r_0^2}{\hbar^2} \tilde{h}_3 + \tfrac{1}{9} a(x, d) [h_4 + 3h_5 + 5h_6] = 0$$

$$\tilde{h}_3'' + (E - \tilde{U}_3) \tilde{h}_3 - \sqrt{8}\, V_T \frac{2\mu r_0^2}{\hbar^2} h_3 - \frac{\sqrt{2}}{9} a(x, d) [h_4 - \tfrac{3}{2} h_5 + \tfrac{1}{2} h_6] = 0$$

$$h_4'' + (E - U_4) h_4 + \qquad 0 \qquad + \tfrac{1}{3} a(x, d) [(h_3 + \sqrt{2} \tilde{h}_7) + \\ - \sqrt{2} (\tilde{h}_3 + \sqrt{2} h_7)] = 0$$

$$h_5'' + (E - U_5) h_5 + \qquad 0 \qquad + \tfrac{1}{3} a(x, d) \left[\left(h_3 - \frac{1}{\sqrt{2}} h_7\right)\right. \\ \left. + \frac{1}{\sqrt{2}} \left(\tilde{h}_3 - \frac{1}{\sqrt{2}} h_7\right) - \tfrac{3}{2} h_8 \right] = 0$$

$$h_6'' + (E - U_6) h_6 + \qquad 0 \qquad + \tfrac{1}{3} a(x, d) \left[\left(h_3 + \frac{1}{5\sqrt{2}} \tilde{h}_7\right) + \\ - \frac{1}{5\sqrt{2}} \left(\tilde{h}_3 + \frac{1}{5\sqrt{2}} h_7\right) - \tfrac{3}{10} h_8 \right] = 0$$

$$h_7'' + (E - U_7) h_7 - \sqrt{8}\, V_T \frac{2\mu r_0^2}{\hbar^2} \tilde{h}_7 - \tfrac{1}{9} a(x, d) [2h_4 + \tfrac{3}{2} h_5 + \tfrac{1}{10} h_6] = 0$$

$$\tilde{h}_7'' + (E - \tilde{U}_7) \tilde{h}_7 - \sqrt{8}\, V_T \frac{2\mu r_0^2}{\hbar^2} h_7 + \frac{\sqrt{2}}{9} a(x, d) [h_4 - \tfrac{3}{2} h_5 + \tfrac{1}{2} h_6] = 0$$

$$h_8'' + (E - U_8) h_8 - \qquad 0 \qquad - \tfrac{1}{10} a(x, d) [3h_5 + h_6] = 0$$

with the following notation

$$^3S_1: \quad r\psi_{00}^{11} = h_3 \qquad ^3P_2: \quad r\psi_{11}^{21} = h_6$$
$$^3\tilde{S}_1: \quad r\psi_{02}^{11} = \tilde{h}_3 \qquad ^3D_1: \quad r\psi_{22}^{11} = h_7$$

$^3P_0: \quad r\psi_{11}^{01} = h_4 \qquad ^3\tilde{D}_1: \quad r\psi_{20}^{11} = \tilde{h}_7$
$^3P_1: \quad r\psi_{11}^{11} = h_5 \qquad ^3D_2: \quad r\psi_{22}^{21} = h_8$

The potentials U_k and $\tilde{U}_k$ are given by

$$U_k = \tfrac{1}{4}a^4 x^2 + \frac{2\mu r_0^2}{\hbar^2} V_k + \frac{l(l+1)}{x^2}$$

$$\tilde{U}_3 = \tfrac{1}{4}a^4 x^2 + \frac{2\mu r_0^2}{\hbar^2} V(^3D_1) + \frac{6}{x^2}$$

$$\tilde{U}_7 = \tfrac{1}{4}a^4 x^2 + \frac{2\mu r_0^2}{\hbar^2} V(^3S_1).$$

The two-body potentials V_k are taken from Reid (1968).

It can be seen that the two sets of equations are intrinsically coupled by the solid-state term. Only two normalization conditions are therefore necessary: one for each set of singlets and triplets. The solution of the $3+8$ differential equations presented us with considerable numerical difficulty in the absence of any Sturm-Liouville type theorems for coupled equations, and the search for the ground state energy of successive shells was a painfully laborious process. The major concern was to make certain that the energy eigenvalues being computed were actually the lowest, i.e., the eigenfunctions h_k's had no nodes.

The energy per particle, E/N, consists of the kinetic energy, which for displaced harmonic oscillator wave functions takes the form

$$E_{\text{K.E.}} = \frac{3}{4} \frac{\hbar^2}{m} \alpha^2 N,$$

together with the potential energy

$$E_{\text{P.E.}} = \sum_{i<j} \frac{\int \Phi(ij) V(ij) \Psi(ij) \, d^3 r_i \, d^3 r_j}{\int \Phi(ij) \Psi(ij) \, d^3 r_i \, d^3 r_j}$$

$$= \tfrac{1}{2} N \sum_{\Delta} N_\Delta \frac{\int \Phi(r) V(r) \Psi(r) \, d^3 r}{\int \Phi(r) \Psi(r) \, d^3 r}$$

where $\Phi(ij)$ is the two-body wave function, i.e., $\Phi = \phi(r_1) \phi(r_2)$ and N_Δ is the number of particles at distance Δ from the one chosen as the origin.

3. Elastic Constants

We must test the system of nucleons arranged in a lattice structure for stability against small deformations. There have been several attempts to derive criteria for the melting of a solid, the best known being Lindemann's rule. This rule gives an empirical criterion that a solid melts when the amplitude of oscillation of a particle becomes a sizable fraction of the nearest neighbor distance. Such a criterion can only be regarded as a convenient 'rule of thumb' designed to test semi-quantitatively the stability of a crystalline structure and indeed, a complete theory of melting must examine the detailed stability of a lattice when it is deformed under shearing stresses.

Any lattice has to satisfy the requirement that its energy density must have a stationary value at equilibrium. However, for the structure to be stable, the energy must have a positive-definite quadratic form, thus increasing its value while undergoing a small strain. Following the notation of Born and Huang (1950), when a cube of side $2a$ is deformed homogeneously, the energy to the second order of deformation in terms of the strain components $e_{\alpha\beta}$ comes out to be

$$E(\delta) = E(0) + \tfrac{1}{2}C_{11}(e_{xx}^2 + e_{yy}^2 + e_{zz}^2) + C_{12}(e_{yy}e_{zz} + e_{zz}e_{xx} + e_{xx}e_{yy}) + \\ + \tfrac{1}{2}C_{44}(e_{yz}^2 + e_{zx}^2 + e_{xy}^2),$$

Here $E(0)$ is the energy of the undeformed cube whose lattice points can be described by $\mathbf{R}_0 = (l_1 a, l_2 a, l_3 a)$, ($l_1, l_2, l_3$ being integers), Φ is the potential, and

$$C_{11} = \frac{2a}{\gamma} \sum D^2 \Phi l_1^4 - P,$$

$$C_{12} = \frac{2a}{\gamma} \sum D^2 \Phi l_1^2 l_2^2 - P,$$

$$C_{44} = \frac{2a}{\gamma} \sum D^2 \Phi l_1^2 l_2^2 - P.$$

and $\gamma = 4$(BCC), $\gamma = 2$(FCC), $V = \gamma a^3$, the summation extending over various shells. The quadratic form in (5) is positive definite provided

$C_{11} + 2C_{12} > 0$ (Sublimation: lattice has no cohesion),
$C_{44} \quad\quad > 0$ (Melting: lattice unstable against shearing stresses),
$C_{11} - C_{12} > 0$ (Gel: elastic resistance against shearing stresses).

For a lattice structure to be stable we must demand the satisfaction of all of the foregoing conditions.

4. Results

We have attempted three different ordered structures: BCC((body-centered cubic), FCC (face-centered cubic) and HCP (hexagonal close packing). A major complication arises because of the possible spin configurations: for a given solid structure, say BCC, one could in principle arrange the nucleons in a large number of ways depending upon the spin configuration. In a BCC lattice one could start with any site with a neutron with its spin up; in the next site on the same cube one can put $N_\uparrow$, $N_\downarrow$, $P_\uparrow$ or $P_\downarrow$. In the next site one has again a similar situation. We have tried several configurations with an arbitrary spin arrangement just to ensure that such arrangements are not

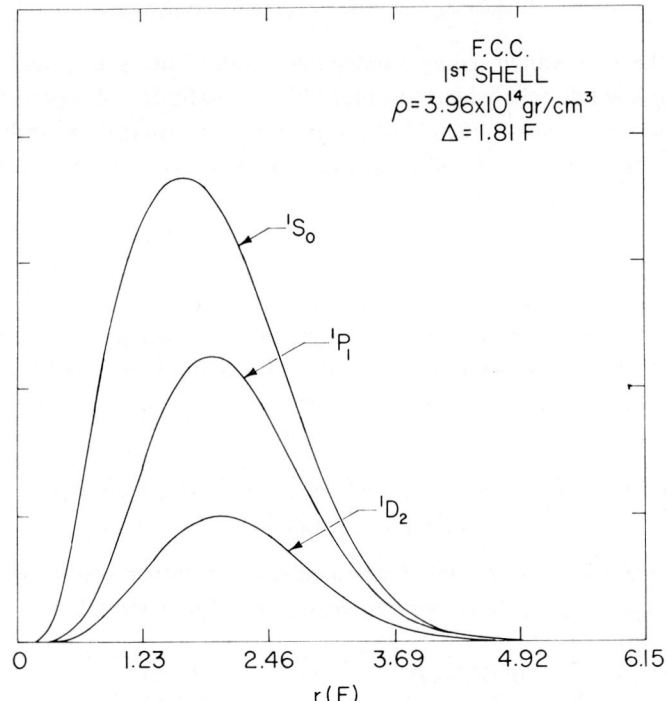

Fig. 3. Singlet state wave functions vs. r at a given density.

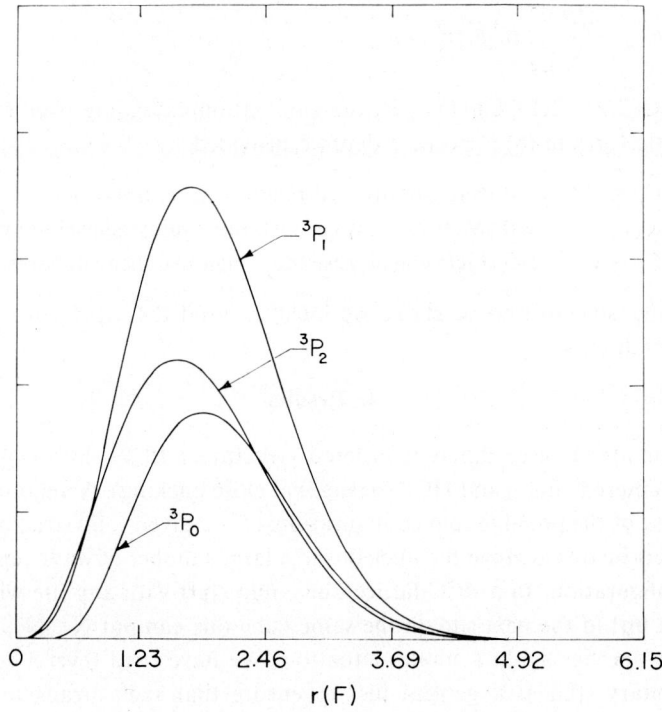

Fig. 4. Triplet state wave functions vs. r at a given density.

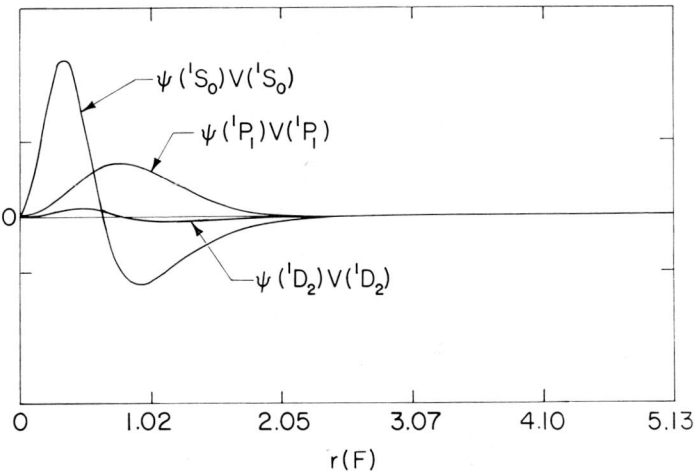

Fig. 5. The product hV vs. r for the singlet states.

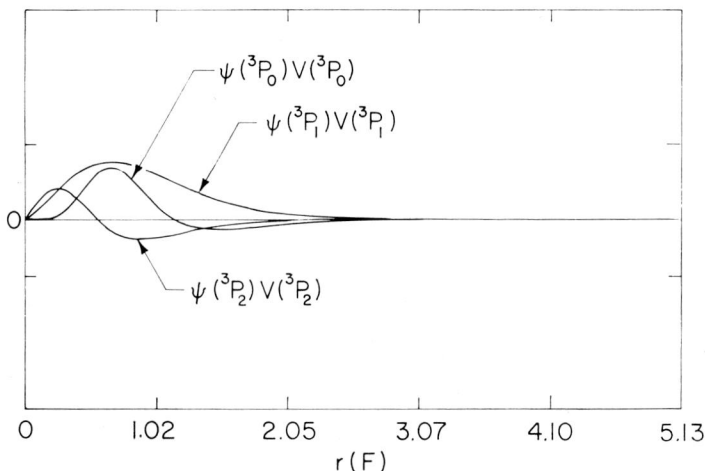

Fig. 6. The product hV vs. r for the triplet states.

TABLE I

Energy per particle (MeV) vs. $10^{-15}\varrho$ (g cm^{-3}) for various configurations with mixed spins

$10^{-15}\varrho$ (g cm^{-3})	FCC N:P = 1:1	BCC Pure neutrons	HCP Pure neutrons	FCC Pure neutrons
1.41	107.4	131.6	118.4	114.5
1.832	181.2	188.7	174.8	171.8
2.596	355.9	310.3	293.7	282.5
3.344	548.9	472.5	441.6	428.5
4.398	911.3	720.5	684.1	652.4
5.237	1194.0	923.2	884.7	864.6

TABLE II
FCC pure neutrons mixed spins

$10^{-15} \varrho$ (g cm^{-3})	Δ (fermi)	α^{-1}(H.F.) (fermi)	K.E. (MeV)	P.E. (MeV)	B/N (MeV)	
1.41	1.188	0.460	146.4	−31.9	114.5	
1.603	1.138	0.435	163.8	−24.2	139.6	Stable
1.832	1.089	0.410	183.9	−12.1	171.8	
2.294	1.010	0.376	219.4	19.6	239.0	
2.596	0.969	0.361	238.3	44.2	282.5	
3.344	0.891	0.323	297.2	131.3	428.5	
3.776	0.855	0.310	322.3	186.9	509.2	
4.398	0.813	0.291	366.0	286.4	652.4	
5.237	0.767	0.271	421.9	442.7	864.6	

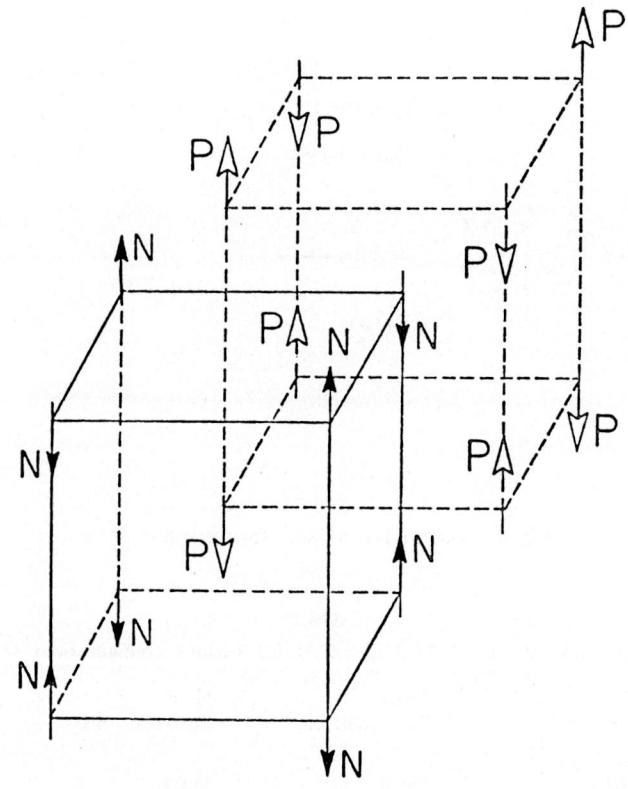

B.C.C.

Fig. 7. BCC configuration for an equal number of neutrons and protons.

TABLE III
FCC pure neutrons (mixed spins) elastic constants

$10^{-15} \times \varrho$ (g cm^{-3})	$10^{-36} \times c_{11}$ (dyne cm^{-2})	$10^{-36} \times c_{12}$ (dyne cm^{-2})	$10^{-36} \times c_{44}$ (dyne cm^{-2})
5.237	38.46	14.40	7.55
4.398	22.625	8.602	4.142
3.776	14.529	6.052	3.212
3.344	9.286	4.065	1.985
2.596	4.164	1.683	0.675
1.832	1.065	0.5414	0.1084
1.603	0.5177	0.3386	0.0326
1.41	0.1883	0.1838	−0.0362

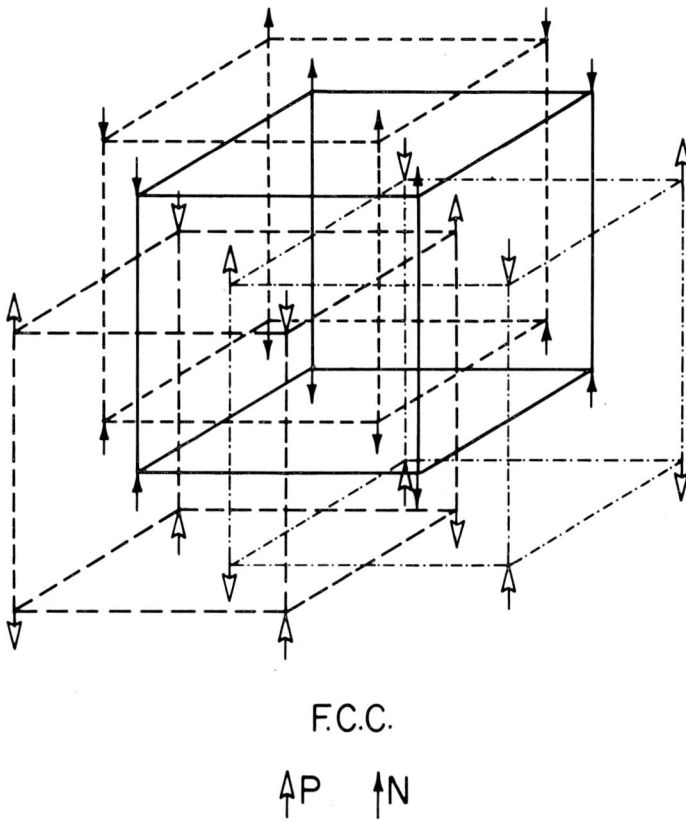

Fig. 8. FCC configuration.

energetically convenient. We were then able to establish that for a given configuration the minimum for energy was achieved whenever the spins of neighboring particles were symmetrically arranged, i.e., in the same cube the particles on neighboring sites should have anti-parallel spins (mixed spins). An analogous situation was found to hold for the other two configurations studied, namely FCC and HCP.

For a given configuration with a prescribed spin arrangement the system of 3+8 equations has to be solved for each shell to yield the lowest eigenvalue. The resulting h_k's are inserted in the energy expression and the computations are carried far enough (approximately up to 24 shells) that contributions of further shells are unimportant. The resulting singlet wave functions $^1S_0$, $^1P_1$, $^1D_2$, and the triplet wave functions $^3P_0$, $^3P_1$, $^3P_2$, are respectively displayed in Figures 3 and 4 and the products hV for the two cases are shown in Figures 5 and 6.

TABLE IV

FCC pure neutrons – mixed spins equation of state

$10^{-15}\varrho$ (g cm^{-3})	E/N (MeV)	$10^{-36} \times \varepsilon$ (erg cm^{-3})	$10^{-36} \times P$ (dyne cm^{-2})	Γ
1.0	75.50	0.969	0.101	1.714
1.2	96.88	1.187	0.150	2.082
1.4	119.33	1.415	0.217	2.330
1.6	143.33	1.654	0.304	2.479
1.8	169.08	1.906	0.412	2.545
2.0	196.08	2.170	0.541	2.567
2.2	225.63	2.448	0.693	2.647
2.4	256.49	2.741	0.877	2.799
2.6	289.42	3.052	1.100	2.888
2.8	324.54	3.381	1.355	2.764
3.0	361.51	3.729	1.622	2.514
3.2	399.71	4.095	1.891	2.326
3.4	438.66	4.477	2.174	2.369
3.6	478.48	4.878	2.505	2.625
3.8	519.71	5.299	2.909	2.883
4.0	562.85	5.743	3.383	2.985
4.2	608.12	6.212	3.905	2.918
4.4	655.31	6.706	4.449	2.751
4.6	704.07	7.225	4.999	2.581
4.8	754.00	7.769	5.557	2.489
5.0	804.90	8.336	6.144	2.514

The geometrical configuration of the nucleons for BCC, FCC and HCP respectively are displayed in Figures 7, 8, and 9 and Table I summarizes the energy per particle as a function of the density for the three configurations. On energetic grounds it is convenient to have only neutrons at the high density end as can be seen from a comparison of the energies computed for an FCC consisting of pure neutrons and one with an equal number of neutrons and protons. The FCC structure made up of neutrons with mixed spins does indeed appear to be the lowest energy configuration and the detailed results for such a system are presented in Table II where the density is given in the first column, the nearest neighbor distance in the second, the spread of the wave function in the third, with the kinetic, potential, and total energy per particle following in order. The region of stability is indicated on the right where it is shown that the lattice is stable only in the density range upwards of $\sim 1.5 \times 10^{15}$ g cm^{-3}. The elastic constants C_{11}, C_{12}, C_{44} are shown in Table III as a function of the density: it can be seen that $C_{44} < 0$ for $\varrho \lesssim 1.5 \times 10^{15}$ g cm^{-3} indicating that the FCC structure made up of neutrons can withstand shearing stresses at a density upwards of this value. The spread of the wave function as measured by α^{-1} (fermi) vs. the density is exhibited in Figure 10 along with the nearest neighbor distance: it can be readily seen that α^{-1} is about a third of the nearest neighbor distance over the density range $1.41 \times 10^{15} \leq \varrho \leq 5.237 \times 10^{15}$ g cm^{-3}.

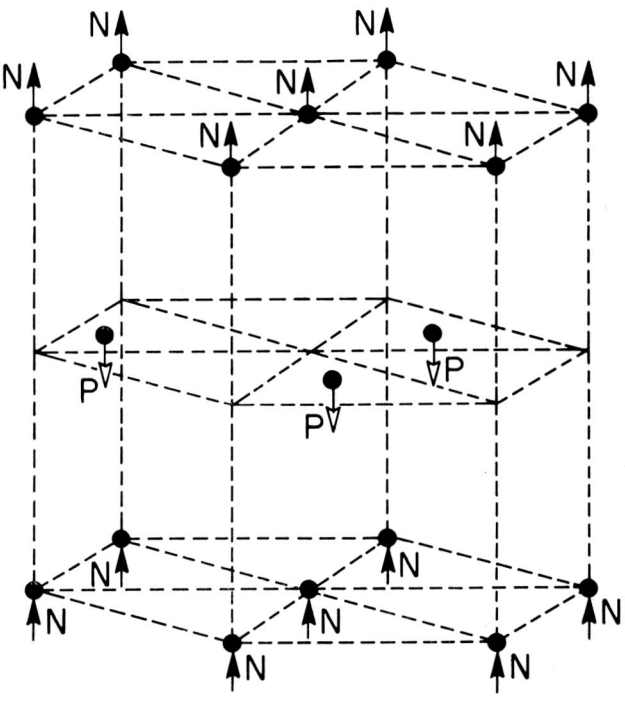

H.C.P.

Fig. 9. HCP configuration.

The equation of state derived by arranging neutrons with their spins mixed in an FCC lattice is summarized in Table IV. Here the first column lists the density, the second the energy per particle, the third the mass-energy density, the fourth the pressure derived by using the relation $P = -(\partial E/\partial V)$ and the last column shows the adiabatic index $\Gamma = (\varepsilon + P) P^{-1} \partial P/\partial \varepsilon$. Figure 11 exhibits a comparison of our energies vs. density (obtained on the assumption that nucleons are arranged in a lattice) and the energies obtained for a neutron gas by Pandharipande. It is clear that at the high density end a lattice made up of neutrons is energetically most convenient. The interesting feature to be noted is that the curves cross around $\varrho \cong 1.5 \times 10^{15}$ g cm^{-3} where the stability analysis shows that $C_{44} < 0$ (i.e., the lattice cannot withstand any shearing stresses at a lower density). Below $\varrho \cong 1.5 \times 10^{15}$ g cm^{-3} the energies obtained by doing a lattice calculation exceed the corresponding energies given by a gas computation. We should therefore like to conclude that our computation seems to indicate a possible solid phase for cold matter at densities exceeding 1.5×10^{15} g cm^{-3}.

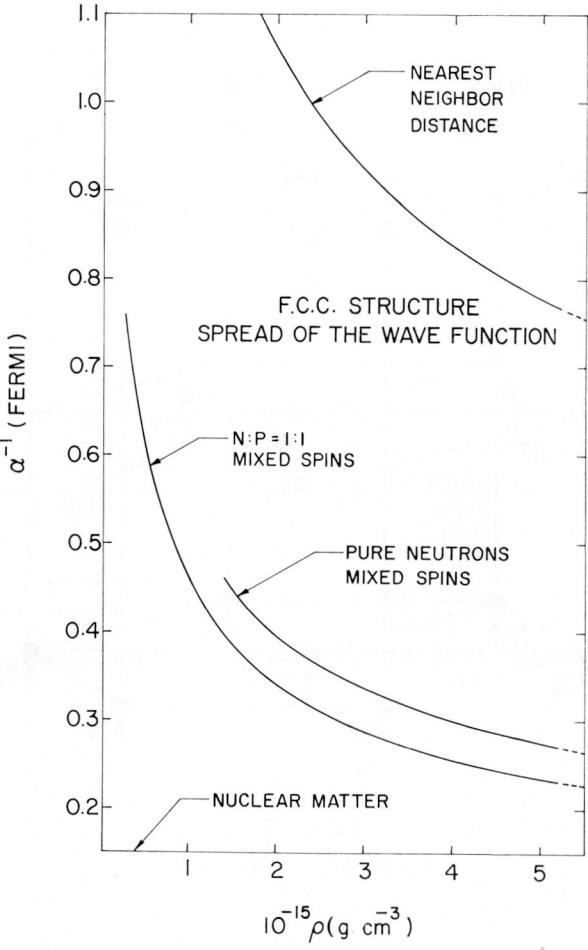

Fig. 10. Spread of the wave function and the nearest neighbor distance as a function of the density.

We have constructed neutron star models composed of cold matter by integrating the equations governing the relativistic stellar structure. For this purpose our equation of state, obtained by assuming a crystalline neutron state, was smoothly joined up with Pandharipande's pure-neutron equation of state at a density of 1.5×10^{15} g cm^{-3}. In Figure 12 we show the resulting neutron star masses in units of the solar mass as a function of the central density. We find a maximum mass of 1.56 $M_\odot$ for a stable neutron star at a central density 3×10^{15} g cm^{-3}, not very different from a maximum mass of 1.66 $M_\odot$ calculated by Baym et al. (1971) at a central density of 4.1×10^{15} g cm^{-3}.

We also attempted a few baryonic crystals by including hyperons in an FCC lattice.

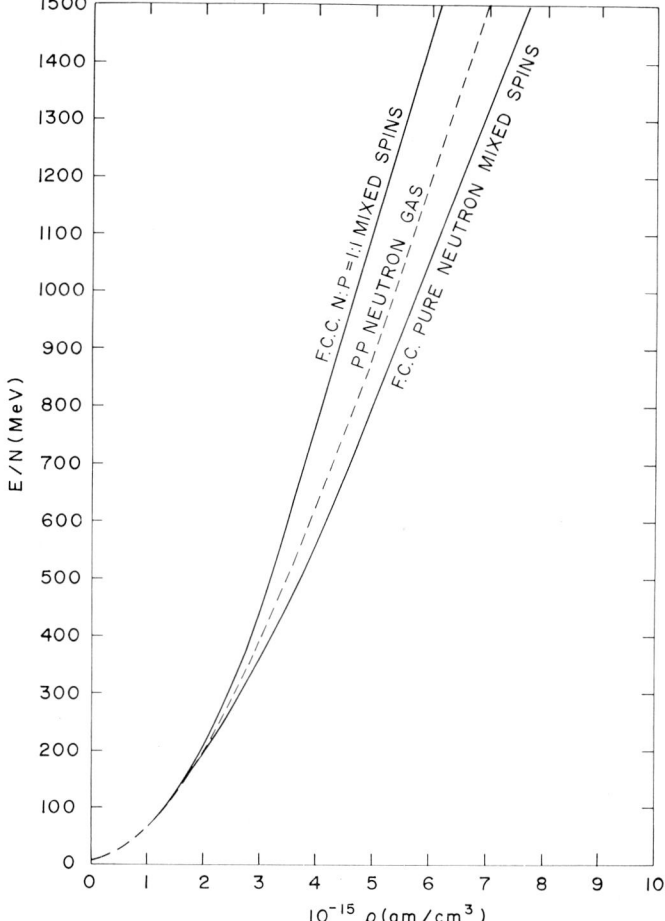

Fig. 11. Energy per particle vs. the density for an FCC lattice made up of pure neutrons and of equal number of neutrons and protons; the dashed line shows Pandharipande's result for neutron gas.

The hyperonic potentials were derived by extending the work of Brown et al. (1970), and a typical hyperonic potential (Λ, N; $T = 1/2$) is displayed in Table V. The resulting energies for various distributions of baryons are shown in Figure 13: it is at once evident that the energies are considerably lowered compared to the pure neutron matter arranged in a lattice. One of the principal reasons for this lowering of energy can be

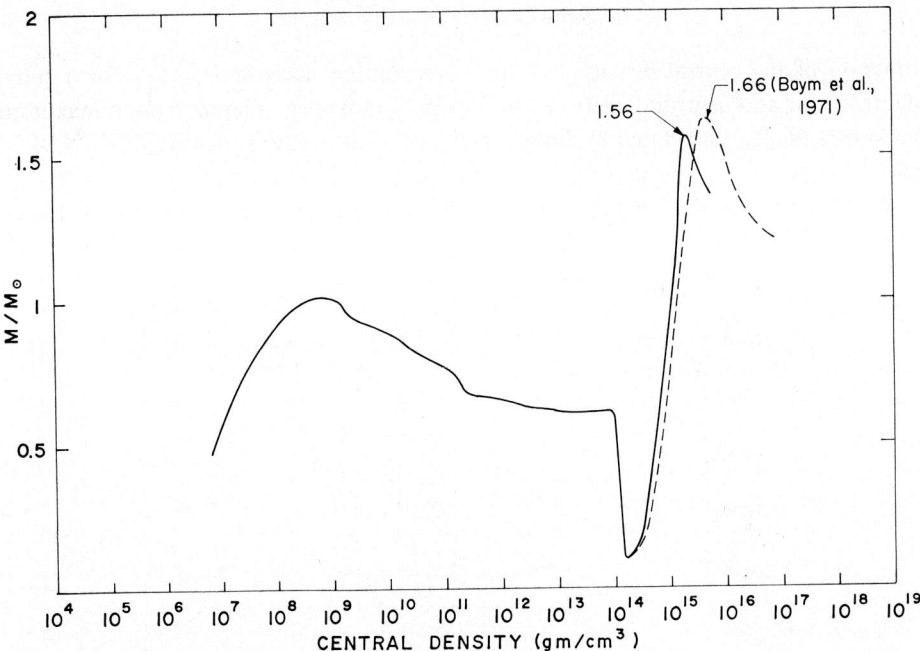

Fig. 12. Neutron star mass in units of the solar mass vs. the central density; the dashed curve shows the results of Baym *et al.*

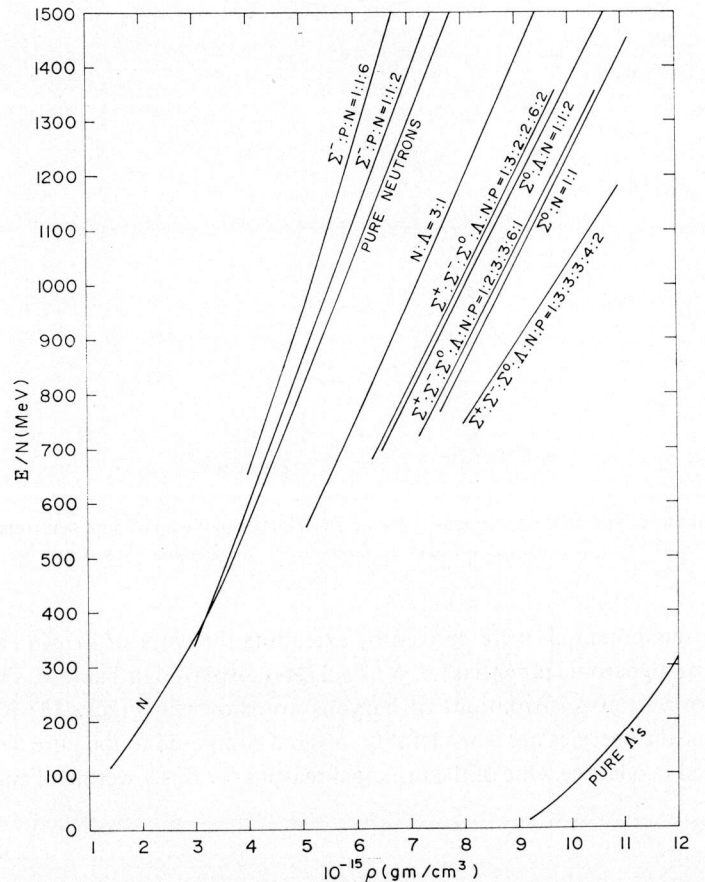

Fig. 13. Energy per particle for a variety of baryonic compositions in an FCC lattice.

TABLE V

Λ, N potential: $T = 1/2$; $V = V_C + V_\sigma \boldsymbol{\sigma}_1 \cdot \boldsymbol{\sigma}_2 + V_T S_{12} + V_{LS} \mathbf{L} \cdot \mathbf{S} + V_Q W_{12}$

	$\pi(139, T = 1, \text{Ps})$	$\eta(548, T = 0, \text{Ps})$	$k(495, T = 1/2, \text{Ps})$
	$\pi \equiv 0.7 \, f^{-1}$	$\eta \equiv 2.7817 \, f^{-1}$	$k \equiv 2.3465 \, f^{-1}$
V_C	0	0	0
V_σ	0	$-19.395 (e^{-\eta r}/r)$	$+(-)^{L+S} 57.743 (e^{-kr}/r)$
V_T	0	$-58.185 f(r) (e^{-\eta r}/r)$ $f(r) = \frac{1}{3} + 0.359/r + 0.129/r^2$	$+(-)^{L+S} 173.288 f(r)(e^{-kr}/r)$ $f(r) = \frac{1}{3} + 0.4261/r +$ $+ 0.1816/r^2$
V_{LS}	0	0	0
V_Q	0	0	0

	$\sigma(490, T = 0, S)$	$\omega(888, T = 0, V)$
	$\sigma \equiv 2.4873 \, f^{-1}$	$\omega \equiv 4.5076 \, f^{-1}$
V_C	$-1131.08 (e^{-\sigma r}/r)$	$+6815.036 (e^{-\omega r}/r)$
V_σ	$+0.9883 f(r)(e^{-\sigma r}/r)$ $f(r) = 0.402/r + 0.3232/r^2 + 0.129/r^3$	$+[616.7849 - 43.6140 f(r)](e^{-\omega r}/r)$ $f(r) = 0.2218/r + 0.0984/r^2 + 0.0218/r^3$
V_T	0	$-925.177 f(r)(e^{-\omega r}/r)$ $f(r) = \frac{1}{3} + 0.2218/r + 0.04929/r^2$
V_{LS}	$-135.9047 f(r)(e^{-\sigma r}/r)$ $f(r) = 0.402/r + 0.1616/r^2$	$-5754.498 f(r)(e^{-\omega r}/r)$ $f(r) = 0.2218/r + 0.04929/r^2$
V_Q	$-1.9766 f(r)(e^{-\sigma r}/r)$ $f(r) = 0.1616/r^2 + 0.1299/r^3$	$+87.228 f(r)(e^{-\omega r}/r)$ $f(r) = 0.0492/r^2 + 0.0218/r^3$

	$\varrho(755, T = 1, V)$	$k^*(890, T = 1/2, V)$
	$\varrho \equiv 3.8324 f^{-1}$	$k^* \equiv 4.4275 f^{-1}$
V_C	0	$(-)^{L+S} 552.4166 (e^{-k^* r}/r)$
V_σ	0	$(-)^{L+S}[353.103 - 147.8237 f(r)](e^{-k^* r}/r)$ $f(r) = 0.2258/r + 0.1020/r^2 + 0.0230/r^3$
V_T	0	$-(-)^{L+S} 529.65 f(r)(e^{-k^* r}/r)$ $f(r) = \frac{1}{3} + 0.2258/r + 0.0510/r^2$
V_{LS}	0	$-(-)^{L+S} 1495.686 f(r)(e^{-k^* r}/r)$ $f(r) = 0.2258/r + 0.0510/r^2$
V_Q	0	$(-)^{L+S} 295.647 f(r)(e^{-k^* r}/r)$ $f(r) = 0.0510/r^2 + 0.0230/r^3$

attributed to the $^3P_1$ wave which in the nucleon-nucleon interaction is purely repulsive, but is modestly attractive in all the hyperon-hyperon and hyperon-nucleon interactions. The question of the hyperonic interactions, however, needs to be examined further before any firm conclusions can be drawn.

Acknowledgement

It is a pleasure to thank Mal Ruderman and Al Cameron for stimulating discussions.

References

Anderson, P. W. and Palmer, R. G.: 1971, *Nature Phys. Sci.* **231**, 145.
Banerjee, B., Chitre, S. M., and Garde, V. K.: 1970, *Phys. Rev. Letters* **25**, 1125.
Baym, G., Pethick, C., and Sutherland, P.: 1971, *Astrophys. J.* **170**, 299.
Born, M. and Huang, K.: 1950, *Dynamical Theory of Crystal Lattices*, Clarendon Press, Oxford.
Brandow, B.: 1972, *Ann. Phys.* **74**, 112.
Brown, J. T., Downs, B. W., and Iddings, C. K.: 1970, *Ann. Phys.* **60**, 148.
Clark, J. W. and Chao, N.-C.: 1972, *Nature Phys. Sci.* **236**, 37.
Guyer, R. A.: 1969, in D. Turnbull (ed.), *Solid State Phys.* **23**, 413.
Guyer, R. A. and Zane, I.: 1969, *Phys. Rev.* **188**, 445.
Pandharipande, V. R.: 1971, *Nucl. Phys.* **A174**, 641.
Reid, R. V.: 1968, *Ann. Phys.* **50**, 411.

DISCUSSION

Bethe: When Johnson and I attempted to calculate a crystal model for neutron matter, we found the coupling between higher angular momentum states very important, even though these states have essentially no potential energy. The reason is that the unperturbed wavefunction in relative coordinates,

$$\phi = \exp - (r - a)^2 \alpha$$

is centered at a, the lattice distance, and therefore has strong components with large l, when expanded around $r = 0$. It is legitimate to stop with $l = 2$?

Canuto: We do not have any reason to believe that the higher waves are unimportant. Since the coupling amongst the waves is very complicated, we have decided to examine results for $l \leq 2$. The higher waves will have a strong centrifugal repulsion and as a consequence the wavefunctions for those waves are expected to be pushed away from the origin where the potential is essentially attractive. We therefore feel that if anything their inclusion will lower the energy.

Bethe: Your conclusion is done with essentially the Brueckner-Goldstone method. When I tried to apply this to high density *gas*, I got much too high binding energy in the 2-body approximation. I realize that many-body clusters are *less* important for the crystal than for the gas, but still I believe that Pandharipande method is more reliable at high density. In any case, gas and crystal should be done with the *same* method.

Canuto: I agree that gas and crystal should be computed with the same method. It remains to be seen if a variational method as the one employed by Pandharipande is preferable. Historically speaking, as far as the quantum crystals are concerned, people have gone from variational computation to the Bruckner-Goldstone method and not vice versa. We have employed what quantum crystal physicists have suggested to us as being the best method available.

Bethe: In light of Ruderman's remark after my paper, I withdraw my statement that neutron crystals can not exist at high density. However, I do not believe that Canuto and Chitre have yet proved their existence by the present calculations.

SUPERFLUIDITY IN NEUTRON STARS

GEORGE GREENSTEIN
Amherst College, Amherst, Mass., U.S.A.

Abstract. We present a short Cook's tour of the possible effects of rotation coupled with superfluid properties of neutron star interiors. A suggestion is made to take advantage of forthcoming lunar occultations of the Crab Nebula in order to search for blackbody X-ray emission from the Crab pulsar.

1. Introduction

Let me begin by summarizing our beliefs (this is the correct term) concerning the internal structure of neutron stars. (Extensive reviews, with references, have been given by Ruderman, 1969a, 1972; Pines, 1970; and Cameron, 1970.) As one penetrates from the surface of a neutron star inwards the density rises rapidly, the electrons quickly become degenerate, and the atoms completely ionized. The nuclei arrange themselves into a rigid lattice. At densities of about 3×10^{11} g cm^{-3} neutrons evaporate from the nuclei. Beneath this point the lattice of nuclei coexists with two uniform gases; a relativistic degenerate gas of electrons and a non-relativistic degenerate gas of neutrons. The number of neutrons outside the nuclei is greater than the number within them. At densities of about 2×10^{14} g cm^{-3} the nuclei dissolve. Beneath this point we have three uniform degenerate gases; relativistic electrons, non-relativistic neutrons and non-relativistic protons. In this regime too the neutrons outnumber the protons. At densities above about 10^{15} g cm^{-3} hyperons appear and rapidly become the dominant component.

We know that at densities below about 2×10^{14} g cm^{-3} the $^1S_0$ phase shift between two neutrons is negative; the interaction of two neutrons of opposite spin, in a state of zero relative angular momentum, is attractive. Precisely the same situation holds for electrons in an ordinary metal, although in this case the origin of the attraction is different (in neutron stars the attraction is due to the strong interaction whereas in metals it arises through the electron-phonon interaction). Within the BCS theory of superconductivity the existence of such an attractive interaction constitutes the necessary condition that the electrons become superconducting at sufficiently low temperatures. This result is independent of the origin of the attraction. By analogy, one then reasons that those neutrons in neutron stars for which the $^1S_0$ phase shift is negative – i.e., those within the crust – form a neutral superconductor – a superfluid.

Near the base of the crust, at 1.6×10^{14} g cm^{-3}, the $^1S_0$ phase shift becomes positive. At higher densities the $^3P_2$ phase shift becomes negative, leading to anisotropic superfluidity – a state that has never been observed in the laboratory. In what follows we will treat this anisotropic superfluid in precisely the same way as the more familiar isotropic superfluid – not necessarily a safe procedure.

The strong interaction between protons is the same as that between neutrons. Thus the protons lying below the crust may form a superconductor (isotropic, since the density of protons always remains relatively small).

The picture, then, is this; a neutron star consists of an outer crust, an inner crust coexisting with an isotropic superfluid, an anisotropic superfluid coexisting with an isotropic superconductor, and finally a hyperon core of largely unknown properties. Most of the moment of inertia of the star resides in the superfluid.

Calculated superfluid energy gaps vary with density by a factor of perhaps ten. Resulting transition temperatures then vary from about 10^9 to about 10^{10} K, depending on position within the star. It seems clear that neutron stars cool below 10^{10} K quite rapidly. It is *not* clear, though it seems likely, that the Crab and Vela pulsars have cooled far below 10^9 K. Thus much, but not necessarily all, of the superfluid in those two pulsars lies far below its transition temperature. All older pulsars lie very far below their transition temperatures.

It is often said that neutron stars of sufficiently low mass are entirely solid. This is true but misleading. Such stars do remain solid all the way down to their cores. However, they *also* contain a superfluid whose moment of inertia exceeds that of the crust. There are, however, neutron stars for which superfluidity is unimportant. These are the most massive stars, which consist almost entirely of hyperons surrounded by thin shells of superfluid and crust. (I am neglecting here the possibility that the neutral hyperons may be superfluid; we will have enough to worry about as it is.)

It is sobering to realize that the whole picture I have drawn has received *no direct observational test whatsoever*, and only two indirect tests. The first indirect test relates to the observed lack of cool white dwarfs. This phenomenon has been successfully interpreted as resulting from the existence, in the cores of these stars, of the same solid lattice we expect to find in the outer regions of neutron stars. As the lattice cools below its Debye temperature its specific heat drops dramatically. The star then rapidly cools to invisibility (Greenstein, 1969). Our only other piece of confirming observational evidence is the very elegant interpretation, which I will describe below, of the post-speedup behavior of the Crab and Vela pulsars. Two indirect tests do not add up to very much. To be conservative I believe we should keep in the back of our minds the possibility that our picture of neutron star interiors is *entirely wrong*. If you don't believe me, consider the solar neutrino experiment.

The intense magnetic field pervading the star (the superconductor is not expected to exhibit a Meissner effect) will force the crust, the electrons, the protons (whether superconducting or not) and the charged hyperons to rotate uniformly. The rotation of this 'charged particle system' determines the rate of ticking of the pulsar. The superfluid, however, need *not* rotate uniformly and, if it does, need not rotate at the same rate as the pulsar. In order to find out what the superfluid is doing we need to understand the ways it has of interacting with the charged particle system.

Because the superfluid is probably far below its transition temperature, these interactions will take place via the superfluid vortex lines. Those neutrons that lie within the normal cores of these lines will scatter against charges. Only one scattering process has been quantitatively considered so far: this is the scattering of neutrons in the isotropic superfluid against electrons. Others include scatterings against phonons in the crust, and protons in the cores of proton flux lines. As yet we do not know which of these is the dominant process.

All of these processes lead to a force per cm³ **f** between superfluid and charges of the following form

$$\mathbf{f} = \frac{\varrho_n \varrho_c}{\varrho \tau} (\mathbf{v}_n - \mathbf{v}_c), \tag{1}$$

where ϱ_n is the mass-energy density of neutrons, ϱ_c that of charges, $\varrho \equiv \varrho_n + \varrho_c$, $(\mathbf{v}_n - \mathbf{v}_c)$ is the relative velocity between neutrons and charges and τ has the dimensions of a time. In the absence of external forces acting on the system velocity differences between neutrons and charges are damped out in this time τ. All the physics of the problem is buried in the calculation of τ. If the protons and neutrons are normal τ is microscopic – something like 10^{-20} s. The rotation of the neutrons is then locked with that of the pulsar. If the protons are normal and the neutrons are superfluid then τ is on the order of fractions of a second. If the neutrons are superfluid and the protons are superconducting then τ is increased enormously – one finds values on the order of years.

The pulsar is slowing down. A torque, presumably electromagnetic in origin, acts to decelerate the charges. They are rotating more slowly than the neutrons. If we multiply both sides of (1) by $r \sin\theta$, where r, θ are the spherical coordinates of the point in question, then the resulting expression gives the torque density acting on the neutrons. Let us assume for now that this is the *only* torque acting on the neutrons. (In general this is not so.) The velocity $\mathbf{v}_c$ is directed in the ϕ-direction. We are not assured that the neutrons are rotating in any such simple way. Let us, however, assume for now that the velocities of any bulk circulation currents within the superfluid are small compared with local rotation velocities. Then $\mathbf{v}_n$ is almost in the ϕ-direction, as is the torque acting on the neutrons. This torque acts to slow their rotation. Equating the torque density to the rate of change of angular momentum yields

$$\dot{\Omega}_n = \frac{\varrho_c}{\varrho \tau} (\Omega_n - \Omega_c), \tag{2}$$

where the Ω's are angular velocities. In the steady state $\dot{\Omega}_n = \dot{\Omega}_c$ and we can solve (2);

$$\frac{\Omega_n - \Omega_c}{\Omega_c} = \frac{\varrho}{\varrho_c} \frac{\tau}{t_c}, \tag{3}$$

where $t_c \equiv |\Omega_c / \dot{\Omega}_c|$. So the superfluid is rotating somewhat faster than the pulsar.

2. The Simplest Possible Model of a Superfluid Neutron Star

Equations (2) and (3) rest on certain assumptions. For now I wish to consider an approximation within which the assumptions are valid. It consists in considering the quantity $(\tau \varrho/\varrho_c)$ constant throughout the star. Then the right-hand side of (3) is constant. Since Ω_c is constant Ω_n is also. Therefore *the neutrons are rotating uniformly*. In this case both assumptions leading to (2) and (3) are valid – the approximation is consistent. There is, however, not the slightest chance that it is *correct*. I consider it now because the resulting picture is easy to work with and guides our thoughts in fruitful directions. Later on, in the interests of masochism and rigor, the approximation will be relaxed.

2.1. Spindown following a period jump

In the first application of these ideas I want to discuss the observed behavior of the Crab and Vela pulsars following their period jumps. So far each of these pulsars has undergone two such jumps and in each case Ω_c and $|\dot{\Omega}_c|$ increased. In a seminal paper Baym *et al.* (1969) were able to show that, if neutron star interiors are superfluid, an increase in Ω_c naturally leads to one in $|\dot{\Omega}_c|$. Their interpretation is elegant and the numerical results too good to be wrong. Their approach is as follows.

Equation (2), after an obvious transformation, now reads

$$\dot{\Omega}_n = \frac{I_c}{I\tau}(\Omega_n - \Omega_c), \tag{4}$$

where the I's are total moments of inertia. *Two* torques act on the charges – an *accelerating* torque equal and opposite to that which decelerates the neutrons and the *decelerating* radiation torque which we write as $I_c M$. The torque equation for the charges is

$$\dot{\Omega}_c = -M + \frac{I_n}{I\tau}(\Omega_n - \Omega_c). \tag{5}$$

Now the pulsar speeds up. The accelerating torque in (5) is now less since it is proportional to $\Omega_n - \Omega_c$. The decelerating torque is unchanged. Therefore, as observed, the charges – which constitute the pulsar – decelerate more rapidly. A trivial manipulation of (5) gives the fractional change in $\dot{\Omega}_c$ divided by that in Ω_c, i.e.,

$$\zeta \equiv \frac{\Delta \dot{\Omega}_c/\dot{\Omega}_c}{\Delta \Omega_c/\Omega_c} = \frac{I_n \, t_c}{I \, \tau}\left|1 - \frac{\Delta \Omega_n}{\Delta \Omega_c}\right|. \tag{6}$$

There are a number of well-known predictions of this model; the change in $\dot{\Omega}_c$ decays exponentially away with time constant τ, the instantaneous change in Ω_c is related to a permanent change in Ω_c through

$$(\Delta \Omega_c)_{\text{permanent}} = (1 - Q)(\Delta \Omega_c)_{\text{initial}}, \tag{7}$$

where

$$Q = \frac{\tau}{t_c}\zeta \tag{8}$$

and

$$\frac{(\Delta\dot{\Omega}_c)^2}{(\Delta\Omega_c)(\Delta\ddot{\Omega}_c)} = Q. \tag{9}$$

Most of these predictions are independent of the mechanism producing the period jump.

Because the observed decay times of the Crab and Vela pulsars are long (days and years, respectively) we have strong evidence that these pulsars possess superfluid interiors. It has been claimed that they must *also* be superconducting. But this is not necessarily so if the coupling time describing the interaction of the superfluid with phonons in the crust is long. If it is then one can imagine that the protons are not superconducting and spin up in fractions of a second following a period jump. The observed period decay would then represent the spin-up of the isotropic superfluid within the crust. To resolve this point we will need a calculation of the phonon-superfluid interaction.

I wish to point out a further testable consequence of this theory. It is that *in every model of what caused the period jump, ξ should be constant from one jump to the next* (though different for different pulsars). This will be true whether successive jumps have the same or different amplitudes, and whether the time between jumps is long or short compared to the spin-down time τ. It is almost certainly also true within the context of the more realistic model to be described in the next section, although other predictions of the theory (most notably, that of a purely exponential decay) will not remain true. The proof depends upon the process producing the jump.

The first process that comes to mind (Börner and Cohen, 1971) is that a flying saucer crashed onto the surface of the star, striking a glancing blow in such a direction as to speed it up. If we don't like flying saucers we can talk in terms of planets – masses on the order of 1/10 that of the Earth are required in order to understand Vela. In this case $\Delta\Omega_n = 0$ and $\xi = I_n t_c / I\tau$, a quantity characteristic of the pulsar and depending in no way on the circumstances of the jump.

A second process in a neutron starquake (Ruderman, 1969b; Baym and Pines, 1971). It is not obvious what to take for $\Delta\Omega_n$ here. Since the mass of the crust is small compared with that of the neutrons $\Delta\Omega_n \ll \Delta\Omega_c$ seems appropriate, leading to an identical conclusion. If $\Delta\Omega_n$ is comparable to $\Delta\Omega_c$ we would require their ratio to be independent of $\Delta\Omega_c$ for the proof to hold.

A third process is a sudden change in the pulsar magnetosphere (Scargle and Pacini, 1971). Here one imagines a massive cloud of plasma to suddenly detach from the corotating magnetic field and escape to infinity. This has two effects. First, it reduces the moment of inertia of the charged particle system – an effect identical to that of a starquake, with the added attraction that $\Delta\Omega_n$ is always zero. Secondly, the explosive release of plasma can produce an impulsive reaction torque on the star – an effect identical to that of a collision with a planet. Again, ξ is constant in time.

Finally, consider a sudden transfer of angular momentum from neutrons to charges. Whether this might conceivably be accomplished will be discussed later but for now we simply note that typical values of $(\Omega_n - \Omega_c)/\Omega_c$ amount to several percent. The neutron superfluid thus represents an enormous storehouse of 'excess' angular momentum relative to the charges. The transfer of even a small fraction of this excess could easily account for the observed jumps.

In this picture the neutrons slow down as the charges speed up and $I_n(\Delta\Omega_n) = -I_c(\Delta\Omega_c)$. Then $\xi = t_c/\tau$, again constant from one jump to the next. Note that this model would predict $Q = 1$.

What is the observational situation? With regard to Vela we do not yet know the value of $\Delta|\dot\Omega_c|$ for the second jump. It does seem to be positive (Reichley and Downes, 1971). The situation with regard to the Crab pulsar is almost equally ambiguous. The Princeton group (Boynton et al., 1972) has published what seems to be the most complete analysis of its first jump. The only published analysis of its second jump (Lohsen, 1972) does not discuss the post-jump behavior in terms of an exponential period decay. The physical meanings of the values of ξ derived from the two analyses are then quite different, and no comparison is possible. (If we neglect this difficulty and baldly compare values we find ξ (first jump) $\cong 10 \times \xi$ (second jump).) It is fortunate that Lohsen's observations do provide a continuous string of data during the period immediately following the jump. Re-analysis of these data should provide a detailed check of the theory.

2.2. Frictional Heating

Cameron (unpublished) has added a further critical element to our picture of superfluid neutron stars. He noted that *because the bulk of the mass of the star is slowed by frictional means, heat must be steadily dissipated in the process*. Consider 1 cm³ of superfluid. The frictional force $\mathbf{f}$ of (1) on this fluid element multiplied by $(\mathbf{v}_n - \mathbf{v}_c)$ is the rate that work is being done on the neutrons. All this work is dissipated as heat. The rate of dissipation is then

$$\dot E_{heat} = \mathbf{f} \cdot (\mathbf{v}_n - \mathbf{v}_c)$$
$$= (\text{torque})(\Omega_n - \Omega_c)$$
$$= \varrho_n (r \sin\theta)^2 \dot\Omega_n (\Omega_n - \Omega_c) \text{ erg cm}^{-3} \text{ s}^{-1}. \tag{10}$$

Neglecting the rotational energy in the charges the rate of loss of rotational energy is $\dot E_{rotation} \cong \varrho_n (r\sin\theta)^2 \dot\Omega_n \Omega_n$ erg cm^{-3} s^{-1} so that

$$\frac{\dot E_{heat}}{\dot E_{rotation}} \cong \frac{\Omega_n - \Omega_c}{\Omega_n} \cong \frac{\Omega_n - \Omega_c}{\Omega_c}. \tag{11}$$

Neutron stars are end points of stellar evolution. As such they are generally thought to possess no internal sources of energy. We now see that this is not so. If superfluid they are able to convert a fraction, given by (11), of their rotational energy into internal heat. (The mechanism outlined here is not the only way this can be done. Henriksen et al. (1972) have shown that rotational energy can also be dissipated into heat by the action of a steady wobble.) An analysis of *where* this takes place (Greenstein, 1971) within the star shows the frictional energy generation to be predominantly confined to a thin shell lying within the lower regions of the crust.

The rate of energy generation integrated throughout the star must equal the rate the star radiates energy. For internal temperatures $T \leq 10^8$ K the cooling mechanism is predominantly photon emission from the surface. Confining attention to this case

$$4\pi R^2 \sigma T_e^4 = \int d^3 r \dot{E}_{\text{heat}}$$

$$\cong \int d^3 r \rho_n (r \sin\theta)^2 \dot{\Omega}_c (\Omega_n - \Omega_c)$$

$$\cong \int d^3 r \rho_n (r \sin\theta)^2 \dot{\Omega}_c \Omega_c \rho \tau / \rho_c t_c, \qquad (12)$$

where R is the stellar radius, σ the Stephan-Boltzmann constant and T_e the surface temperature. The right-hand side of (12) contains τ, which is a function of T. If the relation between T and T_e is known (12) becomes an equation which can be solved for the temperature. A rough treatment (Greenstein, 1971) yields

$$T_e \cong \frac{4 \times 10^7}{[pt_y^2]^{1/6}} \text{ K}, \qquad (13)$$

where p is the pulsar period in seconds and $t_y \equiv p/\dot{p}$ in years. This result gives the lowest surface temperature such a star can attain. Because of the crudeness of the approximations leading to (13), detailed predictions based upon it are subject to doubt. Keeping this in mind I wish to apply it to three specific examples.

Before doing so a general comment should be made. Before 1967 the way one searched for neutron stars was to search for blackbody X-ray emission from their surfaces. No such sources were ever found. Since the discovery of pulsars and their identification with rotating neutron stars this project has been largely dropped. I wish to emphasize that the evidence that pulsars are neutron stars, though strong, is *indirect*. It would be nice to find something direct. The discovery of a point source of blackbody X-ray emission at a position coinciding with that of a pulsar would constitute such evidence.

The first example I want to discuss is the closest pulsar, CP 0950. Equation (13) predicts its surface temperature to be 2×10^5 K. The spectrum peaks at 0.1 keV energies. The X-ray flux at this energy at the top of the Earth's atmosphere is 10^{-28} erg cm^{-2} s^{-1} Hz^{-1} (estimating the pulsar's distance from its dispersion measure and neglecting interstellar absorption). Such a flux, were it concentrated at keV energies, would be detectable. Because it is concentrated at 0.1 keV energies the observations will be harder. Whether they are impossible is not clear.

The next example is the Crab pulsar. Its predicted temperature is 5×10^6 K. The spectrum peaks at 2 keV at which point the flux is 3×10^{-27} erg cm^{-2} s^{-1} Hz^{-1}. Such a flux is eminently detectable. Why, then, has it not been detected? Because it represents a small fraction (about 15%) of the flux from the nebula as a whole at this energy. It seems that the only way to detect this flux would be to observe the nebula during a lunar occultation.

Let me describe a possible observation one might perform during such an occultation. As the limb of the moon sweeps across the nebula the received X-ray flux diminishes steadily. As the limb sweeps across the pulsar the flux decreases discontinuously. Of this sudden kink some is due to the obscuration of the X-ray pulsar. Suppose that one accurately knows the pulsar intensity. The difference between it and the observed kink represents the steady, non-pulsed, X-ray emission from the pulsar. If we *assume* this to be blackbody radiation from a neutron star we can take a canonical radius for the star and calculate its temperature. But there is far more we can do if the *spectrum* of the point source can be determined (by performing the above observation at several energies). If the spectrum is blackbody we will have strong reason to attribute it to thermal radiation from a neutron star. The spectrum determines the temperature. The received flux then determines the radius of the star. If this radius does not agree with expected neutron star radii we will have an interesting contradiction. If it does, and if we trust our models, the radius will determine the mass of the star. Now we can use the formula

$$\dot{p}p \cong \frac{4\pi^2}{c^3} B^2 \frac{R^6}{I}, \tag{14}$$

to determine B, the pulsar's magnetic field.

(Equation [14] is true in those simple magnetospheric models such as those of Ostriker and Gunn (1969), and Goldreich and Julian (1969), which predict a braking index $N=3$. It is presumably not too badly off for the Crab, for which $N \cong 2.5$.) Finally, τ, the relaxation time describing the post-speedup decay of the Crab pulsar, is known to be ≈ 4 days. Theoretically, τ is a function of temperature and mass. A number of fruitful comparisons between theory and observation are possible here too. So, in one fell swoop, we would have gained an enormous amount of information.

I would not have wasted so much time on this were the next lunar occultation some time in the 1980's. In fact *the next lunar occultation is only a year and a half away* on March 29, 1974. From then on the Moon will occult the Crab *every month* till late 1975, a total of 20 occultations. Some details concerning the occultations are given in Table I. The years 1974–75 are going to present golden opportunities, not to return for something like a decade. I believe that reliable observations such as I have described could provide crucial data to the theoreticians. I hope they will be performed. So tell all your friends.

We already have one observational handle on the temperature of the Crab pulsar. This is the historic lunar occultation experiment of Bowyer *et al.* (1964) which showed the bulk of the X-ray emission to be nebular in origin. This experiment just missed detecting the X-ray pulsar. At a temperature predicted by (13) the thermal radiation from the star at 2 keV would be slightly higher than that from the pulsar. Thus, *were its temperature significantly higher than 5×10^6 K, thermal X-rays from the Crab* *neutron star would already have been detected.* Again, this result neglects interstellar (or nebular) absorption.

The last example is the recently discovered X-ray source GX340+0 (Margon *et al.*, 1971). This is the only source we know of whose spectrum has been definitely established to be blackbody. The observed temperature is 1.5×10^7 K – quite hot. From distance estimates (4 kpc) and the observed flux the radius of the emitting body is found to be 8 ± 4 km. It looks exactly like a neutron star. How old is it?

We can get a cooling curve from (13) (Greenstein, 1972) if we assume a magnetospheric model. If we assume the approximate validity of those for which the braking index $N=3$ then $p\dot{p}$ is constant in time. Writing $t_c = p^2/p\dot{p}$ and noting that if $N=3$, $t_c = 2t$ we can find the temperature in terms of the time:

$$T_e \cong \frac{10^8}{\chi^{1/12} t^{5/12}} \text{ K}, \tag{15}$$

where t is the age in years. Here χ has been defined to be $p\dot{p} \times 10^{15}$. In known pulsars χ ranges from 0.06 to 150 with a mean value of order unity. Specializing to the observed temperature of GX340+0 yields an age for this object of 125 $\chi^{-1/5}$ yr, a period of 0.003 $\chi^{2/5}$ s and a luminosity (rate of loss of rotational energy) of (10^{40} to 10^{42}) $\chi^{-3/5}$ ergs s^{-1}. It makes sense to search carefully for a supernova remnant and a pulsar at the location of this source. If we assume the supernova remnant expands at $\approx 10^3$ km s^{-1} its angular diameter would be a few seconds of arc.

TABLE I

Forthcoming lunar occulatations of Crab Nebula*
Adopted position (1950) 05ʰ31ᵐ30ˢ.5
+21°59′01″

Date	Time of Conjunction (nearest hour)	Elongation of Moon from Sun	Area of visibility
1974	(hours)	(degrees)	
March 29	22	77	Arctic and North Scandinavia at low altitude
April 26	04	50	Central and Northern Asia, Arctic, Alaska, Canada and North USA.
May 23	11	23	Northeast America, Greenland, Arctic, North Scotland, Scandinavia, Finland, Central Asia
June 19	21	356	Northeast Asia, Arctic, Alaska, Canada, Northeast of USA
July 17	08	329	Northeast Canada, Greenland, Arctic, N. Scotland, Scandinavia, Finland, Central Asia
Aug. 13	17	302	Japan, E. Asia, Alaska, Arctic, Greenland, N. America
Sept. 10	01	275	N. Africa, Europe (west at low altitude), Central Asia, Japan

Table I (continued)

Date	Time of Conjunction (nearest hour)	Elongation of Moon from Sun	Area of visibility
Oct. 07	07	248	Central and N. America (but not White Sands), N. Atlantic, S. and S. W. Europe, N. Africa
Nov. 03	12	221	Central Pacific, Hawaii, Central America and north of S. America at low altitude
Nov. 30	20	194	Central Africa (west at low altitude), Arabia, India, S. E. Asia, northern cost of Australia at very low altitude.
Dec. 28	06	167	Central Pacific, Hawaii, Central and N. America, north of S. America, central Atlantic
1975			
Jan. 24	17	140	Central Africa, Arabia, India, S. E. Asia, northern coast of Australia at low altitude.
Feb. 21	02	113	S. Pacific, S. America (east at low altitude), Mexico
March 20	10	86	Indian Ocean, East Indies, Australia, New Zealand at low altitude
April 16	15	60	E. coast of S. America at very low altitude, S. Atlantic, S. Africa and Madagascar
May 13	21	33	S. Pacific, S. America, Antarctic
June 10	04	6	Indian Ocean, Australia, New Zealand
July 07	13	339	S. America, Antarctic, very near Cape of Good Hope at low altitude
Aug. 03	23	312	Australia, New Zealand, Antarctic
Aug. 31	08	285	Tip of South America, Antarctic

* *Source:* L. V. Morrison, H. M. Nautical Almanac Office, Royal Greenwich Observatory, Herstmonceux Castle, England.

I am grateful to Dr. Morrison for furnishing me with this information.

3. The Most Complicated Possible Model of Superfluid Neutron Star

None of the results we have derived so far are rigorous. Indeed, they may be off by orders of magnitude. Why?

They were derived under the assumption that $\varrho\tau/\varrho_c$ is constant. In fact it varies by orders of magnitude throughout the star. Thus the right-hand side of (3) is a strong function of density. Since Ω_c cannot be then Ω_n must be. Therefore *the neutrons are in a state of differential rotation*. This does not sound particularly horrible. Were the neutrons to form an ordinary fluid it would not be. But because they form a superfluid it makes all the difference in the world.

3.1. SUPERFLUID TURBULENCE (Greenstein, 1970).

A rotating superfluid must contain vortex lines. These lines are in a state of tension and would like to form a uniform array parallel to the rotation axis. The tension in a line is finite, however, so that a sufficiently strong shearing force can disturb this state. Let us first completely neglect line tension. At every point within the superfluid the lines move with essentially the same velocity as the superfluid (the relative velocity between line and fluid, determined by the Magnus effect, amounts to $\approx 10^{-4}$ cm s^{-1}). Different regions of the superfluid rotate with different angular velocities. Each vortex line is therefore steadily lengthening and twisting about the rotation axis (in the same way that a rubber band, stretched between two cars moving with different speeds, steadily lengthens). Every time one region of the superfluid has lapped another once, the vortex lines passing between them have wrapped once more about the star. Eventually the wrapping is very tight. Along the equator lines of opposite senses are brought near each other. This situation is unstable and eventually two opposing lines will be brought sufficiently close to reconnect, forming a vortex ring and a shorter vortex line. The ring migrates away, the line steadily lengthens, new rings form and the process continues. As a given ring migrates about it collides with others and with lines, exciting vortex waves in them. Eventually a ring will be broken into two smaller rings. Feynman (1955) speculates that the ring of smallest possible diameter is a roton. If so, the breaking up of large rings leads in the end to the heating of the fluid. The whole process constitutes the superfluid version of the dissipation of velocity differentials into heat. In the steady state the superfluid contains tightly wrapped vortex lines, no longer parallel to the rotation axis, plus large numbers of rings. The distribution of lines fluctuates irregularly. The superfluid velocity field, determined by the distribution of lines, also fluctuates irregularly. This state is a superfluid version of fully developed turbulence. The turbulence is *microscopic* with typical eddy sizes being the distance between lines ($\lesssim 10^{-2}$ cm).

Recall that τ, the relaxation time describing the coupling between charges and superfluid, depends on scattering processes taking place within the normal cores of vortex lines (and rings). It is therefore inversely proportional to the total length of vortex line present. If the lines are greatly lengthened τ is proportionately shortened.

It is difficult to estimate the magnitude of this effect. A given line must wrap very many times about the star before opposing ends come sufficiently close to reconnect and form rings. These rings can themselves survive for long periods of time before being broken apart. In the absence of detailed knowledge it seems safe to say that τ is decreased by this effect by orders of magnitude. The problem is we don't know by *how many* orders of magnitude!

Of course vortex lines are in a state of tension. This tension acts to 'rigidify' the rotation of the superfluid. Let us begin with a uniformly rotating superfluid and ask if line tension is sufficient to maintain this state. In uniform rotation each line is parallel to the rotation axis. The scattering force f acting upon it, given by (1), varies by orders of magnitude along its length. The line will be able to maintain the state of uniform rotation if its tension force F_{tension} is greater than the total scattering force $F_{\text{scattering}}$ (equal to f integrated along the line) acting upon it. Numerically

$$F_{\text{tension}} = \pi \varrho_n \left(\frac{\hbar}{2m_n}\right)^2 \ln\left(\frac{b}{a}\right)$$

$$\cong 5 \times 10^8 \text{ dyne} \qquad (16)$$

at $\varrho_n = 10^{14}$ g cm^{-3}. Here m_n is the mass of the neutron, b the distance between vortex lines and a the vortex core radius. We can find $F_{\text{scattering}}$ as follows. In the steady state it adjusts itself until the total torque on the neutrons is sufficient to slow them down at the same rate as the charges. Thus

$$\text{torque} = I_n \dot{\Omega}_n \cong MR^2 \dot{\Omega}_c. \qquad (17)$$

Also

$$\text{torque} \cong \begin{pmatrix} \text{number of} \\ \text{lines in} \\ \text{the star} \end{pmatrix} RF_{\text{scattering}}$$

$$\cong (4 \times 10^{15} \, \Omega_c) \, RR_{\text{scattering}}. \qquad (18)$$

Thus

$$F_{\text{scattering}} = \frac{MR^2 \dot{\Omega}_c}{4 \times 10^{15} \, \Omega_c R} \cong \left(\frac{M}{M_\odot}\right) \frac{10^{17}}{6 t_y} \qquad (19)$$

($t_y \equiv |\Omega_c/\dot{\Omega}_c|$ in years). The condition $F_{\text{tension}} > F_{\text{scattering}}$ then yields

$$t_y > 3 \left(\frac{M}{M_\odot}\right) \times 10^7 \text{ yr}. \qquad (20)$$

So all pulsars whose characteristic age is less than $\approx 3 \times 10^7$ yr will exhibit superfluid turbulence.

There may be a flaw in this proof. A uniform vortex lattice exerts forces that cannot be understood in terms of the properties of isolated lines. These forces are generally (see, e.g., Fetter and Stauffer, 1970) thought to be far weaker than tension forces. Ruderman (1972), however, has argued against this point of view. He will briefly describe his arguments elsewhere in these proceedings.

3.2. Hydrodynamics of the superfluid

If vortex tension is unable to enforce rigid body rotation we are left with the task of finding out what the superfluid is doing. It is well known that fully developed turbulence possesses a turbulent kinematic viscosity $v = $ (mean velocity in an eddy) $\times$ (mean diameter of an eddy). The velocity field of a superfluid about a vortex line is $v = nh/2m_n r$ where $n = 1, 2, 3 \ldots$. Therefore, $v = nh/2m_n$. This estimate is in agreement with observed properties of the turbulent flow of superfluid helium (Vinen, 1961). The existence of this viscosity will result in the formation of a boundary layer just below the base of the crust in which the superfluid corotates with the crust.

The existence of this boundary layer, produced by turbulent viscosity, fulfills the conditions required for Eckman pumping to begin. This process spins down the superfluid in a time

$$t_E \approx R/\sqrt{\nu\Omega_n} \approx 1/\sqrt{n\Omega_c} \text{ yr}.$$

This time scale (one month for the Crab) is always short compared with t_c. The neutron superfluid, then, will almost exactly corotate with the crust. Thus the dominant interaction between superfluid and charges may well be *classical* (viscous) in nature. Whether this is so depends on whether the turbulence develops sufficiently fully for such ideas to be applicable. There is a further problem. Are we correct in regarding the crust as 'containing' the superfluid? It may be that vortex lines interact with the crust quite weakly. If so, no boundary layer would form and, though viscous, the superfluid would not exhibit Eckman pumping.

If not, what does it do? The following ideas are exceedingly tentative. If we look at the rotation curve obtained from (3) we find two regions in the star in which Ω_n decreases outwards. These regions coincide with those in which the superfluid energy gap decreases with decreasing density, i.e., at densities between 10^{14} and 10^{15} g cm^{-3}, and at densities $<2\times 10^{13}$ g cm^{-3}. In these two regions Ω_n may be decreasing outwards sufficiently rapidly for the quantity $r^2\Omega_n$ to decrease outwards. *If it does the flow is hydrodynamically unstable* (the Rayleigh instability) towards the development of classical turbulence. Because only the neutrons undergo these motions *there is no composition gradient acting to stabilize the flow*.

Suppose the fluid is turbulent in this classical sense. Now the turbulent eddies are macroscopic. Perhaps the superfluid resembles a pan of boiling water. We know that such a pan occasionally spits up a blob of water into the air. Suppose the superfluid lying within the inner crust does this. The blob will penetrate into the outer crust and lodge there – but the superfluid is rotating more rapidly than the crust. Therefore, the blob will impart its excess angular momentum to the crust, and will suddenly *speed it up*.

Should we identify this process with the observed *intermittent* large period jumps or with the smaller scale *continuous* fluctuations in the pulse repetition rate that the Princeton group (Boynton *et al.*, 1972) claims to be random in nature? It is hard to be sure. Even the rotation curve derived from (3) cannot be correct, for (3) neglects strictly hydrodynamical forces acting on the superfluid. A rigorous analysis of the stability of the superfluid flow should begin from the full equations, averaged over many vortex lines, of superfluid hydrodynamics. For now we are only able to make a plausibility argument – that it seems reasonable that in those regions in which the coupling between neutrons and charges is weakest Ω_n should be greatest. If so we would expect instabilities to develop.

I should emphasize that the problems we need to solve in order to understand what the superfluid is doing are *soluble* – an unusual state of affairs in neutron star physics. They are all problems in classical hydrodynamics (vortex lines are nineteenth century objects). We do not need to invent any new physics. It may even prove possible to exploit the strong resemblance between the neutron superfluid and superfluid helium to design laboratory experiments with which to gain insight into these phenomena.

References

Baym, G. and Pines, D.: 1971, *Ann. Phys. N.Y.* **66**, 816.
Baym, G., Pethick, C., Pines, D., and Ruderman, M.: 1969, *Nature* **224**, 872.
Börner, G. and Cohen, J.: 1971, *Nature Phys. Sci.* **231**, 146.
Bowyer, S., Byram, E. T., Chubb, T. A., and Friedman, H.: 1964, *Science* **146**, 912.
Boynton, P. E., Groth, E. J., Hutchinson, D. P., Nanos, G. P., Partridge, R. B., and Wilkinson, D. T.: 1972, *Astrophys. J.* **175**, 217.
Cameron, A. G. W.: 1970, *Ann. Rev. Astron. Astrophys.* **8**, 179.
Fetter, A. and Stauffer, B.: 1970, *Nature* **227**, 584.
Feynman, R.: 1955, in C. J. Gorter (ed.), *Prog. Low Temperature Phys.* **1**, 17.
Goldreich, P. and Julian, W.: 1969, *Astrophys. J.* **157**, 869.
Greenstein, G.: 1970, *Nature* **227**, 791.
Greenstein, G.: 1971, *Nature Phys. Sci.* **232**, 117.
Greenstein, G.: 1972, *Nature Phys. Sci.* **238**, 71.
Greenstein, J. L.: 1969, *Comments Astrophys. Space Phys.* **1**, 62.
Henriksen, R. N., Feldman, P. A., and Chau, W. Y.: 1972, *Astrophys. J.* **172**, 717.
Lohsen, E.: 1972, *Nature* **236**, 70.
Margon, B., Bowyer, S., Lampton, M., and Cruddace, R.: 1971, *Astrophys. J. Letters* **169**, L45.
Ostriker, J. P. and Gunn, J. E.: 1969, *Astrophys. J.* **157**, 1395.
Pines, D.: 1970, in E. Kanda (ed.), *Proc. of the 12th International Conference on Low Temperature Phys.*
Reichley, P. and Downes, G.: 1971, *Nature Phys. Sci.* **234**, 48.
Ruderman, M.: 1969a, *J. Phys.* **30**, C3-152.
Ruderman, M.: 1969b, *Nature* **223**, 597.
Ruderman, M.: 1972, *Ann. Rev. Astron. Astrophys.* (in press).
Scargle, J. and Pacini, F.: 1971, *Nature Phys. Sci.* **232**, 144.
Vinen, W. F.: 1961, in *Prog. Low Temperature Phys.* **3**.

DISCUSSION

Ruderman: When the force along an *isolated* superfluid vortex line varies by more than the tension in the line, then the vortex line will behave as described by Prof. Greenstein. It will bend and twist in response to the external force. But an array of vortex lines can behave quite differently and the criterion that if an isolated line twists in a complicated way, then an array will likely become turbulent is generally not valid. From the vortex line point of view a group of N closely spaced vortices has N^2 times the tension of a single vortex but only N times the differential force: the array is very much stiffer relative to the impressed force.

A uniformly rotating neutron star superfluid is described by a dense parallel vortex array of spacing $b \sim 10^{-2}$ cm. The scale of spatial variation of the differential torques on the neutron star superfluid is of order $R \sim 10^6$ cm so that $b/R \sim 10^{-8}$. The limit $b/R \to 0$, almost reached here, should be the same as that in which $\hbar \to 0$, i.e. classical hydrodynamics. This classical problem, the response of a uniformly rotating gravitating, compressible, nonviscous fluid in an axially symmetric container to a paraxial torque which varies with position but not angle (ϕ), can be solved. The results (an extension of the Taylor-Proudman theorem) show a uniform ϕ-directed acceleration on each coaxial cylinder together with a slow circulation in the planes of constant ϕ (i.e. planes through the symmetry axis). In the vortex line description, the vortex lines remain almost rigid and parallel to the rotation axis. The only change with time is a motion which changes their number density so that this density remains proportional to the classical $\nabla \times \mathbf{v}$.

The classical instability which can occur when the angular momentum per unit mass decreases with increasing radius corresponds to $\nabla \times \mathbf{v}$ changing sign. In the superfluid vortex picture, where $\nabla \times \mathbf{v}$ vanishes so does the vortex line density and, consequently, so does the torque upon the fluid in that region. Thus the torque would not convert a stable rotating fluid into an unstable one whatever its spatial distributions. (A sign reversal in $\nabla \times \mathbf{v}$ cannot be obtained merely by moving vortex lines around but would need the spontaneous creation of oppositely directed lines in the body of the fluid.)

Bethe: It seems that opposite statements have been made in the talk by Greenstein and in the discussion of it. Can someone elucidate what we should believe?

Greenstein: Don't believe anything quite yet. The problem with my approach is that I have considered the properties of an isolated vortex line. In fact the vortices form a lattice because there are long-range interactions between them. These interactions are very difficult to understand rigorously. The problem with Ruderman's approach is that it has not yet been formulated sufficiently fully to decide whether or not it represents a satisfactory treatment.

Stauffer has pointed out that, because the Tkachenko vortex lattice has a finite shear modulus, the system may split up into a few cylindrical regions, each rotating rigidly (if Ruderman's argument is correct).

PHASE DIAGRAM OF A CHARGED BOSE GAS*

J. P. HANSEN, B. JANCOVICI, and DANIEL SCHIFF

*Laboratoire de Physique Théorique et Hautes Energies, Bâtiment 211,
Université Paris-Sud, Orsay, France*

Abstract. The phase diagram of a charged Bose gas is drawn. The domains of existence of the solid, fluid and superfluid phases are discussed. It is predicted that superdense helium can be superfluid at densities higher than 10^6 g cm^{-3}.

---

* Presented by B. Jancovici. This paper has been published in *Phys. Rev. Letters* **29**, 991 (1972).

SUPERLUMINAL SOUND AND FERROMAGNETIC TRANSITION IN THE ZELDOVICH MODEL

G. KALMAN and S. T. LAI*

Dept. of Physics, Boston College, Chestnut Hill, Mass. 02167, U.S.A.

Abstract. The implications of the Zeldovich model (baryons interacting through a massive vector field) for the problem of superluminal sound propagation and ferromagnetic transition are examined. In a classical baryon gas at high densities correlation effects lead to the pressure increasing faster than the energy, ultimately resulting in superluminal sound; crystallization phase transition appears however at comparable densities, thus competing with the onset of superluminal sound. For a high density fermi gas the domains of ferromagnetic transition are delineated, indicating a minimal and maximal density below and above which no ferromagnetic transition can be expected. The latter is further affected by relativistic effects requiring a different approach to the calculation of exchange energy and of the ferromagnetic phase.

1. Introduction

The study of matter at ultra-high density poses some questions of fairly general character, the answers to which are expected to be virtually independent of the details of the interaction between the particles. Zeldovich's model has been devised (Zeldovich, 1962) and employed (Bludman and Ruderman, 1968) to investigate such questions. The model describes the interaction between nucleons with the aid of a massive vector field. The resulting repulsive short range potential is taken at its face value (without corrections resulting from the exchange of several mesons) and can be used both classically and quantum mechanically. It has been recently speculated (see the paper by Bethe in this volume) that it is indeed in the domain of very strong coupling where such a model provides a fairly reliable approach and a superposition of Zeldovich-like potentials has been used in several attempts (Leung and Wang in this volume) to describe realistically nuclear matter at ultrahigh ($\varrho > 10^{15}$ g cm^{-3}) densities. Although our justification for using the Zeldovich model is not that we expect it to be a realistic nuclear potential, in order to fix ideas we might stipulate that we are dealing with densities $n > 1$ fm^{-3} ($\varrho > 1.6 \times 10^{15}$ g cm^{-3}) with a vector meson of mass ~ 770 MeV (for ω meson $m_\omega = 784$ MeV) corresponding to a range of interaction $\mu^{-1} = 0.26$ fm and a coupling constant $g^2/\hbar c = 10 \sim 20$.

In this paper we investigate the puzzle of superluminal sound propagation and the problem of ferromagnetic transition; we also comment on the question of high density crystallization. These points we studied in the context of the Zeldovich model: it is hoped that the investigation of the various aspects of a single model is illuminating, even though no immediate connection with concrete physical systems can be established.

2. Superluminal Sound Propagation

2.1. Equation of State

As calculations for the equation of state for neutron matter at supernuclear and near-

* Now at Logicon Inc., False Church, Va. 22044, U.S.A.

relativistic densities led to such high pressures that p (pressure) became comparable to ε (energy density, including rest mass), the question of sound velocity acquired some interest. To be sure, results indicating substantially superluminal sound velocities emerged (as in Barker et al., 1967, for example), but it was argued that they were due to nonrealistic hard core potentials, inconsistent use or complete neglect of relativity, etc., and it was generally held that in the ultrarelativistic limit $p \leqslant \varepsilon/3$ should be obeyed, the equality being reached asymptotically. Correspondingly, since the speed of sound is determined by

$$c_S = c \left(\frac{\partial p}{\partial \varepsilon} \right)^{1/2}_S \tag{1}$$

it follows that the maximum sound speed would have been limited to $c_S \leqslant c/\sqrt{3}$.

It has been demonstrated by Zeldovich (1962) that the consistent relativistic use of his model does not support this contention. Indeed, in the Hartree approximation which was used by Zeldovich both the maximum limiting p/ε and c_S/c ratios approach one, viz.,

$$\begin{aligned} p/\varepsilon &\leqslant 1, \\ c_S/c &\leqslant 1. \end{aligned} \tag{2}$$

This result originates from the presence of the positive Hartree terms for a one-component system with repulsive short range interaction. The Hartree terms are of the order $O(n^2)$ as compared to $O(n)$ for the kinetic terms, and hence at extremely high densities the Hartree terms dominate, resulting in the asymptotic equation of state:

$$\lim p \to \varepsilon. \tag{3}$$

This leads to $\lim_{n\to\infty} c_S \to c$, implying possible existence of luminal sound speed at extreme relativistic densities. Similar results on zero sound in a relativistic quantum gas described by the Zeldovich model were obtained by Kalman (1967).

Bludman and Ruderman (1968) went beyond the Hartree approximation by including correlations among particles in a crystal lattice. They showed that under these circumstances both ratios in (2) exceed unity and ultrabaric equation of state ($p > \varepsilon$) and superluminal sound ($c_S > c$) obtained at sufficiently high densities.

It has been argued by Ruderman (1968) that the source of the acausal behavior – which is at the root of the appearance of superluminal sound propagation – is connected with the inherent instability of a classical one-particle or many-particle system interacting with its own radiation field (cf. the problem of the classical runaway electron discussed in Rohrlich, 1965). In the detailed dynamical description of the system this indeed casts doubt on the validity of choosing and constructing Green functions in the customary way (Bludman and Ruderman, 1968; Ruderman, 1968); it is difficult to see, however, how such an argument will affect direct thermodynamical considerations. (Whether the conventional thermodynamic or statistical approach is valid when the retarded interaction between particles is dominant is, however, questionable. A study of the appropriate kinetic theory would be desirable.)

Here we wish to report on our study of whether the introduction of correlations in the original Zeldovich model, where particles are in a gaseous (plasma) state, leads to superluminal rather than luminal sound propagation. The use of the word 'plasma' is indeed appropriate in the present context, since the behavior and the proper description of a system we are contemplating (a gas of 'charged' classical elementary particles at a finite temperature, interacting through a massive vector field) and that of

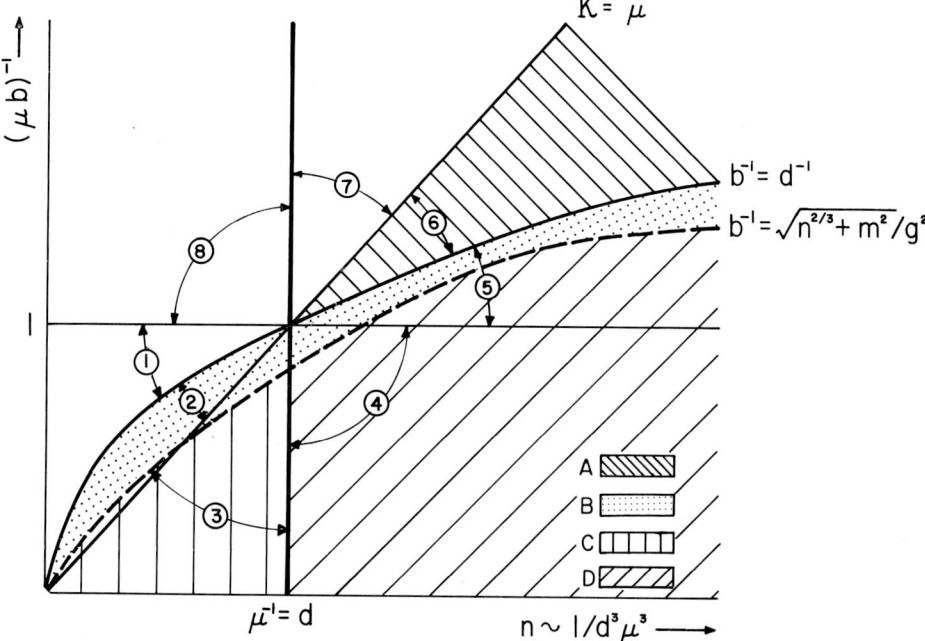

Fig. 1. Various domains in the parameter plane: for fixed values of the coupling constant and of the range of the field the parameter along the horizontal axis designates the density and the one along the vertical axis designates the temperature. The numbers enumerate the domains of different physical behaviors. Different shadings distinguish between the plasma domain (A), the classical lattice domain (B), the domain of fermi gas behavior (C), and the domain of quantum crystal behavior.

a classical electrodynamic plasma have much in common. To be sure, we are dealing with a system with a large coupling constant, but since we are studying the competitive roles of Hartree and correlation effects, the plasma approach provides a good qualitative description of and a reliable physical insight into the fundamental behavior of the system.

Our answer to the question of whether superluminal sound appears in the present model is in the affirmative: we do get $c_s > c$ when correlations become strong enough, although our method is not quantitatively reliable in the domain where this is obtained. We also acquire detailed descriptions of how correlation effects modify the sound speed, the equation of state and the stability of the plasma state.

We now outline the essential ideas and steps of our derivation. The distribution function, $F(\mathbf{x}, \mathbf{p}, t)$, of a plasma in a 'self-consistent field approximation' is governed by the Vlasov equation:

$$\left\{\frac{\partial}{\partial t} + \mathbf{v} \cdot \frac{\partial}{\partial \mathbf{x}} + g\mathbf{E} \cdot \frac{\partial}{\partial \mathbf{p}}\right\} F(\mathbf{x}, \mathbf{p}, t) = 0, \tag{4}$$

where E_μ is derived from the equations

$$\begin{aligned} (\Box - \mu^2) A_\mu &= -4\pi j_\mu, \\ \mathbf{j} &= g \int \mathbf{v} F(\mathbf{x}, \mathbf{p}, t) \, d\mathbf{p}, \\ \varrho &= g \int F(\mathbf{x}, \mathbf{p}, t) \, d\mathbf{p}, \\ \mathbf{E} &= -\dot{\mathbf{A}}. \end{aligned} \tag{5}$$

From this system, the 'dielectric' response-function ε of the system is obtained as:

$$\begin{aligned} \varepsilon(\mathbf{k}, \omega) &= 1 + \mathbf{k} \cdot \frac{\omega^2 - k^2}{\omega^2 - k^2 - \mu^2} \frac{\omega_0^2 m}{k^2} \int d\mathbf{p} \, \frac{\partial F^0/\partial \mathbf{p}}{\omega - \mathbf{k} \cdot \mathbf{v}}, \\ \omega_0^2 &= 4\pi \frac{g^2 n}{m}, \end{aligned} \tag{6}$$

where F^0 is the equilibrium distribution function. By invoking the fluctuation-dissipation theorem (Kubo, 1957; Sitenko, 1967; Golden and Kalman, 1969), the pair-correlation function $G(\mathbf{k})$ becomes related to the imaginary part of the static response function $\varepsilon(\mathbf{k}, 0)$. We obtain (Kalman and Lai, 1972b)

$$\begin{aligned} G(\mathbf{k}) &= -\frac{\kappa^4}{4\pi\beta} \frac{1}{k^2 + \kappa^2 + \mu^2}, \\ \kappa^2 &= 4\pi g^2 n \beta, \\ \beta &= 1/kT. \end{aligned} \tag{7}$$

The pair-correlation function leads to expressions for all the relevant thermodynamic quantities. Independently, the grand partition function is calculated by the cluster ring summation as in Brout and Carruthers (1963). This independent method verifies the thermodynamic results obtained earlier. The equation of state is given ultimately by

$$\varepsilon(n) = \frac{1}{\beta}[3 + \beta m G_0(\beta m)] n + \frac{1}{2}\left(\frac{4\pi g^2}{\mu^2}\right) n^2 - \frac{\kappa^2}{8\pi\beta}[(\kappa^2 + \mu^2)^{1/2} - \mu],$$

and

$$\begin{aligned} p(n) = \frac{n}{\beta} &+ \frac{1}{2}\left(\frac{4\pi g^2}{\mu^2}\right) n^2 - \frac{1}{3}\frac{\kappa^2}{8\pi\beta}[(\kappa^2 + \mu^2)^{1/2} - \mu] + \\ &- \frac{\mu}{24\pi\beta}[(\kappa^2 + \mu^2)^{1/2} - \mu]^2. \end{aligned} \tag{9}$$

Expansions in density n reveal ultrabasic behavior ($p > \varepsilon$) at high densities. At infinite densities, the Zeldovich result is recovered.

To calculate sound speed c_S, one can handle the partial derivatives by the Jacobian method (Landau and Lifshiftz, 1959), which leads to

$$\left(\frac{\partial p}{\partial \varepsilon}\right)_S = \frac{\left[\left(\frac{\partial p}{\partial T}\right)_V^2 \bigg/ \left(\frac{\partial S}{\partial T}\right)_V - \left(\frac{\partial p}{\partial V}\right)_T\right]}{\frac{1}{V}\left(p + \frac{E}{V}\right)}. \tag{10}$$

Alternatively, $(\partial p/\partial \varepsilon)_S$ can be calculated from the thermodynamic functions $p(n, \beta)$, $\varepsilon(n, \beta)$, and $S(n, \beta)$ by expressing β in terms of n at constant entropy S. Then, keeping S constant, we have

$$\left(\frac{\partial p}{\partial \varepsilon}\right)_S = \left\{\frac{\mathrm{d}p\,[n, \beta(n, S)]}{\mathrm{d}n}\right\} \bigg/ \left\{\frac{\mathrm{d}\varepsilon\,[n, \beta(n, S)]}{\mathrm{d}n}\right\}. \tag{11}$$

At high densities, superluminal behavior emerges, because (10) or (11) yield

$$\left(\frac{\partial p}{\partial \varepsilon}\right)_S = 1 + \tfrac{2}{9}\mu^2 g \beta^{1/2} \pi^{-1/2} n^{-1/2} + 0(n^{-1}). \tag{12}$$

As $n \to \infty$, Zeldovich's result is again recovered, viz.,

$$\lim_{n \to \infty} \left(\frac{\partial p}{\partial \varepsilon}\right)_S \to 1. \tag{13}$$

These results are qualitatively similar to those obtained by Bludman and Ruderman (1968).

As the density in the plasma state is increased beyond the plasma domain, symptoms of van der Waals type phase transition manifest themselves through the non-monotonic behavior of the pressure and energy density as functions of density. Whether the ensuing phase transition leads to a liquid or crystal structure cannot, of course, be determined at this point. Once the phase transition occurs the plasma approximation scheme evidently breaks down and no further prediction concerning the value of the sound velocity can be made. In the 'plasma domain' where our results are rigorously justified, neither ultrabaric equation of state nor superluminal sound result from (10), (11), and (12).

2.2. Higher order correlations

In order to somewhat refine our estimates, we consider the higher-order correlations by taking into account the 'watermelon' diagrams in the expansion of the partition function (Abe, 1959). These diagrams are of the order $\beta^{3/2} n^{1/2} g^3$. The higher-order correlations contribute to the free energy F the correction term F_c given by

$$F_c = \frac{n^2}{2\beta} \sum_{m=3}^{\infty} \int dr_1\, dr_2 \left\{ -\beta g^2 \frac{\exp[-(\kappa^2 + \mu^2)^{1/2} r]}{r} \right\}^m \frac{1}{m!}. \tag{14}$$

Explicit evaluation of this infinite series of cluster integrals leads to corrections of pressure and energy density,

$$p_c = \frac{n\lambda^2}{\beta\,12} \left(\log 3\Lambda + \frac{\gamma}{2} - \frac{11}{6} + \frac{1}{2} \frac{\mu^2}{\kappa^2 + \mu^2} \right),$$

$$\varepsilon_c = 3p_c - \frac{(\kappa^2 \mu g)^2}{48\pi(\kappa^2 + \mu^2)}, \tag{15}$$

$$\lambda \equiv \kappa\beta g^2 \quad \text{and} \quad \Lambda \equiv (\kappa^2 + \mu^2)^{1/2} \beta g^2, \quad \gamma = \text{Euler constant}.$$

Since higher-order correlation terms are small refinements to the ring results, they do not cause any significant alteration in our previous finding concerning the physical behavior of the system.

The correction term to the sound speed is obtained as

$$\Delta_c \left(\frac{dp}{d\varepsilon} \right)_s = \frac{1}{4\pi} \left[\frac{5}{6} \left(\frac{\log 3}{2} + \gamma - \frac{2}{3} \right) + \frac{13}{72} \right] \lambda \frac{\kappa^3}{n} + 0 \left(\lambda \frac{\kappa^3}{n} \log \lambda \right), \tag{16}$$

where γ is Euler's constant. This result shows that as long as $(\kappa\beta g^2)\kappa^3 n^{-1} < 1$, the correction is an insignificantly small negative term and the sound speed is overestimated in the ring approximation. There is no substantial modification of the phase transition criterion either.

2.3. COMMENTS AND CONCLUSIONS

As demonstrated in our model calculations, the emergence of ultrabaric behavior and superluminal sound stems from the fact that while correlations reduce both the pressure and energy density, the former is less affected thus increasing the value of $dp/d\varepsilon$. As the density and thus correlations are increased, phase-transition competes with the emergence of superluminal sound. Ultrabaric behavior occurs after superluminal sound, since at low densities energy density is always greater than pressure p, and hence the $p(n)$ curve has to climb at a steeper rate than $\varepsilon(n)$ in order to catch up with the latter.

It is instructive now to compare the densities where the superluminal and the ultrabaric behaviors occur with the critical density for the phase transition. For (9) and (11) we find that $p/\varepsilon = 1$ for

$$n_{\text{ultra}} = \frac{9}{\pi} \frac{1}{g^6 \beta^3} + \cdots, \tag{17}$$

and $\partial p/\partial\varepsilon=1$ for

$$n_{\text{super}} = \frac{9}{\pi} \frac{1}{g^6 \beta^3}. \tag{18}$$

So far as the phase transition is concerned, a somewhat more detailed examination of (9) leads to

$$n_{\text{crit}} = \frac{1}{12\sqrt{3\pi}} \mu^3, \tag{19}$$

where n_{crit} is the density at the critical point on the phase transition curve (Figure 2). The more detailed behavior of these quantities is illustrated in Figure 2. The general

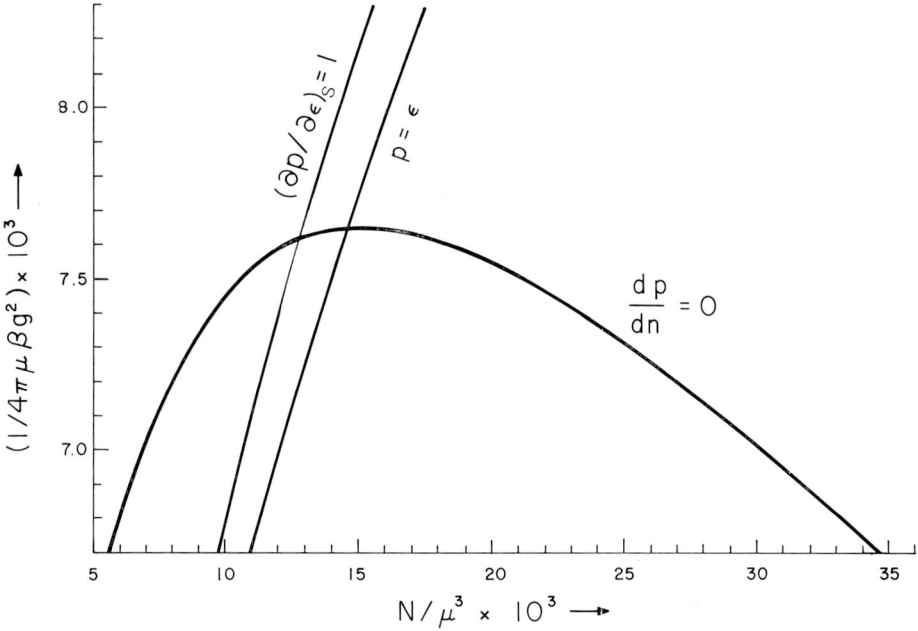

Fig. 2. Domains of phase transition ultrabaric and superluminal behavior. Note that they occur at comparable densities, but there are ultrabaric and superluminal domains outside the phase transition region at higher temperatures.

conclusion is that the ultrabaric and superluminal behaviors at high temperatures occur *outside* the phase transition domain, but in the region where short range correlations are important; thus the 'plasma' approximation is not really appropriate in this region. The qualitative conclusion, however, that the superluminal and ultrabaric behaviors *do* take place in the density and temperature domains indicated should not be affected by the more accurate description of correlations.

3. Ferromagnetic Transition

3.1. Nonrelativistic

The Zeldovich model provides an economic approach by which we can study the qualitative features of and the conditions for the ferromagnetic phase transition in a fairly realistic interaction model for nuclear matter at high densities. In particular, as the density approaches the relativistic region the spin-spin and transverse interaction (inherent in the vector character of the model) substantially modify the role of the exchange energy.

At non-relativistic densities the one-particle Hartree-Fock (HF) energy $\eta(\mathbf{p})$ is given by (Lai, 1970; Kalman and Lai, 1971)

$$\eta(\mathbf{k}) = \frac{\hbar^2 k^2}{2m} + \frac{2\pi g^2}{\mu^2} - \int_{|\mathbf{k}'|<k_F} \frac{d\mathbf{k}'}{(2\pi)^3} \frac{4\pi g^2}{|\mathbf{k}-\mathbf{k}'|^2 + \mu'^2}. \tag{20}$$

The Hartree term is spin independent. The competition between the kinetic and exchange terms determines whether the system favors a ferromagnetic state. If the spin populations $N_\pm$ are unequal, the system is ferromagnetic with magnetization $\zeta = (N_+ - N_-)/(N_+ + N_-)$. By comparing the average energy

$$E(\zeta) = \sum_{\uparrow\downarrow} \int \frac{d\mathbf{k}}{(2\pi)^3} \eta(\mathbf{k};\zeta)$$

per particle for $\zeta=0$ and $\zeta\neq 0$, the lowest energy state at a given density can be determined.

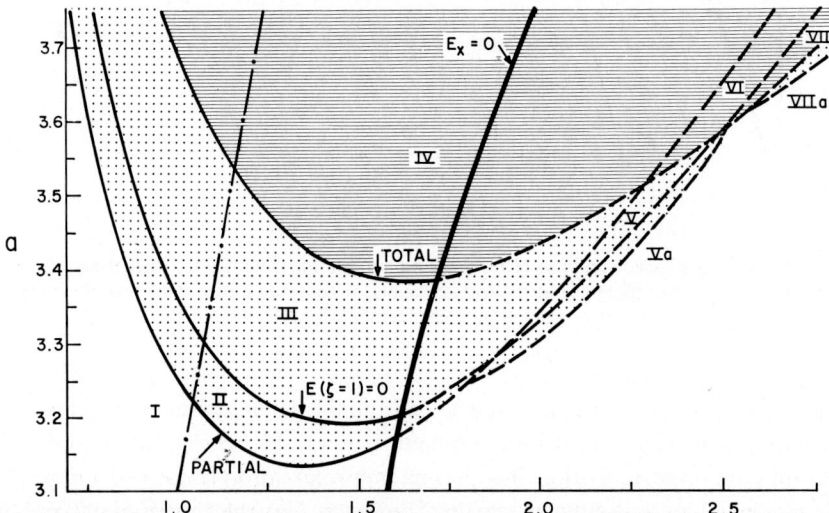

Fig. 3. Various boundaries of ferromagnetic transitions plotted against $a \equiv 2mg^2/\mu\pi\hbar^2$ and $x \equiv 2k_F/\mu$. The dash-dot line corresponds to the quantum crystallization density as estimated on the basis of the calculation by Canuto and Chitre (in this volume). The heavy solid line corresponds to $E_x = 0$.

Starting with low densities the lowest energy state is at $\zeta=0$ until the density is sufficiently high so that the location of min$[E(\zeta)]$ moves away from $\zeta=0$. Thus the condition for the onset of partial magnetization is $\partial E/\partial \zeta\,|_{\zeta=0}=0$ and $\partial^2 E/\partial \zeta^2\,|_{\zeta=0}=0$. Total magnetization is attained when min$[E(\zeta)]$ is located at $\zeta=1$, i.e.,

$$\partial E/\partial \zeta|_{\zeta=1} = 0 \quad \text{and} \quad \partial^2 E/\partial \zeta^2|_{\zeta=1} > 0.$$

The results are shown in Figure 3. Ferromagnetism is favored at intermediate densities because in the low density regime where μ^{-1} is small compared to the interparticle distance d, the system behaves like one with δ-function interaction while at high densities where $\mu^{-1} > d$ the system resembles a Coulomb gas.

Correlation induced phase transitions are obviously absent in the Hartree-Fock (HF) model. A HF-type phase transition (Gartenhaus and Stranahan, 1965a, b) is also absent in this case, since the presence of the Hartree terms maintains dp/dn positive, both in the normal and ferromagnetic state.

3.2. Relativistic Ferromagnetism

At relativistic densities ($\gtrsim 10^{15}$ g cm^{-3}) the problem of exchange energy in a fermi gas exhibits some unconventional features. The new physical aspects of the problem are the 'retardation,' i.e., the explicit energy dependence of the interaction potential and the importance of the transverse interaction; in addition, certain formal difficulties arise because of the description of the system in terms of helicity rather than spin states (Salpeter, 1961; Zapolsky, 1960).

As a result, the nonrelativistic exchange part of the HF Hamiltonian H_X^{NR} is modified to H_X^R as follows (Kalman and Lai, 1972a):

$$H_X^{NR} = \begin{vmatrix} \mathbf{p}_1 S_1 \\ \mathbf{p}_2 S_2 \end{vmatrix} \Big\langle \begin{matrix} \mathbf{p}_1 S_1 \\ \mathbf{p}_2 S_2 \end{matrix} \Big| \phi(\mathbf{p}_1, \mathbf{p}_2) \begin{vmatrix} \mathbf{p}_2 S_2 \\ \mathbf{p}_1 S_1 \end{vmatrix} \Big\langle \begin{matrix} \mathbf{p}_2 S_2 \\ \mathbf{p}_1 S_1 \end{matrix} \Big|,$$

$$H_X^R = \begin{vmatrix} \mathbf{p}_1 h_1 \\ \mathbf{p}_2 h_2 \end{vmatrix} \Big\langle \begin{matrix} \mathbf{p}_1 h_1 \\ \mathbf{p}_2 h_2 \end{matrix} \Big| \phi^R(\mathbf{p}_1, \mathbf{p}_2) \begin{vmatrix} \mathbf{p}_2 h_2 \\ \mathbf{p}_1 h_1 \end{vmatrix} \Big\langle \begin{matrix} \mathbf{p}_2 h_2 \\ \mathbf{p}_1 h_1 \end{matrix} \Big|,$$

where

$$\phi(\mathbf{p}_1, \mathbf{p}_2) = \frac{g^2}{|\mathbf{p}_1 - \mathbf{p}_2|^2 + \mu^2} \tag{21}$$

$$\phi^R(\mathbf{p}_1, \mathbf{p}_2) = \frac{g^2(1 - \boldsymbol{\alpha}_1 \cdot \boldsymbol{\alpha}_2)}{|\mathbf{p}_1 - \mathbf{p}_2|^2 - [\eta(\mathbf{p}_1) - \eta(\mathbf{p}_2)]^2 + \mu^2} \tag{22}$$

and S stands for spin, h for helicity.

The properties of the relativistic exchange integrals are displayed in Figure 4. In the extreme relativistic region the exchange energy becomes positive in contrast to the negative value it has in the non-relativistic region.

In Figure 5 we have indicated the boundary where the exchange energy changes sign. No ferromagnetic behavior in the region to the right of the boundary can be expected and the likelihood for the ferromagnetic state corresponding to a lower energy diminishes as one approaches the boundary. The strong range-dependence of the

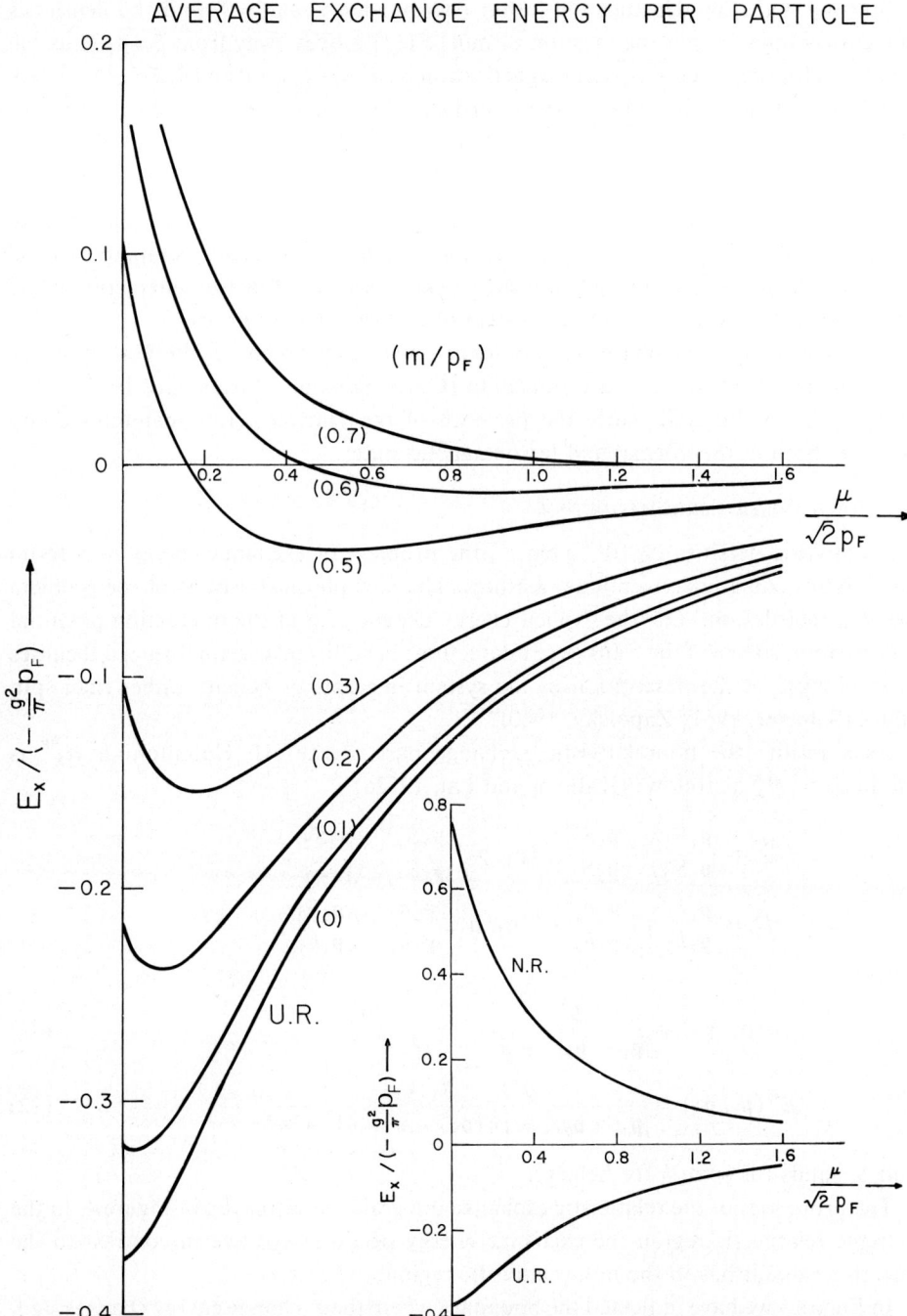

Fig. 4. Dependence of exchange energy E_x on the range μ^{-1} of the field and the fermi momentum p_F. At sufficiently high densities, exchange energy becomes positive. The small insert depicts the transition boundary of E_x.

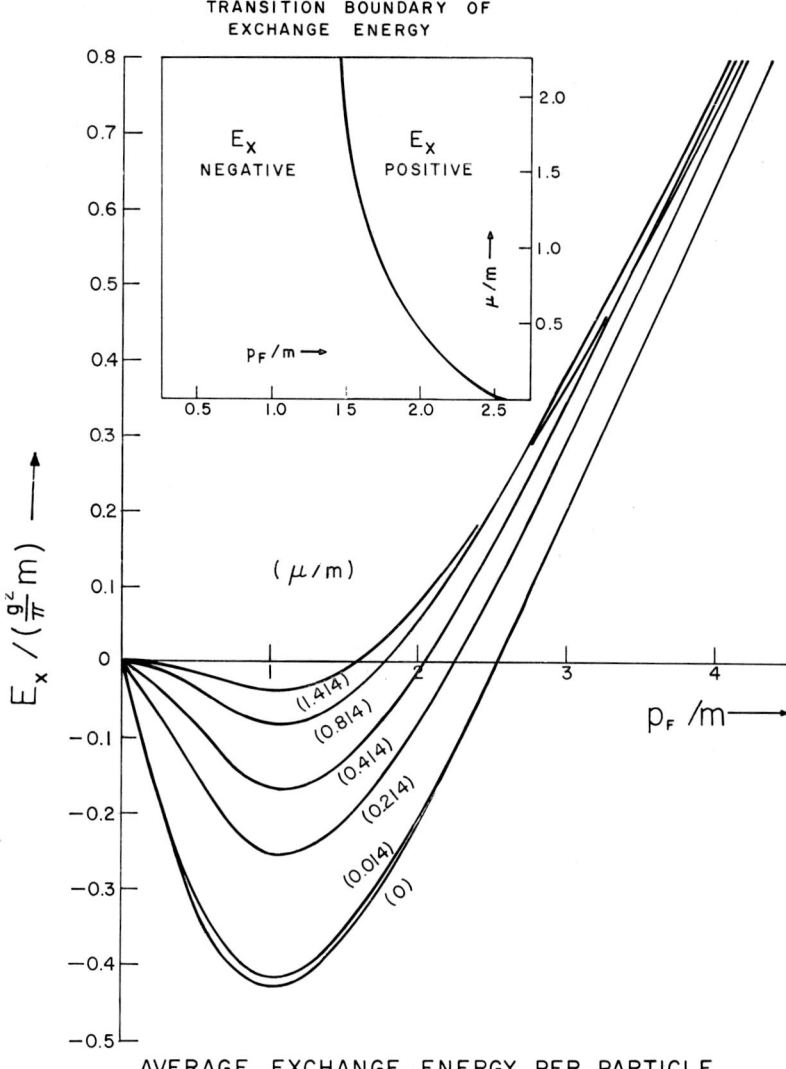

Fig. 5. Dependence of exchange energy on the range μ^{-1} of the field and the mass m of baryon. The small insert illustrates the comparison of the non-relativistic and the ultrarelativistic behaviors.

density value where the transition occurs should be noted. For example, for an electron gas the exchange energy changes sign at about 15.6 n_c ($n_c = m^3c^3/3\pi^2\hbar^3$, the critical density for relativistic degeneracy) while for $\mu/m \simeq 0.8$ this takes place around 5.4 n_c ($n_c = 5.7 \times 10^{15}$ g cm^{-3} for a neutron gas); however around 5×10^{15} g cm^{-3} the relativistic decrease amounts to about 60% of the non-relativistically calculated value.

3.3. METHODOLOGICAL QUESTIONS

In the relativistic domain the system is properly described in terms of helicity eigenstates: the ferromagnetic state which requires the spin rather than the helicities to take up a certain value is not an eigenstate of the HF Hamiltonian (within positive energy states). The proper definition of the ferromagnetic state now can be accomplished by considering a coherent superposition $|F\rangle$ of helicity states which maximizes the expectation value $\langle F|\Sigma_3|F\rangle$. Setting

$$|F\rangle = \frac{1}{(1+b^2)^{1/2}}\{|p\uparrow\rangle + b|p\downarrow\rangle\} \tag{23}$$

one finds

$$b = -\frac{\varepsilon}{m}\cot\theta + \left(\frac{\varepsilon^2}{m^2}\cos^2\theta + 1\right)^{1/2} \tag{24}$$

with $\varepsilon = (p^2+m^2)^{1/2}$ and θ being the angle between $\mathbf{p}$ and the z-axis. The maximum magnetic moment per particle now becomes

$$\langle\mu\rangle_{\max} = \frac{e\hbar}{mc}\frac{m^2}{2\varepsilon^2}\left(1+\frac{\varepsilon}{m}\right)\left(\frac{\varepsilon^2}{m^2}\cos^2\theta + \sin^2\theta\right)^{1/2}. \tag{25}$$

A further problem arises if one wishes to apply the HF approximation in a consistent fashion, which requires that the one-particle Hamiltonian (including the exchange term) be diagonalized. This is automatically satisfied in the nonrelativistic approximation where the $|\mathbf{p}S\rangle$ base states diagonalize the exchange term [cf. (20)]. The latter however, is not diagonalized by the relativistic $|\mathbf{p}hr\rangle$ (r designates the signature of the energy) base states: therefore a transformation to a new set of states, say $|\mathbf{p}\chi\varrho\rangle$, is required,

$$|\mathbf{p}\chi\varrho\rangle = \sum_{\text{exch}} C_{\chi\varrho,hr}|\mathbf{p}hr\rangle, \tag{26}$$

in such a manner that the expansion coefficients C are to be determined self-consistently. This can be done by expressing the exchange Hamiltonian as

$$H^{\text{exch}}_{phr,\bar{p}\bar{h}\bar{r}} = \sum_{\mathbf{p}'}\phi(\mathbf{p},\mathbf{p}')\delta_{\mathbf{p}\bar{\mathbf{p}}}n_{\mathbf{p}'}X_{hr,\bar{h}\bar{r}}(\mathbf{p},\mathbf{p}') \tag{27}$$

$$X_{hr,\bar{h}\bar{r}} = \sum_{\substack{h'r'\\h''h''}}\left\langle \begin{matrix}hr\\h'r'\end{matrix}\right|(1-\boldsymbol{\alpha}_1\cdot\boldsymbol{\alpha}_2)\left|\begin{matrix}h''r''\\hr\end{matrix}\right\rangle \Omega_{h'r',h''r''} \tag{28}$$

with the aid of the density matrix Ω

$$\Omega_{h'r',h''r''} = \sum_{\chi\varrho} C^*_{\chi\varrho,h'r'}n_\chi n_\varrho C_{\chi\varrho,h''\varrho''}. \tag{29}$$

The diagonalization of X provides the new base states $|\chi\varrho\rangle$ and thus the expansion coefficients C; since X is already an implicit function of the C-s through (28) and (29) this procedure provides the self-consistency condition for the C-s. This work is in progress: at the time of the writing of this paper we have no definitive result to report on.

4. Remarks on the Question of Crystallization

The question of crystallization of matter at very high densities in the context of Zeldovich model is worth some further exploration. Some order-of-magnitude estimates can easily be given. The temperature-density region for a classical plasma state is such that the kinetic energy exceeds the potential energy and the particles behave collectively despite the short range of the field; these lead to the condition: $(\beta g^2)^{-1} < d < \kappa^{-1} < \mu^{-1}$. As shown in Figure 1, this condition can be satisfied at very high densities provided that the temperature is sufficiently high. On the other hand, the conditions for the existence of a classical crystal are such that the kinetic energy is less than the potential energy and the thermal energy is greater than the fermi energy; these lead to the conditions: $(\beta g^2)^{-1} < n^{1/3}$ and $(\beta g^2)^{-1} > (n^{2/3} + m^2)^{1/2} (\hbar c/g^2)$. If the last inequality is reversed, the system becomes a quantum crystal. The appropriate temperature-density regions for these conditions are shown in Figure 1. The criterion for the existence of a quantum crystal can be given as

$$\frac{\mu c}{\hbar} < n^{1/3} < \frac{g^2 m}{\hbar^2},$$

i.e., the system should be in the 'Coulomb' region, but the interparticle distance should not be smaller than the 'Bohr radius.' These conditions are not contradictory for strong coupling $(g^2/\hbar c > \mu/m)$ and will not be violated even for extreme relativistic regions.

As the system approaches the critical density for phase transition the strong correlations will diminish the role of the exchange energy and in the crystalline state the fermi gas approach is obviously invalid. For the purpose of orientation, we have depicted in Figure 3 the line

$$n_{critical} = \left[3.2 \times 10^{-2} \times \frac{g^2 m}{\hbar^2} \right]^3,$$

suggested by the results obtained for the actual solidification boundary by the recent computations of Canuto and Chitre (as given in this volume) with realistic nuclear potentials.

References

Abe, R.: 1959, *Prog. Theor. Phys. Kyoto* **22**, 213.
Barker, B. M., Bhatia, M. S., and Szamosi, G.: 1967, *Nuovo Cimento* **52B**, 355.
Bludman, S. A. and Ruderman, M. A.: 1968, *Phys. Rev.* **170**, 1176.
Brout, R. and Carruthers, P.: 1963, *Lectures on Many-Electron Problems*, Interscience, New York.
Gartenhaus, S. and Stranahan, G.: 1965a, *Phys. Rev. Letters* **14**, 341.
Gartenhaus, S. and Stranahan, G.: 1965b, *Phys. Rev. Letters* **14**, 621.
Golden, K. J. and Kalman, G.: 1969, *J. Stat. Phys.* **1**, 415.
Kalman, G.: 1967, *Phys. Rev.* **158**, 144.
Kalman, G. and Lai, S. T.: 1971, *Phys. Letters* **34A**, 75.
Kalman, G. and Lai, S. T.: 1972a, *The Problem of Exchange Energy in a Relativistic Electron Gas*, paper presented at the Spring Meeting of The American Physical Society, New England Section.

Kalman, G. and Lai, S. T.: 1972b, *Ann. Phys. (N.Y.)* **73**, 19.
Kubo, R.: 1957, *J. Phys. Soc. Japan* **12**, 570.
Lai, S. T.: 1970, Brandeis University Ph.D. Thesis.
Landau, L. D. and Lifshitz, E. M.: 1959, *Statistical Physics*, Pergamon Press, New York.
Rohrlich, F.: 1965, *Classical Charged Particles*, Addison Wesley, Reading.
Ruderman, M. A.: 1968, *Phys. Rev.* **172**, 1286.
Salpeter, E. E.: 1961, *Astrophys. J.* **134**, 669.
Sitenko, A. G.: 1967, *Electromagnetic Fluctuations in Plasmas*, Academic Press, New York.
Zapolsky, H.: 1960, Cornell University Report (unpublished).
Zeldovich, Ya. B.: 1962, *Soviet Phys. JETP* **14**, 1143.

SUPERNOVAE AND NEUTRON STARS

STIRLING A. COLGATE

New Mexico Institute of Mining and Technology, Socorro, N.M. 87801, U.S.A.

Abstract. The formation of a neutron star and the event observed as a supernova are presumed by most astrophysicists to be simultaneous. The theoretical difficulty of understanding this is compounded by a hierarchy of conflicting phenomena dominated by the possibility of effective neutrino transport of the heat of the neutron star formation to the outer envelope. A thermonuclear detonation of a carbon core may take place at just the right density such that both neutron star formation and neutrino assisted envelope ejection can occur together.

1. Supernova and Neutron Star: The Same Event?

This paper reviews one particular problem of supernova theory; namely, how can the birth of a neutron star and a supernova be one and the same event and yet how can they not be?

The theories of supernovae are faced with a major dichotomy: on the one hand, we see a neutron star remnant at the site of one supernova, the Crab, and infer that many other cases are similar – yet on the other hand, serious doubt has been cast on the numerical theoretical calculations that might explain such events.

The problem centers on the requirement of simultaneously ejecting something like a solar mass at the same time that a similar mass collapses to a neutron star. First of all, what is the justification for both 'a solar mass' and 'near simultaneity'? The collapse to a neutron star requires that sufficient mass be assembled at a low enough temperature such that the available specific gravitational energy is greater than the specific energy required to convert a proton-electron pair into a neutron, when that proton is at the nuclear binding corresponding to that in helium. At the stellar densities where collapse to a neutron star is imminent, there is little question that stellar nucleosynthesis has proceeded to the point of at least helium – and more likely carbon-oxygen and even to the end point – iron. Ninety percent of the nuclear binding occurs in helium and so whatever additional synthesis has occurred will not significantly change the approximate result. The minimum energy required to convert He to free neutrons is about 10^{19} erg g^{-1}. This must be equated to MG/r, the gravitational potential. If we want to minimize M, we must minimize the radius. The minimum radius in turn is set by the condition that we do not want the degenerate electron pressure to require more work in collapse than the conversion of protons to neutrons – or put another way, protons will be converted to neutrons whenever the electron Fermi level becomes greater than the nuclear energy difference. This occurs at $\varrho \simeq 2 \times 10^{11}$ g cm^{-3} ($E_{\text{Fermi}} \simeq 10$ MeV) which is therefore the density of incipient collapse. Using this density results in $r \simeq 10^7$ cm and $M \simeq M_\odot$. If the temperature is raised (compared with $E_F \simeq 10$ MeV) the minimum mass will be increased. The result is comfortably assuring in that we know that it requires a mass greater than the

Chandrasekhar limit 1.13 $M_\odot$ to overcome electron degeneracy pressure and reach this density in the first place, which, anyhow, is a similar way of looking at the condition for collapse.

The estimate of the ejection of a solar mass is an observational estimate derived from the light curves. One observes a peak luminosity and temperature; granted that these observations have some considerable error and even variation as to class of supernova, but for a typical type I supernova the peak luminosity may be 2×10^{43} erg s^{-1} for a surface temperature $\simeq 10^4$ K. The surface area required, $A(c/4)aT^4 = L$, becomes 2×10^{31} cm^2. The absolute minimum mass would be this area divided by the optical opacity. At the necessarily low density, this is Compton opacity, 10^{32} g for the surface layer. Models describing the mechanism of producing the optical radiation require some finite depth of the expanding (observed by Doppler shift) matter which at a minimum may be 10 mean free paths thick (in order that the diffusive release time of the radiation be as long as the observed width of one week) implying that at least 1/2 $M_\odot$ is ejected.

Finally, we are presuming that these events occur nearly simultaneously. Extrapolation of the slowing down of the Crab pulsar leads to an origin within years of the supernova event. One could assume ejection took place (say 10 years?) before or after collapse; namely, that the events are only secondarily related, but the theoretical difficulties become even more staggering.

If ejection comes after collapse, then the matter must be lifted from the neutron star surface at the bottom of a gravitational well of greater than 100 MeV per nucleon yet be observed 'dribbling' out into space at 1 to 2 MeV per nucleon (maximum bulk expansion velocity $\simeq 2 \times 10^9$ cm s^{-1}). No energy source has been proposed that could possibly achieve this. Ejection before collapse is more difficult to exclude. Conceivably, a thermonuclear detonation of a carbon-oxygen layer could take place at the surface of a more highly evolved, say, silicon core. The outer layer might be ejected and the smaller energy release in the core might be too small to disrupt the core and leave a large enough mass that later evolution would lead to collapse. The subtlety of conditions that would lead to thermonuclear detonation and partial disruption without simultaneously triggering collapse would seem hard to arrange – almost contrived – and so unless there is overwhelming evidence to the contrary, most would agree that despite current difficulties, collapse and ejection must occur in the same dynamical event, but it is difficult to manage both at the current sophistication of theoretical analysis.

The original concept of achieving both events simultaneously depends upon utilizing the binding energy of the neutron star ($\simeq 100$ MeV per nucleon) to eject an outer, less strongly bound (1 MeV per nucleon) mass fraction. We emphasize net binding energy, the difference between the internal and gravitational energies.

If comparable masses are ejected and collapsed, it would seem that the large energy difference would be more than adequate to achieve partial ejection. However, the difficulty arises in the inefficiency of energy transfer. If initially cold matter of relatively small binding energy were to collapse to a neutron star, the difference in binding energy would have to appear in internal degrees of freedom – oscillations and ultimately heat. Because of the propensity for the 'cold' fluid to form a shock wave during collapse, the formation process is relatively inelastic; the 'bounce' is weak, and essentially all the binding energy appears as heat. This heat, in turn, can be conducted to the outer layer by 'radiation' transport, heating the outer layers to the point where they blow off. The problem is that ordinary Planck radiation diffuses so slowly (the characteristic time is the Helmholtz contraction time of 10^5 yr) that the outer layers would have long since collapsed in seconds onto the neutron star core and possibly exceeding the stable mass limit and forming a black hole. Fortunately, for the theoretician, at the high temperature of the just-formed neutron star of $\simeq 30$ MeV (3×10^{11} K or 100 MeV per nucleon) radiation must include neutrinos as leptons in thermodynamic equilibrium and this equilibrium occurs in times of 10^{-8} to 10^{-6} s depending upon density, temperature and which kind of neutrino is involved, electron or muon. The thickness of the imploding matter onto the forming neutron star is variously one to ten mean free paths for the electron neutrinos and much less than one mean free path for the muon neutrinos. Therefore, only the electron neutrinos can transport heat to the outer layers and two conditions can prevent or greatly reduce this heat transport, and hence prevent sizeable mass ejection. If the imploding matter is thick enough (about 10 mean free paths) so that the neutrino radiation diffusion velocity ($\simeq c/10$) is less than the implosion velocity, then the diffusing heat will be continuously carried back to the neutron star and the outer layers will not be heated. Similarly, if the muon emission is great enough, it will allow the escape of the heat without interaction with the outer layers. Fortunately, one effect argues strongly in favor of some partial heat transport. As the temperature falls, so also does the neutrino electron scattering cross section, $\sigma_v \propto E^2$, and with nuclear thresholds and degeneracy $\sigma_v \propto E^{4-5}$ so that decreasing temperature allows a more rapid (longer mean free path) diffusion of heat. In addition, a lower temperature favors electron neutrino emission as opposed to muon neutrino emission. The actual temperature and ratio of electron to muon neutrinos depend upon details of the hydrodynamics, but some of these details become crucial to the final outcome.

Initially, a simple transport of the neutrino heat from the standing shock in the imploding matter on the neutron star core appeared to give ample energy to eject the envelope at several MeV per nucleon. However, later calculations with more detailed neutrino transport negated this conclusion and only stars of mass greater than 4 to 6 solar masses ejected an envelope, and even so the residual core masses were greater than 2.5 $M_\odot$. A hierarchy of effects could alter this unhappy state of affairs:

(i) The standing shock on the neutron core star preferentially emits the 'binding energy' in electron neutrinos in several meters equilibrium distance before muon neutrinos are emitted in several hundred meters distance.

(ii) In larger mass stars, the temperature immediately prior to collapse at $\varrho \sim 2 \times 10^{11}$ g cm^{-3} makes a sensitive difference to the hydrodynamics during the near free fall to the neutron star surface. This matter is cooled just before collapse by thermal decomposition to proton rich nuclei with subsequent electron capture, neutrino emission and hence cooling to a lower adiabat during collapse. The lower adiabat leads to a stronger standing shock, lower neutrino opacity in the collapsing matter and hence a larger heating of the external layers.

Finally, to confound the issue further, collapse calculations have been usually performed with relatively large and finite mass zoning so that not only is the detail of the various shocks neglected, i.e., electron vs. muon neutrino emission, but in the expanding phase after the envelope is heated and starts to expand some one inner mass zone develops a large ratio of inner and outer boundary radii. Further collapse onto the neutron star is halted due to this artifact independent of further neutrino cooling or what-have-you. An analytical approximation of the explosion phase indicates that for a very wide range of explosion conditions, wider than the most optimistic explosion advocates would claim, one-half the mass of matter initially ejected will fall back onto the neutron star. This is because in a spherical explosion, the inner regions perforce must expand relatively slowly. When the gravitational and thermal sink of a neutron star is introduced at the center, collapse of even the hottest matter back onto the neutron star is assured. Although this secondary collapse of matter, which might have experienced transient densities of $10^9 \leqslant \varrho \leqslant 10^{12}$ g cm^{-3} is comforting to those who worry about the possibility of producing too much neutron rich matter for the Galaxy, nevertheless, the fact that only one-half the mass is really ejected compared to what we used to think, makes simultaneous ejection of a solar mass and collapse of less than 2 $M_\odot$ even more difficult.

2. Thermonuclear Detonations

In recent years, as well as at the start of all theoretical considerations of supernova mechanisms, the detonation of an evolving carbon star has been an attractive way to make supernova, heavy elements, etc. However, for awhile it seemed that calculations indicated that despite the relatively modest overpressure ($\simeq 10\%$) created by the thermonuclear burning of carbon to completion (a mixture of all nuclear species up to about iron), still the star seemed to disrupt itself completely leaving no neutron star remnant! In order to conceivably form a neutron star, the matter had to reach several $\times 10^{10}$ g cm^{-3} before detonation so that beta decays could take place fast enough (due to the high electron Fermi level) so that collapse took place before explosion. Such a high density seemed impossible to reach without prior detonation of carbon at several $\times 10^9$ g cm^{-3}. Again, a new theory has saved the day – one which predicts convective URCA shells with sufficiently steep temperature dependence as to overcome the carbon-carbon reaction sensitivity.

It now appears that a star can evolve stably through carbon burning up to the end point of iron and hence collapse with no nuclear energy available. A more attractive viewpoint is that collapse is initiated when there is still potentially explosive fuel, carbon-oxygen in the outer layers of the presupernova core, so that detonation and collapse to a neutron star occur nearly simultaneously. Eureka! But not quite. Long ago this was tried numerically where 5 $M_\odot$ of equivalent carbon-carbon burning energy were dumped into the outer layers of a 10 $M_\odot$ star at the time when the inner part started collapsing to a neutron star. It *all* fell in unless neutrino transport was added. In other words, the overpressure created by a carbon detonation is so small (because of the high specific heat of the many nuclear species and radiation) that the rarefaction created by an inner collapse completely dominates the explosion and turns the matter around to an implosion. There the matter currently rests – in neutron stars, black holes – wherever. Perhaps a subtle combination of detonation in the outer layers, retarded implosion and hence lower density in the matter imploding onto the neutron star will cause a strong enough standing shock such that the electron neutrino heat can escape fast enough to boost the carbon-carbon detonation in the outer layers sufficiently to eject part of these detonating outer layers – and form us.

Acknowledgements

I am deeply indebted to the many original and creative scientists that have conceived and written the original papers from which this discussion was drawn. Foremost among them are: William Fowler, Fred Hoyle, Geoffrey Burbidge, Margaret Burbidge, Al Cameron, Jim Truran, Dave Arnett, Bob Schwartz, B. Paczynski, G. Rakavy, G. Shaviv, Jim Fraley, Don Clayton, Dave Schramm, Dick White, Jim Wilson, M. LeBlanc, Carl Hansen, Craig Wheeler, Zalman Barkat, J.-R. Buchler, Steve Bruenn, Chester McKee, and others.

DISCUSSION

Van Horn: Since the problem of black holes has been raised I would like to ask the following question: situations must arise in which the core implosion of a supernova results in the creation of a black hole instead of a neutron star. Has this shown up in any of your calculations and, if so, what were the consequences?

Colgate: In the calculation of Dick White and myself the mass of the neutron star formed in the explosion was always significantly less than the currently accepted critical neutron star mass. As a consequence the Newtonian hydrodynamics used was justified. In the calculations of Arnett, on the other hand, more mass accumulated on the neutron star than would be stable in general relativity. The implication was that a black hole would have been formed had general relativistic hydrodynamics been used. Wilson has made general relativistic calculations that indeed show the formation of a black hole. To the extent that some mass is ejected there would be negligible observable difference for an outside observer. Of course the lack of a pulsar would be evident at some later date.

NEUTRON STAR STRUCTURE FROM PULSAR OBSERVATIONS*

DAVID PINES and JACOB SHAHAM**

Dept. of Physics, University of Illinois at Urbana-Champaign

and

MALVIN A. RUDERMAN

Dept. of Physics, Columbia University, New York, N.Y., U.S.A.

Abstract. We examine what inferences can be made regarding neutron star structure from observations of micro- and macroglitch behavior. After considering various theories it seems plausible that crustquakes offer an explanation for the Crab microglitches, while corequakes can explain the Vela macroglitches. It is concluded that the Crab pulsar has a mass of less than 0.5 $M_\odot$ and is ~90% superfluid neutrons while the Vela pulsar may possess a solid neutron core and have a mass of ~0.7 $M_\odot$ with a superfluid neutron abundance of ~15%.

1. Introductory Remarks

Following the identification of pulsars as rotating neutron stars, interest in the calculation of the stellar structure of neutron stars has increased appreciably. As we have heard at this meeting from Bethe and Negele, the structure and composition of the outer portion of a neutron star ($\varrho \lesssim 2 \times 10^{14}$ g cm^{-3}, say) is by now comparatively well understood theoretically; however, for densities which lie between 2×10^{14} g cm^{-3} and 10^{16} g cm^{-3} there have been a number of different proposals for the stellar composition (Sawyer, 1972; Scalapino, 1972; Sawyer and Scalapino 1973; and the papers of Pandharipande and Canuto in this volume) with concomitant predictions for various aspects of stellar behavior. It may well be some time before there exists a theoretical 'consensus' on the behavior of neutron star matter in this density region, so that it is natural to consider to what extent *observations* of pulsar behavior provide confirmation of stellar structure calculations.

The clues which pulsar observation provide concerning neutron star structure are not exactly numerous. To be sure, the long term stability of pulsar signals strongly suggest that the outer crust of a pulsar is solid, in accord with theoretical predictions (Ruderman, 1969). Moreover, for one pulsar, that in the Crab nebula, energy balance considerations provide, in principle, a way of determining the stellar moment of inertia; however, as we shall see, this approach presently offers little more than an order of magnitude estimate of this quantity. The remaining current observational clues come from the Crab and Vela pulsars, both of which have been observed to speed up suddenly on more than one occasion (Richards *et al.*, 1969; Lohsen *et al.*, 1971; Papliolios *et al.*, 1970, 1971; Lohsen, 1972; Reichley and Downs 1969, 1971; Radhakrishnan and Manchester 1969) and both of which display a generally restless behavior between speedups (Boynton *et al.*, 1972; Reichley and Downs, 1970).

---

* Presented by D. Pines.
** Present address: Racah Institute of Physics, Hebrew University, Jerusalem, Israel.

We learn from these observations in three ways:

(i) Examination of the behavior of the pulse frequency in the period following a large speedup, or *macroglitch*. The spin-up of the outer crust, and those parts of the interior which are strongly coupled to it, acts as an external probe of the remaining interior matter (Baym et al., 1969b).

The observation of macroscopic relaxation times (~ 1.2 yr for Vela, ~ 4 days for the Crab pulsar) for the transfer of the shift in angular velocity from the crust to the interior neutrons provides strong evidence for the presence of superfluid neutrons and protons inside these stars. Moreover, from the fraction of the frequency jump which relaxes we can estimate the stellar abundance of the neutron superfluid.

(ii) Consideration of the origin of macroglitches. To the extent that these result from processes inside the neutron star, an understanding of their origin, magnitude, and frequency may confirm existing stellar structure calculations or suggest possible inconsistencies therein.

(iii) Consideration of the origin of the restless behavior of the Crab pulsar. The resulting noise in the rotational frequency spectrum may be attributed to frequency microglitches (of either sign); again, an understanding of its origin provides a further test of the applicability of present stellar models to pulsar behavior.

The major portion of this talk will be devoted to the above three problems; in the concluding sections we examine other ways of deducing stellar structure from pulsar observation, and consider what future observations might be especially relevant to the determination of the structure of neutron stars.

2. After a Macroglitch

After the observations of the sudden Vela speedup in 1969 a two component description of the dynamics of a neutron star was proposed to explain quantitatively the two prominent features of the pulsar's postglitch behavior (Baym et al., 1969b).

(a) the tendency for a substantial part of the sudden increase in Ω to relax in a roughly exponential way.

(b) the observation that immediately after the glitch the fractional jump in the slowing down rate is much greater than that in $\Omega \, (\Delta \dot{\Omega}/\dot{\Omega} \gg \Delta \Omega / \Omega)$.

The two component model assumes that all the conducting components of the star, which are presumably tied together by a uniform magnetic field, spin up together during a glitch – whatever the glitch origin. This is reasonable since the Alfvén velocities in the magnetosphere and within the star communicate any spin-up to all such conducting components in a time ($\sim 10^3$ s) too short to be observed. The moment of inertia of this fraction of the star, I_c, includes the electrons and nuclei of the crust, the electrons and protons of the core, possible charged hyperons in a superdense central core of a heavy neutron star, the magnetosphere, and finally any other component sufficiently strongly coupled to these that it shares the increased angular velocity of the glitch in a time too short to be resolved (a few days in Vela, a few hours in the Crab pulsar). The rest of the star, which is weakly enough coupled to the charged components to respond to them on a longer time scale, has a moment of inertia I_n. This slowly responding component is the neutron superfluid. Because of the weak coupling, the crystal spin-up thus effectively acts as an external probe of the neutron superfluid; study of the superfluid response therefore provides a measure of both the strength of the coupling and of the stellar abundance of the superfluid.

The simplest version of the two component theory assumes that the neutron superfluid can be described by a single angular velocity, $\Omega_n \neq \Omega$, and that the coupling between the charged and superfluid components is given by:

$$I_c \dot{\Omega} = -\alpha - \frac{I_c}{\tau_c}(\Omega - \Omega_n), \tag{1}$$

$$I_n \dot{\Omega}_n = \frac{I_c}{\tau_c}(\Omega - \Omega_n). \tag{2}$$

(The external torque α, and τ_c also depend on Ω.) After a sudden initial jump $(\Delta\Omega)_0$ in Ω, the post-glitch behavior described by (1) and (2) is

$$\Omega(t) = \Omega_0(t) + (\Delta\Omega)_0 [Qe^{-t/\tau} + (1-Q)], \tag{3}$$

where $\Omega_0(t)$ is the extrapolated frequency in the absence of the glitch,

$$Q = \frac{I_n}{I}\left(1 - \frac{\Delta\Omega_n}{(\Delta\Omega)_0}\right), \tag{4}$$

and

$$\tau = \tau_c \frac{I_n}{I}. \tag{5}$$

$\Delta\Omega_n$ is the initial jump in Ω_n, and I is the total stellar moment of inertia.

The form of the 'post-glitch function' given in (3) is in rough accord with the Vela observations and also, but with different Q and τ, with reported observations of the Crab glitches. A key test of the model is whether or not all post-glitch functions in a given pulsar have the same Q and τ. Period noise intrinsic to the pulsar introduces some ambiguity into the separation between post-glitch function and noise fluctuations in the reduction of the data. At present there is no reported inconsistency with the post-glitch function hypothesis. From a quantitative fit based on (3), to the initial speedups for both pulsars, one obtains the glitch parameters given in Table I.

The observed magnitudes of Q and τ inform us about certain properties of the neutron star interior. Q furnishes a direct measure of the extent to which the glitch reduces the total stellar moment of inertia, since the relative change in I is

$$\frac{\Delta I}{I} = -\frac{(\Delta\Omega)_\infty}{\Omega} = -(1-Q)\frac{(\Delta\Omega)_0}{\Omega}, \tag{6}$$

where $(\Delta\Omega)_\infty$ is the glitch-induced long term increase in the crustal rotation frequency. Q likewise gives an indication of the stellar abundance of neutron superfluid, since as long as $I_c \ll I_n$, and/or $\Delta I_n \ll \Delta I_c$ (where ΔI_n and ΔI_c are the changes in the moments of inertia of the two components at the glitch), one has

$$Q \simeq I_n/I. \qquad (7)$$

In a lighter neutron star ($M \lesssim 0.5\ M_\odot$) almost all of the matter beneath the crust is expected to be protons, electrons, and neutron superfluid. The theoretical Q is then ~ 0.90–0.95, just in the range of that inferred from the Crab pulsar glitches. This is an encouraging numerical agreement. In a heavier neutron star (corresponding to a central density, say, of greater than 10^{15} g cm^{-3}), one expects Q to be smaller, since in such stars the easily spun-up charged hyperon core and/or solid neutron core will contribute to I_c, producing a corresponding decrease in the fraction of neutron superfluid. However, no matter how heavy the star, one expects a non-vanishing value of

TABLE I

Speed-up observations for the Vela and Crab pulsars

	Vela	Crab
Ω (rad/sec)	70.5	190
T(yr)	2.4×10^4	2.4×10^3
$\Delta\Omega/\Omega$	2.34×10^{-6}	$(6.9 \pm 0.7) \times 10^{-9}$
$\Delta\dot\Omega/\dot\Omega$	6.8×10^{-3}	$(8.5 \pm 3.5) \times 10^{-4}$
τ	1.2 yr	(7.7 ± 3) days
Q	0.145	0.96 ± 0.08

Q, since there will always be some neutron superfluid present; an approximate minimum value for Q is ~ 0.05. For the Vela pulsar, the inferred $Q \sim 0.15$ suggests that it has a mass $\gtrsim 0.7\ M_\odot$ (Pines et al., 1972).

The crust-superfluid coupling is characterized by τ, essentially the time for the I_c components to come to rest if the core were suddenly to stop spinning. That this is a macroscopic time strongly suggests the presence of superfluid neutrons in the core of the star, as well as the superfluidity of those protons which interpenetrate the neutrons (Baym et al., 1969a, 1969b); it is incompatible with the interior neutrons forming a 'normal' degenerate quantum liquid. Assuming the core protons to be superfluid, the dominant coupling to the neutrons comes through electron-neutron interaction; this interaction will spin up a normal Fermi liquid of neutrons in a time $\sim 10^{-11}$ s. On the other hand, if the neutron liquid is superfluid, only that part of the neutron fluid within vortex cores can interact in a normal way with the core electrons. Even for these, the interaction is suppressed by a factor $\exp(-\pi\Delta^2/4E_F kT)$ where Δ is the superfluid energy gap and E_F is the neutron Fermi energy (Feibelman, 1971). The net electron-superfluid neutron interaction is proportional to the total length of vortex line in the superfluid. The minimum length of vortex line in a sphere with radius R of uniformly rotating superfluid with angular frequency Ω is $\sim \Omega m_n R^3/\hbar$. The radius of a vortex core $\sim 10^{-12}$ cm. The minimum fraction of the rotating superfluid contained in the quasi-normal vortex cores of the Crab pulsar is thus 10^{-18}. With this fraction of superfluid neutrons,

$$\frac{1}{\tau_c} \sim \frac{\Omega}{40} \left(\frac{\Delta}{1\,\mathrm{MeV}} \right) \left(\frac{kT}{E_F} \right) \exp\left(-\frac{\pi \Delta^2}{4 E_F k T} \right) \mathrm{s}. \tag{8}$$

The estimated magnitude for τ_c is quite reasonable; for a temperature of 10^8 K and typical pulsar densities, τ_c is of the order of days for $\Delta = 1.7$ MeV and of the order of a year for $\Delta = 2.4$ MeV; since, moreover, τ is very sensitive to small changes in temperature, one can easily understand the different τ's observed for the Vela and Crab pulsars.

A longer post-glitch healing time would be expected for the Vela pulsar even if the mutual friction torque were the same as the Crab pulsar, since the healing time is roughly proportional to $IQ(1-Q)$. However, the torques are likely not identical, since Vela is older (and hence colder) and rotating more slowly (and hence possesses a smaller number of vortex lines): both these latter effects are in the right direction.

We note that if the protons, which interpenetrate the neutron superfluid, form a normal Fermi liquid, rather than a superfluid, the above times would be appreciably decreased, since the proton-neutron coupling strength is some 10^3 times that of the magnetic electron-neutron interaction and the proton Fermi energy is much smaller than that of the electrons. Given the above parameters, the resulting τ's would then be microscopic in contradiction with observation.

We have thus far assumed the neutron superfluid is not turbulent. If, as suggested by Greenstein (1970), there exists any appreciable degree of turbulence, the total length of vortex line is greatly extended beyond its normal length. For the theorists this is literally opening a can of worms, and the questions he must answer are: how many 'worms' (squirming vortex lines) are there, and what is their total length? Any vortex line extension by a factor f would decrease the τ by $1/f$. A huge f (complete turbulence on a microscopic scale would make $f \sim 10^{18}$) would therefore predict a τ_c much too small to be compatible with observation. We conclude there cannot be any very large turbulence present in the rotating neutron superfluid.

3. Origin of Macroglitches

Some fifty-odd papers which seek to explain the macroglitches observed for the Vela and Crab pulsars have by now appeared in print, and yet another fifty may well appear where the problem is regarded as satisfactorily resolved by the astrophysical community. In preparing this talk it seemed to us that a critical review of macroglitch theories might be useful in sorting out whether certain classes of theories could be rejected at this time and whether any theory or combination of theories could be regarded as providing a plausible or even possible explanation of the glitch origin; by relating theories in the latter group to observation on the one hand, and to stellar structure on the other, one might hope to gain further perspective on the nature of these two neutron stars, and on neutron stars in general.

Let us define a macroglitch as a frequency jump such that

$$\frac{\Delta\Omega}{\Omega} \gtrsim 10^{-9}.$$

The present observational facts which a macroglitch theory must explain include the following:

(i) Macroglitch size vs. pulsar period. Why have macroglitches been seen only in the Crab and Vela pulsars, and why are those observed in the Vela pulsar ($\Delta\Omega/\Omega \sim 2 \times 10^{-6}$) some two or more orders of magnitude larger than those seen in the Crab ($10^{-9} \lesssim \Delta\Omega/\Omega \lesssim 10^{-8}$)? In other words, why does one have

$$\left(\frac{\Delta\Omega}{\Omega}\right)_{\text{Vela}} \gg \left(\frac{\Delta\Omega}{\Omega}\right)_{\text{Crab}} \gg \left(\frac{\Delta\Omega}{\Omega}\right)_{\text{other}} ?$$

(ii) Macroglitch sign. Why do all macroglitches thus far observed have the same sign?

(iii) Macroglitch repetition frequency. The time, τ_g, between Vela macroglitches is of the order of a few years, while that between Crab macroglitches is months to years.

(iv) $\tau_g \neq \tau$. The time between glitches is certainly not the same as the healing time, τ, in the case of the Crab pulsar, and most likely is not the same either for the Vela pulsar, since $(\tau_g)_{\text{Vela}} \sim 2$ yr, $(\tau)_{\text{Vela}} \sim 1.2$ yr.

(v) Macroglitches are 'sudden' events. For the Crab pulsar (Lohsen, 1972) one sees that a macroglitch takes place in no more than a few hours; for Vela the corresponding limit is at present a few days.

(vi) $(Q)_{\text{Vela}} \ll 1 \neq (Q)_{\text{Crab}} \simeq 1$. The fraction of the glitch which relaxes is appreciably less than unity in the case of the Vela pulsar.

(vii) No appreciable change in pulse shape is seen following the speedup of either the Crab or Vela pulsars.

The classes of macroglitch theories may be conveniently specified in terms of the hypothesized physical location of the macroglitch; from way out to far in, we have:

Planets
Magnetospheric instabilities
Accretion
Crustquakes
Hydrodynamic instabilities associated with the superfluid neutron core
Corequakes

and it is in this order we shall consider the theories.

3.1. PLANETARY PERTURBATIONS (Michel, 1970; Rees *et al.*, 1970)

It seems rather difficult to assume that the macroglitches and microglitches are a result of a linear motion of the pulsar caused by a system of planets surrounding it.

Even if one accepts their existence in the immediate vicinity of a supernova remnant, there are considerable difficulties in fitting such a model to the observations. For one, it seems impossible to fit the sharp rise of a glitch together with its much slower subsequent slowing down to a near passage of a planet: indeed, no reasonably smooth function can fit the observation. In the Crab pulsar, an imitation of the restless behavior by planets presents even a more formidable project. The Princeton group (Groth, 1971) at one stage obtained a reasonable fit for their 1969–1970 observing season of the Crab pulsar, by postulating the existence of three orbiting planets. Together with their four cubic polynomial parameters, this was a 19 parameter fit. However, they could not then predict correctly their 1970–1971 observations; it is likely that one needs an ever increasing number of planetary companions as the fit period is increased.

On the other hand, a planetary passage hypothesis is consistent with the Vela observations, in the sense that it can explain both the magnitude of the apparent speedup, its duration, and its repetition frequency, provided one assumes a highly eccentric orbit with a near passage at $\sim 4.5 \times 10^{12}$ cm (Michel, 1970). (This distance is sufficiently far from the pulsar that tidal effects will not cause a major perturbation.) However, since the two Vela glitches were not of identical magnitude, one needs at least two planets in orbit about the Vela pulsar to fit the data. Clearly more observational data and a more detailed theoretical analysis is required before one can accept or reject this proposal for Vela. If the Vela macroglitches are nothing but a linear motion-induced Doppler shift, then no conclusions concerning the superfluidity of the neutron core can be drawn from analysis of post-glitch behavior. However one can still measure a $Q \equiv 1 - (\Delta\Omega)_\infty/(\Delta\Omega)_0$; so defined, Q should be equal to unity, since any spinup is necessarily followed by a spindown. However, only the 'rapid' spin up is easily detected above the intrinsic pulsar noise, so that the apparent measured Q could turn out to be less than unity. (Whether it could be as small as 0.15, however, is not clear.)

3.2. MAGNETOSPHERIC INSTABILITIES (Pacini and Scargle, 1971; Sturrock, 1971)

The two chief arguments in favor of a magnetospheric instability as an origin of macroglitches – the apparent observation of wisp motion or flaring plus a change in the pulsar dispersion measure following the Crab macroglitch of September, 1969 – have become the two principle observational arguments against this explanation; no similar correlations have been found with other Crab macroglitches. However, those still interested in pursuing this possibility (unstable-magnetospheric-theorists?) must explain:

(i) Why only these two pulsars choose to have unstable magnetospheres and why, in these cases, the instability is such that in the glitch an appreciable fraction of the magnetosphere is blown away? Indeed the hypothesized sudden change in the plasma moment of inertia required for the Vela pulsar is of the same order as the maximally allowable $I_{\text{plasma}} \lesssim B_s^2 R^3/6\Omega^2$ (Rosenbluth, 1972), where B_s is the surface magnetic field, R the pulsar radius.

(ii) The characteristic time to fill the pulsar magnetosphere is of the order of seconds; why then should it take months, or years, for an instability to develop?

(iii) Why, if the entire magnetosphere is involved in the instability, is there no change in pulse shape following a macroglitch?

(iv) The behavior of terrestrial plasmas near an instability is generally to avoid a gigantic instability; such plasmas tend rather to fluctuate about some 'minimal' instability. Why should pulsar plasmas be different?

3.3. Accretion

There are likewise a number of problems with attributing macroglitches to accretion (see the paper by Börner and Cohen in this volume).

(i) Where does the accreting matter come from? One needs $\sim 10^{-10} M_\odot$ per macroglitch for the Crab pulsar, and a thousand times more for Vela. One knows, for example, that the accretion rate for a neutron star which forms a compact X-ray source is $\sim 10^{-9} M_\odot$ per year, (Lamb et al., 1973), and that to get this much accreting matter easily, one needs to postulate an accompanying close companion star. The constancy of the pulsars' pulse periods enables one to readily discard any such companions for Crab and Vela. Moreover, the accreting matter could not represent a 'fallback' of matter from the supernova explosion which created the pulsar; any such fallback would take place within the first year of the supernova (Colgate, 1972).

(ii) Assuming that one invents a source of accreting matter, could $10^{-9} M_\odot$ reach the stellar surface sufficiently rapidly to produce a macroglitch? Consider a large piece of stellar (or rather planetary) matter incident on the pulsar. It will first of all be ripped apart by tidal forces. To see this, assume the matter to be in the form of an infalling homogeneous sphere of radius R_s, density ϱ (~ 5 g cm^{-3}) and rigidity μ ($\sim 10^{12}$ dyne cm^{-2}); the shear angle ϕ of the sphere at distance R from the pulsar will be

$$\phi \sim \frac{kM_p}{4\varrho R^3}, \tag{9}$$

where M_p is the pulsar's mass (2×10^{33} g). The Love number k is roughly given by $G\varrho^2 R_s^2/2.5\mu$ for $R_s \ll (1/\varrho)\sqrt{\mu/G} \sim 8 \times 10^8$ cm, and is ~ 1 for $R_s \gg (1/\varrho)\sqrt{\mu/G}$. For a body with $R_s \gg 8 \times 10^8$ cm, we must therefore have

$$\phi \sim \frac{10^{32}}{R^3} \lesssim \phi_c, \tag{10}$$

where ϕ_c is the critical angle for break-up.

If $\phi_c \sim 10^{-4}$, a typical value for such objects, than at a distance of $R \sim 10^{12}$ cm, the infalling chunk will start breaking up. As the pieces continue to fall in, we must eventually have

$$\frac{G\varrho R_s^2 M_p}{10 R^3 \mu} \simeq 7 \times 10^{13} \frac{R_s^2}{R^3} \lesssim \phi_c \tag{11}$$

so that, when the pieces finally reach the pulsar's surface, at $R \sim 10^6$ cm, they have a radius of roughly 1 cm!

We may also get a rough idea as to when the torn pieces will start to move independently in the pulsar's gravitational field. For the pulsar's gravitational pull to be more important than that of a neighboring piece, we must have

$$\frac{GM_p}{R^2} \gg \frac{(4\pi/3)G\varrho R_s^3}{R_s^2}$$

or

$$R \ll \frac{10^{16}}{\sqrt{R_s}}. \tag{12}$$

Therefore, at about $R \sim 3 \times 10^{10}$ cm one may expect that the chunk will not only be torn to bits, but that the bits will also start to move independently of one another. Also, an incoming chunk of material is likely to carry a large amount of angular momentum with it. This means that the large infalling chunk goes first into orbit, and, because of the tidal tearing, forms a disc of particles around the pulsar. The disc will indeed fall in gradually, because of friction in the pulsar's 'atmosphere' and the tidal forces, but it is very difficult to see why this will not be a continuous long process rather than a sharp 'glitch.'

We also note in passing, that with the required rate of infalling material, there should be substantial X-radiation produced (which is not observed), and, indeed, the neutron star may not function as a pulsar at all (Shvartsman, 1971; Lamb *et al.*, 1972).

3.4. Crustquakes

Crustquakes – the sudden release of elastic energy in the solid outer crust of a neutron star – were one of the first mechanisms suggested to explain the Vela macroglitch (Ruderman, 1969). What is appealing about crustquakes is that one is dealing with a physical process which has a clear terrestrial analogue, for which the time between macroglitches, all of which will have the same spin, varies from one to another (and is not correlated with τ), that crustquakes are expected to be a common phenomenon only in comparatively young pulsars, and that there exists a wide variety of mechanisms, many of them plausible, for inducing critical strains in the stellar crust. While, as shall see, they continue to be a likely mechanism for the Crab-pulsar macroglitches, it is no longer likely that they provide a mechanism for those observed in the Vela pulsar.

The mechanism which has been considered in most detail is rotational-induced strain arising from the gradual spin-down of the star as a result of the emission of electromagnetic radiation and charged particles. The crust of the star, formed when the star is spinning comparatively fast, is subject to increasing gravitationally induced stresses as the star slows down; when these stresses exceed the yield point, the crust will crack (a starquake). In the process, some stress is suddenly relieved, the crustal moment of inertia is suddenly reduced, and, by conservation of angular momentum, its rotation rate is suddenly increased, hence a speedup.

A simple description of such crustquakes may be given in terms of a quadrupolar stellar deformation described by a single time-dependent distortion parameter, the crustal oblateness (Baym and Pines, 1971). In this description, if appreciable plastic flow does not take place, the time to the next quake is proportional to the stress relieved in the preceding quake, and may be estimated for a given model of neutron star. To make quantitative estimates of the relevant parameters, we assume the star is axially symmetric and define the oblateness ε according to $I = I_0(1+\varepsilon)$ where I_0 is the moment of inertial for a non-rotating spherical star. The time varying portion of the mechanical energy may be written as

$$E = -\frac{L^2}{2I_0}\varepsilon + A\varepsilon^2 + B(\varepsilon_0 - \varepsilon)^2, \tag{13}$$

where ε_0 is a reference oblateness (which changes only as a result of plastic flow or crustquakes), and the coefficients A and B measure the gravitational and elastic energy stored in the star as a result of rotation. An order of magnitude estimate of A and B may be obtained from the expressions appropriate to a self-gravitating incompressible homogeneous sphere of radius R and crustal volume V_{cr}, $A = (3/25\ GM^2/R,)$ $B = (57/50)\mu V_{cr}$, where μ is the shear modulus of the crustal material. On minimizing the energy, (13), at fixed L and ε_0, we have

$$\varepsilon = \frac{I_0\Omega^2}{4(A+B)} + \frac{B}{A+B}\varepsilon_0 \tag{14}$$

and, since $B \ll A$ for all stable neutron stars of this type (Baym and Pines, 1971), $\varepsilon \simeq I_0\Omega^2/4A$, its perfect fluid value. In this one parameter description, which may be appropriate to 'astrologically young' pulsars, a quake takes place when the mean stress in the crust, $\sigma = (1/V_{cr})(\partial E_{el}/\partial \varepsilon) = \mu(\varepsilon_0 - \varepsilon)$ exceeds some critical value, σ_c. In the quake, both the oblateness and reference oblateness decrease according to

$$\Delta\varepsilon = [B/(A+B)]\Delta\varepsilon_0 \simeq (B/A)\Delta\varepsilon_0, \tag{15}$$

where $\Delta\varepsilon$ is directly observable, since

$$\Delta\varepsilon = \Delta I/I = -(\Delta\Omega)_\infty/\Omega = -(1-Q)(\Delta\Omega)_0/\Omega. \tag{16}$$

After the quake, the stress will start to build up once more, and the time to the next quake is given by

$$\tau_q^\varepsilon = T(\omega_q^2/\Omega^2)|\Delta\Omega_\infty/\Omega|, \tag{17}$$

where T is the slowing-down time of the pulsar $(T = |\Omega/\dot\Omega|)$, and effects of stellar structure are described through the parameter,

$$\omega_q^2 = 2A^2/BI. \tag{18}$$

Results of microscopic stellar model calculations of ω_q^2 are given in Table II, together with the predicted time between quakes, for a speedup involving a relative jump in the moment of inertia of one part in 10^9 [recall that for the Crab pulsar, if $Q \simeq 0.9$, this corresponds to an initial speedup of one part in 10^8, since $\Delta I/I = -+(\Delta\Omega_\infty)/\Omega = -(1-Q)(\Delta\Omega)_0/\Omega$ according to (6)]. We see that for a model star of mass $\sim 0.3\ M_\odot$, the predicted interval between macroglitches of initial relative magnitude 10^{-9} to 10^{-8}, is of the order of months and years as observed, and further note that the corresponding critical strain angle, $\phi_c \sim \sigma_c/\mu \sim 10^{-4}$, a reasonable *ad hoc* value. For the Vela pulsar, however, the corresponding calculated interval between macroglitches of relative magnitude $(\Delta\Omega)_0/\Omega \sim 10^{-6}$, while depending on the assumed mass, is at least a few centuries, and more probably many millenia. This is because the released strain is much larger than in the Crab, while both the slowing down rate, T, and the rate at which strain is replenished ($\sim \Omega^{-2}$) are an order of magnitude smaller. The observed two-year interval between macroglitches, if typical, is therefore not compatible with the treatment of Vela as an 'astrologically young' pulsar in which the strain energy released in one quake is replenished before the next.

TABLE II

Dynamic stellar parameters for the Crab pulsar

Mass ($M_\odot$)	$\omega_q^2 (s^{-2})$	τ_q^ε (yr)	$2\pi/\Omega_W$ (days)	$\varepsilon/10^{-4}$
0.10	2.7×10^8	0.018	3.8×10^{-3}	14
0.15	1.7×10^9	0.11	2.4×10^{-2}	5
0.20	6.3×10^9	0.42	8.8×10^{-2}	3
0.25	2.4×10^{10}	1.6	3.3×10^{-1}	2
0.30	4.7×10^{10}	3.1	6.5×10^{-1}	2
0.46	5.2×10^{11}	35	7.3	1
0.80	7.0×10^{12}	470	9.8×10^1	0.9
1.08	3.7×10^{13}	2400	5.0×10^2	0.7
1.41	4.0×10^{14}	27000	5.5×10^3	0.5

A decrease in ε is, of course, not the only way to build up strain energy in the stellar crust. Pines and Shaham (1972a) have considered the buildup of strain energy as a result of the misalignment of the rotation axis and the elastic reference axis in a star subject to a radiation torque nearly perpendicular to the axis of rotation. The elastic energy then contains an angular contribution which can play an important role (indeed, as we shall see, it is likely responsible for the macroglitches observed in the Crab pulsar); however, although this angular term modifies the above estimate of τ_q^ε, it does not seem capable of doing so sufficiently to explain in plausible fashion the two-year interval between Vela macroglitches.

Still another mechanism for inducing crustquakes has been proposed by Dyson (1969, 1970) who has considered the possible existence of volcanoes on the pulsar surface, through which matter pours out until sufficient material has built up that one gets a starquake. The mountain building process can be a comparatively slow one ($\sim$ months to years) and the mountains are not especially high. (Indeed, one can show that the maximum height, h, of a mountain on the stellar surface is $h/R \simeq 3\phi_c(B/A)$; since $\phi_c \sim 10^{-4}$ and $B/A \sim 3 \times 10^{-3}$, one concludes that mountains on the pulsar surface will not greatly exceed 1 cm in elevation.) The volcano mechanism, while appealing, suffers from a certain lack of plausibility in that to get volcanoes one needs substantial internal temperature gradients, and these are not likely found in the extraordinarily highly (thermally) conducting superdense interior matter. Moreover, in common with the above glitch theories, it cannot easily explain the *size* of the Vela macroglitches. Vela, assuming it has a liquid interior, has a fluid oblateness $\varepsilon \sim I_0 \Omega^2/4\,A \sim 10^{-5}$; hence, in macroglitches the fractional oblateness change approaches the disconcertaingly large value of $\sim 10\%$.

3.5. Hydrodynamic Instabilities Associated with the Fluid Neutron Core

Cameron and Greenstein (1969) have suggested that the neutron core fluid may become classically unstable, as the slowing down torques set up a rotational flow in which the angular momentum per unit mass decreases with distance from the rotation axis. This model has the great virtue that it is based upon an instability that is known to exist. However, the evidence against such a glitch model includes:

(i) In such models the interval between glitches should be approximately the relaxation time for the fluid to return to its preglitch state. But the observed τ in the case of the Crab is two orders of magnitude less than the glitch interval while in the case of Vela the pulsar did not nearly finish its relaxation when the second glitch occurred.

(ii) It may not be possible to set up the unstable flow in a non-turbulent superfluid. Wherever the external torque causes the angular momentum per unit mass to become independent of radius – the limit of stability – the vortex density, and thus the push of the external torque, have to vanish.

(iii) Why are such glitches not seen in the somewhat slower pulsars?

Packard (1972) has proposed that glitches may be manifestations of crust-pinned vortex lines tearing loose. Such a phenomenon is known for individual vortex lines in the laboratory. Arguments against this model include:

(i) As in the Cameron-Greenstein model, the observed relationship between relaxation time and glitch interval is then paradoxical.

(ii) The tension from vortex line bundles and even vortex lines considered as if isolated (a great underestimate of the tension) is so great that for the estimated parameters for pulsar superfluid and crust nuclei no pinning should occur.

(iii) Why, when pinning is easier in slower pulsars, are no glitches observed for them?

3.6. Corequakes

Starting with the recent plausible suggestion that the heavier neutron stars may possess a solid inner neutron core, we have attempted to explain the Vela macroglitches as arising from corequakes which represent the sudden release of elastic energy stored in the solid *inner* neutron lattice (Pines *et al.*, 1972). This neutron solid likely occurs at stellar densities $\geqslant 1.5 \times 10^{15}$ g cm^{-3} (Canuto and Chitre, 1972); because its shear modulus will be some five orders of magnitude larger than that of the crustal material ($\mu \sim 10^{35}$ dyne cm^{-2} instead of $\sim 10^{30}$ dyne cm^{-2}), the core possesses a substantial reservoir of elastic and gravitational energy which can be released in starquakes. Thus one has $B \sim 10\, A$, and the possibility of a quite brittle crustal material, while $\varepsilon \simeq \varepsilon_0$, so that as the star slows down, ε, following ε_0, will change only discontinuously in starquakes. There is thus sufficient elastic energy present that the crust does not have to build up the strain energy released in the previous quake, before quaking anew [and in this respect a solid neutron core resembles the earth (Pines and Shaham, 1972b)], so that there is no difficulty in principle in understanding the appearance of macroglitches every few years in a star rotating as slowly as the Vela pulsar. Moreover the size of the glitches is no longer a major problem because an initial core oblateness of order 10^{-2} would have been reduced by little less than an order of magnitude by macroglitches of magnitude $\Delta\varepsilon \sim 10^{-6}$ occurring over a period of 10^4 yr. Moreover, such macroglitches can be regarded as involving only a minute fraction of the equatorial bulge, rather than the appreciable fraction required for models which yield a current oblateness of $\varepsilon \sim 10^{-5}$.

We further note that the presence of a solid core reduces appreciably the structure factor, $Q \simeq I_n/I$, since the solid neutron core will corotate rigidly with the crust and interior charged particles within microscopic times. The fraction of neutron superfluid can easily be 0.15; indeed, an I_n/I of that magnitude is characteristic of neutron stars with a mass $\sim 0.7\, M_\odot$.

4. Microglitches

In addition to the overall slowing down – polynomial behavior – and the various macroglitches, there is observational evidence that the Crab pulsar – and, possibly, the Vela pulsar as well – display a 'restless' or noisy behaviour, which manifests itself in erratic small variations in arrival times. A detailed analysis of this restless behavior by the Princeton group (Boynton et al., 1972), shows that no reasonably smooth function can fit the corresponding phase residuals over a large observation period. Rather, these behave like shot noise, corresponding to many minute microglitches with possibly both spin downs as well as spin-ups present. From a Fourier analysis of that shot noise, the Princeton group could determine the average value of the rate of jumps (r) times their magnitude squared, $\langle r(\Delta\Omega)^2 \rangle$. It can also be concluded from their analysis, that the noise before the September 1969 macroglitch was larger than the noise after that event; recently Lohsen (1972) found further evidence of that in connection with the October 1971 macroglitch. A shot noise interpretation of the restless behavior explains as well the earlier reported periodicities of the order of months, since shot noise produces apparent periodicities of the order of the time span of the data.

Nelson et al. (1970) have interpreted the restless Crab behavior in terms of larger, less frequent, frequency jumps, which are manifestly of both signs; the average value of the small jumps cannot be obtained from the Princeton analysis, since it is lost in the overall slowing down of the pulsar. Further, a grouping of very frequent, small, frequency jumps can occur, to provide the transition from the Princeton interpretation to that of Nelson et al. (1970). Clearly, longer periods of observation as well as higher temporal resolution are required to determine definitely the microscopic structure of this restless behavior.

Let us define a microglitch as a frequency jump of either sign of relative magnitude, $\Delta\Omega/\Omega \lesssim 10^{-10}$. In principle most of the macroglitch mechanisms discussed earlier can be scaled down to explain microglitches; to some extent the various anti-macroglitch arguments we have used are buried in the observational noise, in that one can no longer observe postglitch behavior if one has microglitches which take place hourly, or even daily. We consider briefly three 'new' microglitch mechanisms: microquakes induced by angular strains in the crust, relaxation of magnetic field stresses, and superfluid 'vacillation.'

4.1. Microquakes

As shown by Pines and Shaham (1972a), one can expect in general that there is an angular contribution to the elastic energy which arises from a misalignment of the rotation axis and the elastic reference axis. However, a crustquake induced by purely angular strains is, as a rule, smaller than an 'ε' quake, since it involves a smaller area of the stellar surface (Pines and Shaham, 1972b). Also, it is a spin-down, since it involves a crustal motion which tends to orient the crustal axis towards the direction of the instantaneous axis of rotation. The coexistence of oblateness strains has the effect of producing either spin-ups or spin-downs, depending on the specific geometry of the quake and the relative importance of the two kinds of strains. When misalignment has a much faster characteristic time than that of the slowing down, then most of the time strain is relieved by microquakes; eventually, however, these are ineffective in relieving oblateness strains and a 'macroquake' must occur. Detailed considerations show that as a result pulsars will be noisier before a macroglitch than after it.

4.1.1. Relaxation of Pulsar Magnetic Fields

If a pulsar is born in a violent event, it is possible that its crust begins to solidify before all of its conducting fluid components have finished moving to maximally relax the stellar magnetic field stresses (Ruderman, 1972). If so, the present crust may sustain considerable magnetic stresses up to $B^2/8\pi \sim 10^{23}$ dyne cm^{-2}. Such stresses might cause a continual crumbling in weaker parts of the crust which could manifest itself in 'noise' in the pulsar spin frequency. This model would not account for a reported increase in timing noise before the last major glitch in the Crab.

4.1.2. Superfluid Vacillation

After a glitch, angular momentum is transferred to the neutron superfluid in the (observed) relaxation time τ. Because the torque which communicates this to the superfluid core is not spatially uniform, a differential angular velocity is induced in the superfluid in addition to any differential rotation the core fluid may have in its usual spinning down state. The slight increase in angular momentum induced locally by the crust is spread throughout the superfluid by vortex-vortex interactions even though there is no viscosity. (Typical estimates for this so-called Tykachenko-wave angular momentum redistribution suggest a wave velocity of order 0.1 cm s^{-1}.) The added angular momentum is shared among a large number of incommensurate normal modes of the superfluid which cause it to oscillate back and forth about its average motion as the angular momentum continually redistributes itself (Ruderman, 1972). There is no viscosity and, in the low temperature limit, no way of damping this added differential rotation 'vacillation,' except by rubbing back on the crust whose sudden 'glitch' initially stirred the superfluid. An estimate of the pseudorandom motions of the crust caused by this underlying superfluid vacillation gives amplitudes comparable to those observed in the Crab pulsar. Again, however, this model would not account for an increase in timing noise *before* a macroglitch.

4.1.3. *Other Ways of Determining Stellar Structure*

We discuss three other possible observational handles on neutron star structure. First, as we have mentioned, energy balance considerations, together with the assumption that the rotational energy of the Crab pulsar is the sole source of power for the Crab nebula, in principle provides a determination of the stellar moment of inertia (and hence the mass). There are two problems (see Ruderman, 1972):

(i) One cannot be certain of the absolute luminosity of the Crab nebula because its distance is not known to within a factor of two or so.

(ii) One is not sure what fraction of the nebular luminosity must be supplied by the pulsar – is it only the energy radiated by the highest energy and the shortest lived electrons in the nebula ($\sim 5 \times 10^{37}$ erg s^{-1} on the basis of current distance estimates) or does one have to supply as well nebular radiation and the kinetic energy of expansion (which could require as much as 4×10^{38} erg s^{-1}). The best one can therefore conclude is that $I = (3.6 \pm 2.8) \times 10^{44}$ gm cm^2, which translates into pulsar masses as (Baym *et al.*, 1971)

$$M_{\text{crab pulsar}} \sim (0.8 \pm 0.6) M_\odot.$$

One may note that this covers almost the whole range of stable neutron stars!

Obviously, a direct measurement of pulsar mass would be highly desirable. The observation of a wobble of the star (analogous to the Chandler wobble of the Earth), which results from the misalignment of the rotational axis and principle reference inertial axis discussed earlier, provides such a determination under certain circumstances. The wobble frequency is given by (Pines and Shaham, 1972b)

$$\Omega_W = \frac{3}{2} \left(\frac{B}{A+B} \right) \varepsilon_0 \Omega. \tag{19}$$

For a star with a liquid interior (the Crab pulsar?), $\varepsilon_0 \simeq \varepsilon \simeq I_0 \Omega^2 / 4A$, and one has (Pines and Shaham, 1972a)

$$\Omega_W \simeq \frac{3}{4} \frac{\Omega^3}{\omega_q^2} \tag{20}$$

so that measurement of the wobble frequency provides a direct measurement of ω_q^2, from which the mass can be inferred from theoretical calculations of B and A. Indeed, it is at least as likely that in this fashion one will be able to determine the distance to the Crab pulsar (since I is 'directly' determined) as it is that better limits on I can be obtained through better distance determinations of the conventional sort. For a star with a solid neutron core (the Vela pulsar?) on the other hand, observation of Ω_W provides a direct measure of ε_0, but gives no information on the stellar mass.

Further information on stellar structure is in principle contained in the fine structure of a macroglitch (Lohsen, 1972). For example, if one interprets a macroglitch as a crustquake, the fine structure provides information on a possible sequence of related crustquakes, one acting to trigger the next, in a fashion which may be sensitive to the pulsar mass.

4.1.4. Observational Tests

We should like to emphasize once more the importance of carrying out a more detailed analysis of existing timing data on both the Crab and Vela pulsars, in order to obtain the following information:

(i) The magnitude and timing of successive macroglitches.

(ii) Post macroglitch behavior (does the two-component theory provide an adequate description?)

(iii) Nature of the residual 'restless' behavior (frequency noise, larger but less frequent microglitches of both signs, or?).

For the Crab pulsar, all three aspects of the data are interrelated, so that, as we have mentioned, it is non-trivial to make an analysis which distinguishes in unambiguous fashion between the above phenomena. (Indeed, the distinction we have made between macroglitches $(\Delta\Omega/\Omega \gtrsim 10^{-9})$ and microglitches $(\Delta\Omega/\Omega) \lesssim 10^{-10})$ is itself an arbitrary one.) For the Crab pulsar, it is in principle possible to combine the results obtained by different groups of observers in order to obtain this information and indeed a start has been made in that direction; however, the present 'state of the art' is such that one cannot yet give a 'Glitch table' which lists the magnitude and time of occurrence of the macroglitches which have taken place since September, 1969 (and it is for this reason that one is not included here). For the Vela pulsar, essentially all the relevant observations have been made by Reichley and Downs with the Goldstone array, and we can only encourage them in the difficult task of data analysis, and hope that an early answer will be provided to the questions we have posed.

Discovery of pulsar wobble would likewise represent a significant forward step. To the extent that one can decide on the existence of a liquid core, through theory or observation, observation of pulsar wobble will provide a direct determination of either the pulsar mass or current oblateness, both quantities of fundamental interest. [It is, we suppose, a measure of the difference between astronomy and particle physics that one does not have six competing teams (and proposals) currently attempting this fundamental observation.]

4.1.5. Concluding Remarks

At first sight, two pulsars provide a singularly slender observational base on which to construct an elaborate theoretical superstructure. However, we are fortunate in that

TABLE III

Present status of macroglitch theories[a]

Observational Property / Theory	Magnitude	Rapid rise time	Sign	$(\Delta\Omega/\Omega)_{Vela} \gg (\Delta\Omega/\Omega)_{Crab} \gg (\Delta\Omega/\Omega)$ and other	Macro-glitch function	Q values	Frequency	$t_q \neq \tau$	Persistence of pulse shape	Explain microglitches as well?	Physically plausible for neutron stars
Planets: Crab	1	3	2	1	3	2	2	2	1	3	1
Planets: Vela	1	3	2	1	2	2	1	2	1	2	1
Magnetospheric instability	2	2	1	2	2	1	1	1	3	1	2
Accretion	2	3	2	2	2	1	2	1	1	2	3
Crustquakes: Crab	1	1	1	1	1	1	1	1	1	1	1
Crustquakes: Vela	2	1	1	2	1	1	3	1	1	1	1
C. G. Hydrodynamic instability	1	1	2	3	3	3	1	3	1	1	1
Corequakes: Crab	1	1	1	1	1	3	2	1	1	1	1
Corequakes: Vela	1	1	1	1	1	1	1	1	1	1	1

[a] 1 ≡ Not inconsistent with observation
2 ≡ May be inconsistent with observation
3 ≡ Inconsistent with observation

there is now good reason to believe that the Crab and Vela pulsars correspond to different classes of neutron stars, in that they display different internal structure. We have discussed at some length the various theories for the origin of macroglitches; our discussion may be summarized in tabular form, and such a summary is presented in Table III. At this stage it would seem that crustquakes offer a plausible explanation for the Crab macroglitches, while corequakes can explain the Vela macroglitches; it may be that suitable modification of many of the other theories would render them equally plausible, with the exception of the accretion mechanism which seems out of the question for either the Crab or Vela pulsars.

On the basis of the evidence in at this time (both observational and theoretical) we conclude that the Crab pulsar has a mass $\lesssim 0.5\ M_\odot$, and is $\sim 90\%$ superfluid neutrons, while the Vela pulsar may well possess a solid neutron core, have a mass of $\sim 0.7\ M_\odot$, with a stellar superfluid neutron abundance of $\sim 15\%$. It will be illuminating to see whether future observations and theoretical developments confirm this preliminary identification.

References

Baym, G., Pethick, C. J., and Pines, D.: 1969a, *Nature* **223**, 673.
Baym, G., Pethick, C. J., Pines, D., and Ruderman, M.: 1969b, *Nature* **224**, 872.
Baym, G., Pethick, C. J., and Sutherland, P.: 1971, *Astrophys. J.* **170**, 299.
Baym, G. and Pines, D.: 1971, *Ann. Phys.* **66**, 816.
Boynton, P. E., Groth, E. J., Hutchinson, D. P., Nanos Jr., G. P., Partridge, R. B., and Wilkinson, D. T.: 1972, *Astrophys. J.* **175**, 217.
Colgate, S.: 1972 (private communication).
Dyson, F. J.: 1969, *Nature* **223**, 486.
Dyson, F. J.: 1970, Lectures at the Scuola Normale Superiore, Pisa.
Feibelman, P. J.: 1971, *Phys. Rev.* **4**, 1589D.
Greenstein, G. S.: 1970, *Nature* **227**, 791.
Greenstein, G. S. and Cameron, A. G. W.: 1969, *Nature* **222**, 862.
Groth, E.: 1971, Ph.D Thesis, Princeton University.
Lamb, F., Pethick, C. J., and Pines, D.: 1973, *Astrophys. J.* **184**, 271.
Lohsen, E.: 1971, *IAU Circ.* No. 2368.
Lohsen, E.: 1972, *Nature Phys. Sci.* **236**, 70.
Michel, F. C.: 1970, *Astrophys. J.* **159**, 225.
Nelson, J., Hills, R., Cudaback, D., and Wampler, J.: 1970, *Astrophys. J. Letters* **161**, L235.
Packard, R. E.: 1972, *Phys. Rev. Letters* **28**, 1080.
Papliolios, C., Carleton, N. P., and Horowitz, P.: 1970, *Nature* **228**, 445.
Papliolios, C., Carleton, N. P., and Horowitz, P.: 1971, *IAU Circ.* No. 2368.
Pines, D. and Shaham, J.: 1972a, *Nature Phys. Sci.* **235**, 43.
Pines, D. and Shaham, J.: 1972b, *Phys. Earth Planetary Interiors* **6**, 103.
Pines, D., Shaham, J., and Ruderman, M.: 1972, *Nature Phys. Sci.* **237**, 83.
Radhakrishnan, V. and Manchester, R. N.: 1969, *Nature* **222**, 228.
Rees, M. J., Trimble, V. L., and Cohen, J. M.: 1971, *Nature* **229**, 395.
Reichley, P. E.: 1970, talk at URSI Meeting, Washington.
Reichley, P. E. and Downs, G. S.: 1969, *Nature* **222**, 229.
Reichley, P. E. and Downs, G. S.: 1971, *Nature Phys. Sci.* **234**, 48.
Richards, D. W., Pettengill, G. H., Roberts, J. A., Counselman III, C. C., and Rankin, J. M.: 1969, *IAU Circ.* No. 2181.
Roberts, D. H. and Sturrock, P. A.: 1971, *Bull. Am. Astron. Soc.* **3**, 463.
Rosenbluth, M.: 1972 (private communication).
Ruderman, M.: 1969, *Nature* **223**, 597.
Ruderman, M.: 1972, *Ann. Rev. Astron. Astrophys.* **10**, 427.

Sawyer, R. F.: 1972, *Phys. Rev. Letters* **29**, 382.
Sawyer, R. F. and Scalapino, P. J.: 1973, *Phys. Rev.* **D7**, 1579.
Scalapino, P. J.: 1972, *Phys. Rev. Letters* **29**, 386.
Scargle, J. D. and Pacini, F.: 1971, *Nature Phys. Sci.* **232**, 144.
Shvartzman, V. F.: 1971, *Soviet Astron.* **15**, 342.

COOLING OF DENSE STARS

SACHIKO TSURUTA*

NASA, Goddard Space Flight Center, Greenbelt, Md., U.S.A.

Abstract. Cooling rates are first calculated for neutron stars of about 1 $M_\odot$ and 10 km radius, with magnetic fields from zero to about 10^{14} G, for two extreme cases of maximum and no superfluidity. The results show that most pulsars are so cold that thermal ionization of surface atoms would be negligible. Next, nucleon superfluidity and crystallization of heavy nuclei are treated more quantitatively, and more realistic hadron star models are chosen. Cooling rates are thus calculated for a stable hyperon star near the maximum mass limit, a medium weight neutron star, and a light neutron star with neutron-rich heavy nuclei near the minimum mass limit. Results show that cooling rates are a sensitive function of density. The lightest star is cooler than others in earlier stages but the trend is reversed later. The medium weight star is generally the coldest of all in lower temperature regions where the effect of superfluidity becomes significant. However, if a heavy star contains pions, its cooling will be even faster. The Crab pulsar and Vela pulsar, expected to be the two youngest, can be as hot as $(2 \sim 4) \times 10^6$ K (on the surface), comparable with the results obtained from internal frictional heating by Greenstein, if they are medium weight to heavy hadron stars. However, older pulsars are cold. In fact, at about a few million years, the age of average radio pulsars, the surface temperature becomes only several hundred to several thousand degrees. Thus, the earlier conclusions about cold pulsars are still valid. Cooling of a massive white dwarf star is also shown.

1. Introduction

In recent years considerable progress has been made in the studies of cold dense matter. In the present paper, we wish to report some recent work on thermal properties of dense stars.

Some years ago when the first few of the galactic X-ray sources were discovered, it was suggested that these X-ray sources might be neutron stars. The detailed account was reviewed, for instance, in the paper by Tsuruta and Cameron (1966a). The conclusion was that the cooling rate alone could not exclude the possibility of detecting neutron stars as X-ray emitters, but that the observed spectra of X-rays from these sources were inconsistent with those of black-body radiation expected from the surface of neutron stars. This and other considerations led us to doubt the prospect of ever observing neutron stars. The discovery of pulsars several years ago, however, changed the whole picture. Now it is generally believed that pulsars are rotating, magnetic neutron stars (Gold, 1968). Moreover, theoretical considerations and some observational evidence (such as the speed-ups of the Crab and Vela pulsars) suggest the presence of superfluids in neutron stars (Ginzburg, 1971; Borner and Cohen, 1972; and the paper by Pines in this volume). In the earlier cooling calculations the effect of magnetic fields and superfluidity was not taken into account. In our recent work, therefore, emphasis was placed on the effect of these new factors, which were expected to reduce cooling rates significantly. The new outcome may prove valuable for the understanding of pulsar and X-ray star problems.

* National Academy of Science-NRC Senior Research Associate at NASA, Goddard Space Flight Center.

In Section 2 the basic equations are introduced. In Section 3.1 we summarize the general effect of strong magnetic fields and superfluidity on neutron star cooling – a joint work of Ruderman, Canuto, Lodenquai and myself, published in Tsuruta *et al.* (1972). In recent months I have tried to treat the problem more quantitatively using realistic hadron star models and other newly available information. The results are reported in Section 3.2. Cooling of white dwarfs is discussed in Section 3.3.

2. Basic Equations

The cooling time t is defined by

$$t = \int_{U_0}^{U_t} \frac{dU}{\bar{L}(U)}, \tag{1}$$

where U_0 and U_t are the energies of the star at $t=0$ and at time t, respectively, and $\bar{L}(U)$ is the average luminosity over the interval dU. The total energy is mainly a function of the internal temperature of the star T_i and is expressed as

$$U = U(T_i) = \sum_k U_k^D + U_{\text{ion}}^N \quad (\text{with } k = \text{n, p, e, m, h}). \tag{2}$$

The sum is over the thermal tails of degenerate particles k and U_{ion}^N is the energy of the non-degenerate heavy ions. Here k labels neutrons (n), protons (p), electrons (e), mesons (m), and hyperons (h).

The total energy loss rate L is

$$\begin{aligned} L &= L_\gamma + L_\nu, \\ L_\gamma &= L_\gamma(T_e) = 4\pi\sigma R^2 T_e^4, \\ L_\nu &= L_\nu(T_i) = L_\nu^u + L_\nu^B + L_\nu^{p1} + L_\nu(\text{others}), \end{aligned} \tag{3}$$

L_γ is the photon luminosity and L_ν is the neutrino luminosity. The superscripts u, B, and pl stand for the neutrino URCA, bremsstrahlung, and plasmon processes, respectively. L_ν(others) accounts for photoneutrino, pair annihilation and neutrino synchrotron losses. In the earlier stages of cooling, when temperatures are high, neutrino emission dominates, while later, at lower temperatures, photon emission is more significant. In the absence of magnetic fields and superfluidity, the URCA process is dominant among neutrino processes though the plasmon process competes with it at higher temperatures and bremsstrahlung at lower temperatures. At extremely high temperatures ($T > 10^{10}$ K) pair annihilation becomes significant, but contributions by other process are relatively minor.

The total energy and neutrino luminosity depend on the internal temperature, whereas the photon luminosity is a function of surface temperature (and radius). Thus, the first step is to find the relation between T_i and T_e. For this purpose the energy transfer equation and hydrostatic equations are integrated over the outer layers of the star from the surface to the point where the temperature gradient disappears. Recent studies suggest that a neutron star has a distinct boundary where $\varrho_s \simeq 10^4$ g cm^{-3} and a sharp, discontinuous density drop thereafter (see, for example Ruderman's paper in this volume). Therefore, this integration was performed both from the photosphere and from the boundary with $\varrho_s \simeq 10^4$ g cm^{-3}. For magnetic neutron stars with $H \gtrsim 10^{12}$ G both methods led to similar results. Convection is very unlikely in neutron stars (Tsuruta, 1964), so that energy transfer will be governed by radiation. The basic equations are therefore expressed as (Tsuruta and Cameron, 1966a)

$$\frac{dT}{dr} = -\frac{1}{4\pi r^2} \frac{3}{4ac} \frac{\kappa\varrho}{T^3} L(r),$$

$$\frac{dP}{dr} = -\frac{G(P/c^2 + \varrho)(4\pi r^3 P/c^2 + m)}{r(r - 2mG/c^2)} \quad (4)$$

$$\frac{dm}{dr} = 4\pi\varrho r^2,$$

along with the equation of state

$$P = P(\varrho, T),$$

which becomes degenerate at higher densities ($\varrho \gtrsim 10^4$ to 10^6 g cm^{-3} depending on temperature). The opacity which appears in the transfer equation is expressed as

$$\frac{1}{\kappa} = \frac{1}{\kappa_R} + \frac{1}{\kappa_C}, \quad (5)$$

where κ_R and κ_C are the radiative and conductive opacities, respectively.

3. Results and Conclusions

3.1. General effects of magnetic fields and superfluidity on cooling

For this purpose we shall use the '$V_\gamma(\text{II})$' medium mass neutron star model of Tsuruta and Cameron (1966b) (see also Hartle and Thorne, 1968). The equation of state for this model is derived partially from the V_γ potential of Levinger and Simmons. Properties of the model are: $M = 1.07\ M_\odot$, $R = 12.33$ km, and ϱ_c (central density) = $= 7.39 \times 10^{14}$ g cm^{-3}.

The major effect of strong magnetic fields is to drastically reduce photon opacities in certain spatial directions. It also reduces significantly the URCA neutrino luminosity. The effect of the presence of superfluidity is mainly to reduce the total energy of superfluid particles at lower temperatures. It also suppresses the URCA luminosity.

First, consider the effect of magnetic fields on opacity. We have found that the following approximation is valid for the radiative opacity in superstrong magnetic fields with $H \gtrsim 10^{12}$ G (see Tsuruta et al., 1972; Lodenquai et al., 1974; Canuto et al., 1971; Canuto, 1970):

$$\sigma_\omega(H) = \left(\frac{\omega}{\omega_H}\right)^2 \sigma_\omega(0), \quad \text{for } \omega \ll \omega_H, \quad (6)$$

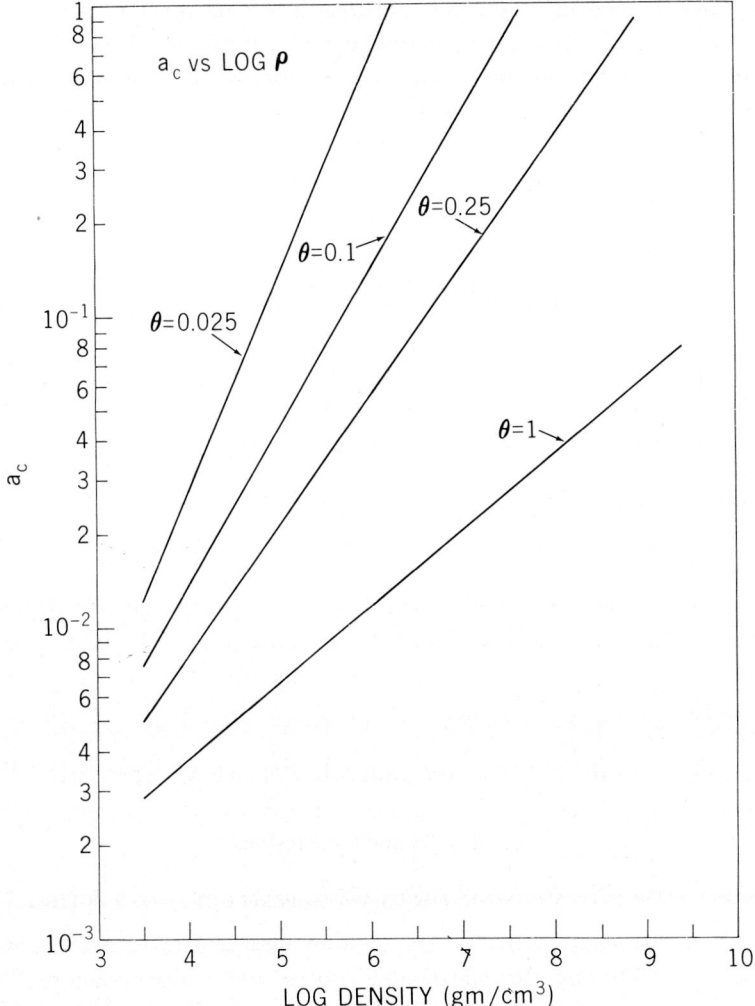

Fig. 1. The factor a_c as a function of density. The a_c is the ratio of conductive opacity with magnetic fields to that without magnetic fields, $\kappa_c(H)/\kappa_c(0)$. $\theta = H/H_q$, and $H_q = 4.41 \times 10^{13}$ G.

where ω is the radiation frequency, $\omega_H = eH/m_e c$ is the electron cyclotron frequency, and $\sigma_\omega(H)$ and $\sigma_\omega(0)$ are the photon cross sections with and without a magnetic field, respectively. The Rosseland mean was used to take account of the frequency dependence. The radiative opacity, thus obtained, can be expressed as

$$\kappa_R(H) = a_R \kappa_R(0); \quad \text{with} \quad a_R \propto \left(\frac{T}{H}\right)^2 \leqslant 1. \tag{7}$$

The conductive opacity is similarly expressed as

$$\kappa_c(H) = a_c \kappa_c(0), \quad a_c \leqslant 1. \tag{8}$$

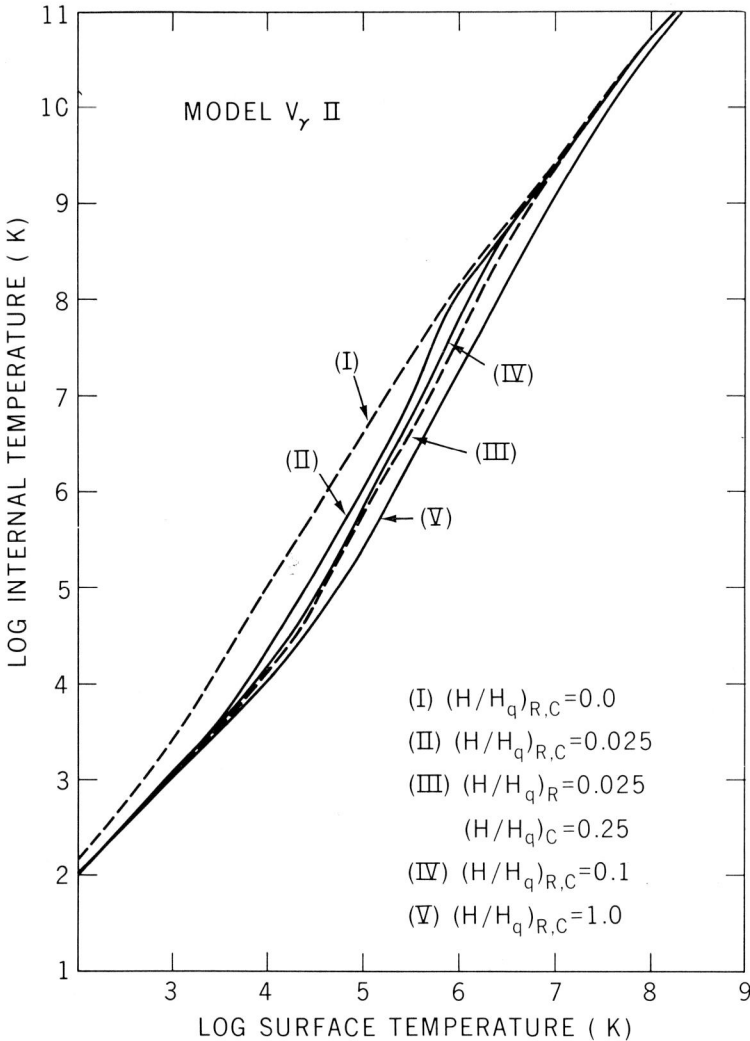

Fig. 2. Internal temperature, as a function of surface temperature at different magnetic field strengths, for Model V_γII as described in Section 3.1. Curves (I), (II), (IV) and (V) stand for a uniform magnetic field of strengths $(H/H_q) = 0$, 0.025, 0.1 and 1, respectively. In the curve (III), the surface field strength is the same as in the curve (II), but the internal field strength is increased by a factor of 10.

The correction factor a_c is generally a complicated function of the magnetic field, density and temperature. In the inner degenerate layers where the conductive opacity plays the dominant role, the temperature dependence drops out and the factor a_c decreases with increasing magnetic field strengths and decreasing density. In Figure 1 the density dependence of the factor a_c is shown for different magnetic field values in the regions of interest. Here, $\theta = H/H_q$, with $H_q = m_e^2 c^3/\hbar e = 4.41 \times 10^{13}$ G. (Similar calculations were made by Canuto and Chiu (1969) for $\theta = 1$ and 0.1 in the vicinity of

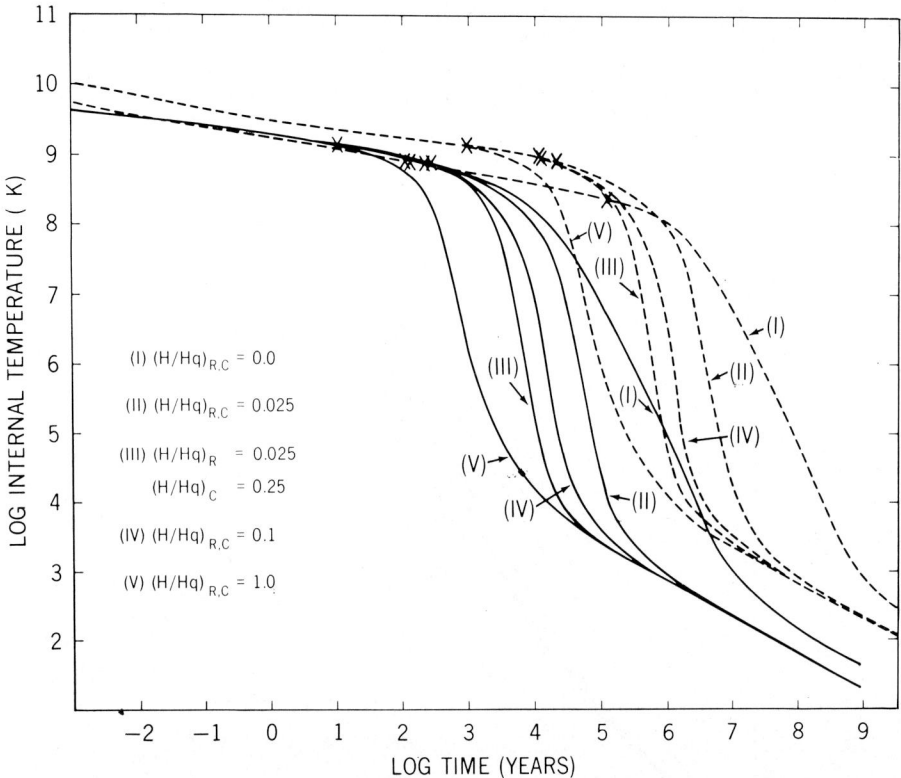

Fig. 3. Internal temperature as a function of time for Model V_γII at different magnetic field strengths. Notation is the same as in Figure 2. The points where the major cooling mechanism shifts from neutrino emission to photon emission are indicated by the crosses. The solid curves and dashed curves represent two extreme cases of maximum and no superfluidity, respectively, as explained in the text.

$\varrho \simeq 10^6$ g cm^{-3}.) We have applied these correction factors to the zero field opacities previously obtained by Tsuruta and Cameron (1966a). It may be noted that electrons become relativistic for densities $\varrho \gtrsim 10^6$ g cm^{-3}. In these high density regions, therefore, we used the formula for conductive opacity (which applies to both relativistic and non-relativistic cases), given by Schatzman (1958). (In the earlier work where the Las Alamos opacity code was used for zero field opacities, the relativity effect was not included. I thank Professor Hayashi for pointing out this to me.)

In Figure 2, the relation between T_i and T_e thus obtained is shown for varying strengths of magnetic fields. In the curve (I), $H = 0$. The curve (II) corresponds to a star of constant field strength of about 10^{12} G. In the curve (III), the surface field is the same as in (II) but the internal field is increased by a factor of 10. We note that the difference between the internal temperature and surface temperature decreases with increasing H and decreasing T. When $H \gtrsim 10^{12}$ G, this difference almost vanishes for $T \lesssim 1000$ K.

In order to study the effect of superfluidity, the following two extreme cases are

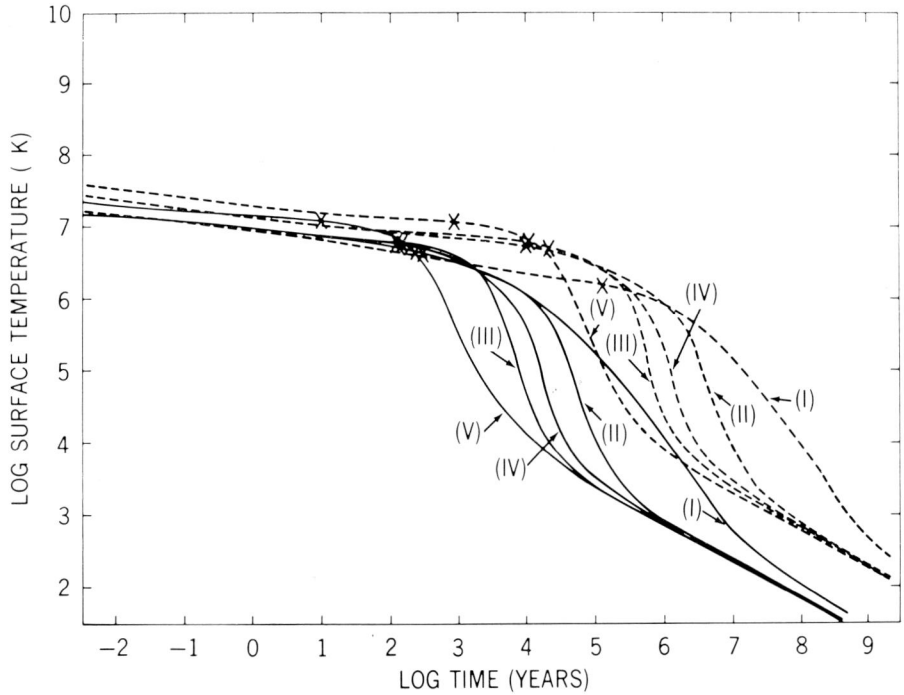

Fig. 4. Surface temperature as a function of time (cooling curves) for Model V_γII at different magnetic field strengths. Notation is the same as in Figure 3.

studied. In Case (S) (maximum superfluidity), the energy of superfluid particles (neutrons and protons) is set equal to zero, and thus degenerate electrons are the main contributors to the total energy. In Case (N) (normal state), superfluidity is completely neglected, and thus degenerate neutrons are the main contributors to the total energy. These effects on the URCA luminosity are also maximized. Thus, we set $L_\nu^u = 0$ for all models in Case (S) and for $H \geqslant 10^{12}$ G in Case (N). The results are shown in Figures 3 and 4, where internal temperatures and surface temperatures, respectively, are plotted as functions of time for different magnetic field strengths. The solid curves stand for Case (S) and the dashed curves represent Case (N).

We note that the effect of both strong magnetic fields and superfluidity are important for neutrons stars older than about 100 yr. For instance, at $t = 1000$ yr (which corresponds to the approximate age of the Crab pulsar NP 0532) the surface temperature is anywhere between $\sim 10^5$ K and a few times 10^7 K. For the Vela pulsar (with $t \simeq 10^4$ years) it is between 10^4 K and 10^7 K. However, for typical older pulsars of a few million years (so far observed only as radio pulsars), the star seems very cold (with the surface temperature anywhere between about 10^2 K and 10^5 K).

3.2. APPLICATION TO SPECIFIC MODELS

In view of the large discrepancy found in the last section between the two extreme

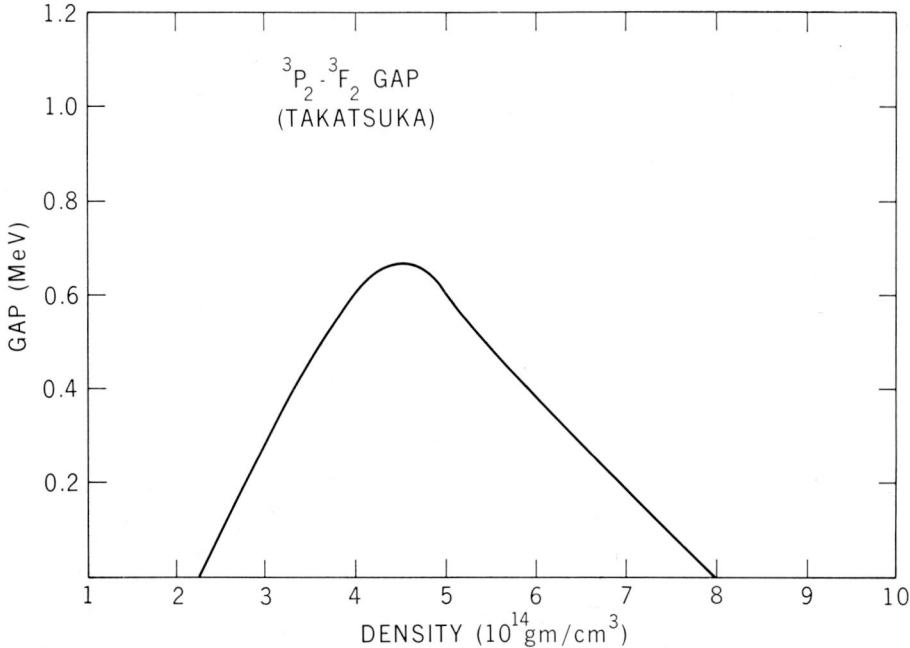

Fig. 5. The density dependence of $\Delta(^3P_2)$ calculated by Takatsuka (1972) including the $^3P_2$–$^3F_2$ tensor coupling.

cases (of maximum and no superfluidity), it might be worthwhile to treat more exactly the effect of superfluidity. Fortunately, the large uncertainty concerning the superfluid energy gap was greatly reduced according to the latest report from the Kyoto group (Takatsuka, 1972; Tamagaki, 1972). There, the $^1S_0$-gap and the $^3P_2$-gap, $\Delta(^1S_0)$ and $\Delta(^3P_2)$, are calculated as a function of density by making use of realistic values of the effective mass, $m^*(\varrho)$. Due to a slight decrease of the value m^*, the gap $\Delta(^1S_0)$ is somewhat reduced but otherwise its general behavior is similar to previous results. There are, however, significant changes in the behavior of the gap $\Delta(^3P_2)$. Taking into account the attractive effect of the $^3P_2$–$^3F_2$ coupling due to the two-nucleon tensor force, Takatsuka (1972) shows that the gap $\Delta(^3P_2)$ behaves as shown in Figure 5. (If the $^3P_2$–$^3F_2$ coupling is neglected, the $^3P_2$-gap becomes negligibly small for realistic values of m^*.) The conclusion is; (i) the $^1S_0$-gap (with maximum $\Delta \simeq 2.5$ MeV) exists in the density region $\varrho \simeq (10^{11} \sim 1.5 \times 10^{14})$ g cm^{-3} and, (ii) the $^3P_2$-gap (with maximum $\Delta \simeq 0.65$ MeV) exists in the range $\varrho \simeq (2 \sim 8) \times 10^{14}$ g cm^{-3}.

It seems clear from the above that the appearance of superfluidity depends critically on the stellar density. If the density is as high as $\sim 10^{16}$ g cm^{-3} for stars near the maximum mass limit (as some neutron star models indicate), the effect of superfluidity will be small. (These heavy stars may possibly contain high concentrations of hyperons and may better be called 'hyperon' or 'hadron stars.') On the other hand, lighter stars near the minimum mass limit may have quite different internal structures; namely the presence of heavy nuclei in the whole interior (Baym et al., 1971b). Therefore, three typical models are chosen for the present investigation. These are: Model (I) a heavy hyperon star near the maximum mass limit, Model (II) a medium weight neutron star, and Model (III) a light neutron star near the minimum mass limit whose central core contains heavy nuclei. Their characteristics are summarized in Table I. In these models the different density ranges are treated in the following manner:

TABLE I

Characteristics of the three models of hadron stars chosen in Section 3.2

Model	Central density	Mass	Radius	
	ϱ_c(g cm^{-3})	$M/M_\odot$	R(km)	
(I)	6.0×10^{15}	1.413	7.10	Heavy
(II)	8.0×10^{14}	0.476	10.9	Medium
(III)	1.1×10^{14}	0.105	76.7	Light

Note: ϱ_c is the central (energy) density in cgs units, $M/M_\odot$ is the mass in solar mass units, and R is the stellar radius in km. Model (I) is a heavy hyperon star, Model (II) is a medium weight neutron star, and Model (III) is a light neutron star with neutron-rich heavy nuclei.

For the outermost layers, where the density $\varrho < \varrho_1 = 3 \times 10^{11}$ g cm^{-3}, the temperature dependence of composition cannot be neglected for our present purposes. This is mainly because the neutron abundance is significant at higher temperatures, while neutrons disappear in this region at zero temperatures (Salpeter, 1961; Wheeler, 1966). In fact, it was found that the matter becomes predominantly composed of neutrons (with small percentages of α-particles, protons and electrons) for $T \gtrsim 2 \times 10^{10}$ K. At about 4×10^9 K, where the composition freezes, there are still significant numbers of neutrons present. Also on the high density side ($\varrho \gtrsim 10^{11}$ g cm^{-3}) heavy nuclei are crystallized before the composition freezes. Therefore, we used the results of recent calculations of equilibrium compositions at finite temperatures ($10^9 \sim 5 \times 10^{10}$ K) by Tsuruta et al. (1973).

In the subnuclear density region, where $\varrho_1 \lesssim \varrho < \varrho_2 \equiv \sim 2 \times 10^{14}$ g cm^{-3}, the matter consists of neutron-rich heavy nuclei (A, Z), neutrons and electrons. For the value Z, we used the recent work by Ravenhall et al. (1972). They show that in this region the value Z is a finite and slowly varying function of density. The ratio A/Z at a given density is more insensitive to different models, and we used the values listed in Baym et al. (1971a). The relative abundance of each component was then calculated using standard methods (see the above references).

In the nuclear region, where $\varrho_2 \lesssim \varrho < \varrho_3 = \sim 8 \times 10^{14}$ g cm^{-3}, we assumed the OPEG type potential for nucleon interactions. This is a Gaussian type soft core potential with one pion exchange constructed by Tamagaki (1968). The relative abundances (neutrons, protons and electrons) was found by a standard method (see, for instance, Baym et al., 1971a).

In the ultradense region, where $\varrho \gtrsim \varrho_3$ the presence of hyperons and muons was assumed. Specifically, the hyperonic composition of model (C) by Pandharipande (1971) was adopted. It may be noted that the appearance of hyperons depends critically on the type of strong interactions assumed among baryons. For instance, it was reported (Moszkowski, 1972) that at above 6 times nuclear density the most stable configuration is still a neutron gas in certain cases. This is for a certain baryon-baryon interaction which he calls the 'Modified Delta Interaction.' However, for a Reid soft

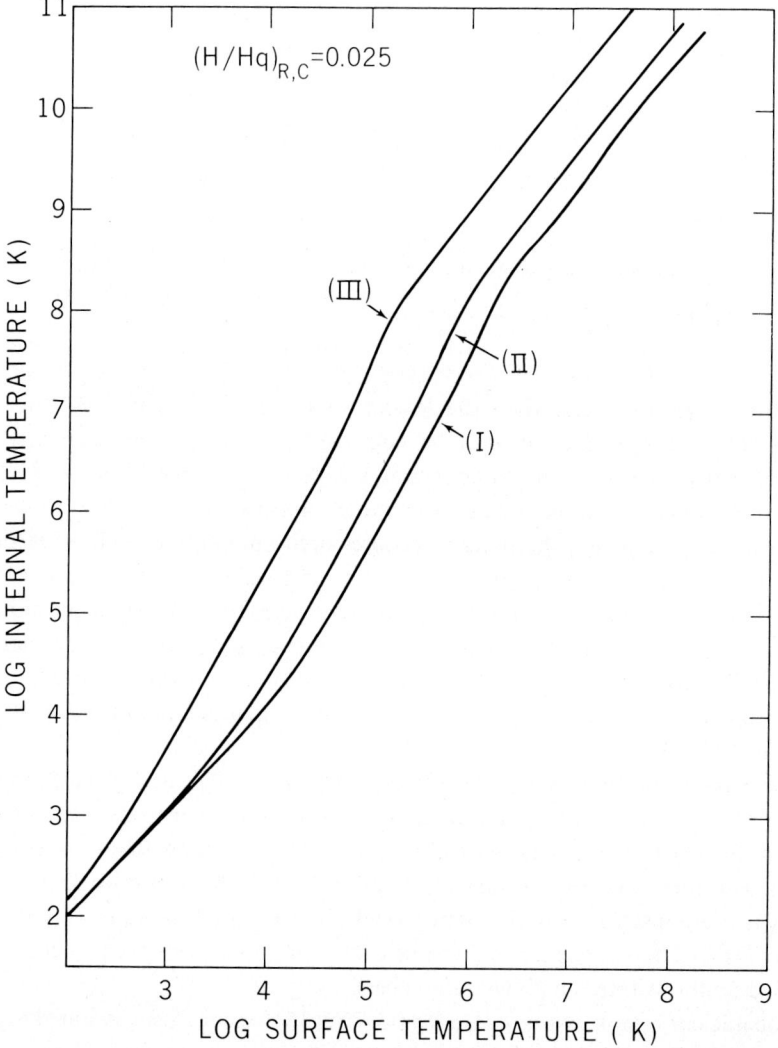

Fig. 6. Internal temperature as a function of surface temperature for the three dense star models (I), (II) and (III) chosen in Section 3.2. A uniform magnetic field of strength $(H/H_q) = 0.025$ is used.

core baryon-baryon interaction with a G-matrix calculated by Sawada and Wong (private communication), there appear significant fractions of hyperons at supernuclear densities (Moszkowski, private communication). (However, both interactions lead to very similar equations of state for pure neutron matter.) In any case, the results of our cooling calculations do not change significantly if our heaviest model is a neutron star instead of a hyperon star.

Using (4) to (8), the relation between internal temperatures and surface temperatures was obtained for the models (I), (II) and (III) described above. The results are shown in Figure 6. In this connection, we may note the angular dependence of the magnetic reduction of opacity. This problem was taken into account by assuming a dipole field and by the estimate that the reduction as expressed by (6) applies in directions within about 10% along both magnetic poles. This approximation is thought to be sufficient in view of greater uncertainties involved in the derivation of (7) and (8).

In the final cooling calculations, the energy and the URCA rates of superfluid nucleons were treated more accurately. We also included the crystallization of heavy ions. Thus, as the energy of degenerate particles k in (2), we used the following formulae:

$$U_k^D = \int C_k^D \, dT,$$

$$C_k^D = \int (C_0)_k^D \, Y_s n_k 4\pi r^2 \, dr,$$

$$(C_0)_k^D = \left(\frac{\pi^2 k_B^2 \, (x_k^2 + 1)^{1/2}}{m_k c^2 \, x_k^2} \right) T,$$

$$x_k = P_k^F/(m_k c),$$

$$Y_s = \frac{8.5 \, T_c}{T} \exp\left(-1.44 \, \frac{T_c}{T} \right), \quad \text{for} \quad T \ll T_c,$$

$$k_B T_c = 0.57 \times \Delta \, (^3P_2)/\sqrt{2} \, \Gamma_0 \quad \text{(with } \ln \Gamma_0 = 1.22)$$

(9)

or

$$k_B T_c = 0.57 \times \Delta \, (^1S_0).$$

Y_s is a superfluid correction factor which becomes unity in the absence of superfluidity. The intermediate region between $T = T_c$ and $T \ll T_c$ was interpolated. The thermal energy of heavy ions is expressed as

$$U_{\text{ion}}^N = \left[\int c_v n_{\text{ion}} 4\pi r^2 \, dr \right] T,$$

where

$$c_v = a_i k_B \mathscr{D} \left(\frac{\theta}{T} \right)$$

$$a_i = \frac{3}{2} \quad \text{for} \quad T \geqslant T_g$$

(10)

and

$$a_i = 3 \quad \text{for} \quad T \leqslant T_m.$$

T_g is the temperature above which ions become a gas and T_m is the melting temperature. In the intermediate region between T_g and T_m, the ions form a liquid and the constant a_i takes on values intermediate between 1.5 and 3. The expression $\mathscr{D}(\theta/T)$ is the Debye function for crystals, which approaches zero as $(T/\theta)^3$ for $T \ll T_m$ (Landau and Lifshitz, 1958). The values T_m and T_g were estimated from the work of Van Horn (1968) and Mestel and Ruderman (1967). In the above equations, k_B is Boltzmann's constant, and n_k and n_{ion} are the number densities of particles k and ions, respectively.

The suppression of the URCA rates due to superfluidity was calculated by using the results of Itoh (1971), Itoh and Tsuneto (1972), Wolf (1966) and Bahcall and Wolf (1965a). Itoh and Tsuneto (1972) also considered the URCA rates in subnuclear regions, and they were taken into account in our models. The recent work by Tsuruta et al. (1973) was used to take account of the URCA rates in the outermost layers (where $\varrho < \varrho_1$). These effects for heavy ions are generally negligible for heavier stars, but they become important for lighter stars.

The solid curve in Figure 7 shows the final, more realistic cooling curve for Model (II) which was obtained in the manner described above. To compare this result with the simplified version of the previous section, the dashed curves in Figure 7 show the two extreme cases, Case (S) (exaggerated superfluidity) and Case (N) (no superfluidity), for the same model.

Figure 8 shows the final cooling curves for our models (I), (II) and (III). We see that the lightest star (III) is cooler than others in earlier stages ($t \lesssim 10^5$ yr) but the trend is reversed when the star becomes older. Model (II) is the coldest star in the

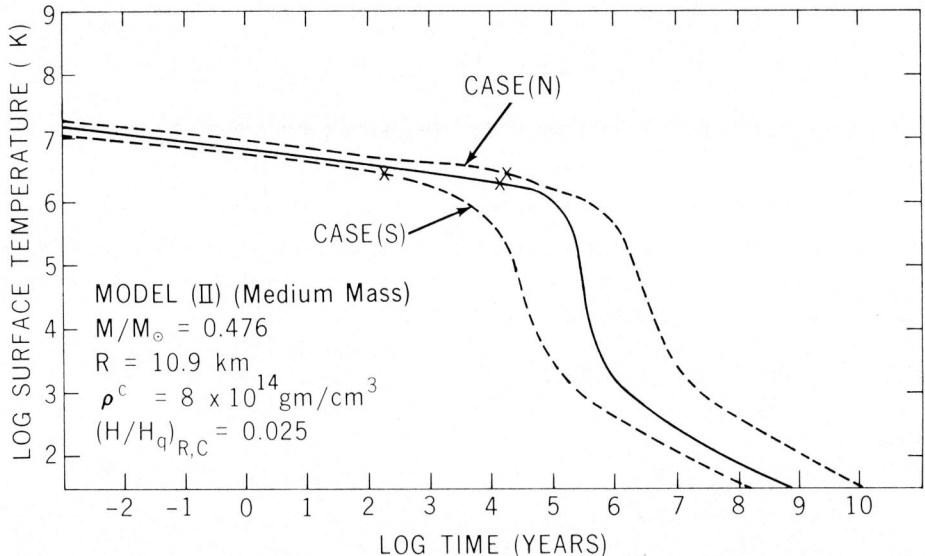

Fig. 7. Cooling curves (surface temperature vs time) for Model (II) of Section 3. The magnetic field strength is the same as in Figure 6. The dashed curves (S) and (N) are for maximum superfluidity and no superfluidity, respectively, as defined in Section 3.1. The solid curve is the final cooling curve obtained by the method described in Section 3.2. The crosses have the same meaning as in Figure 3.

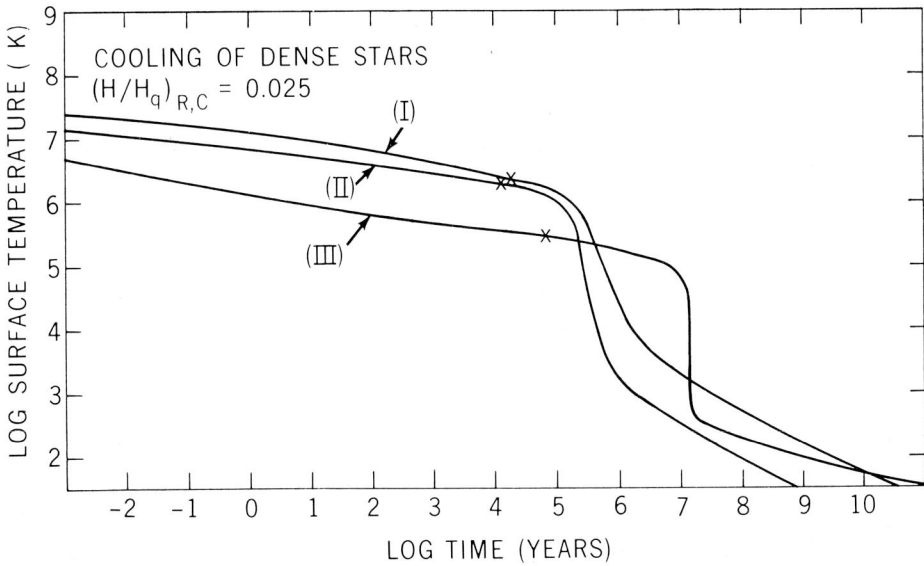

Fig. 8. Final cooling curves (surface temperature vs time) for the three models chosen in Section 3.2, with the same magnetic field strength as in Figure 6. The crosses have the same meaning as in Figure 3.

lower temperature regions where the effect of superfluidity becomes significant. At this point it may be pointed out that the major difference in cooling behavior comes from differences in density rather than mass differences. For instance, Model (II) of the present section and the V_γ(II) model of the previous section, due to their similar densities, have similar cooling behaviors even though the latter is more than twice as massive as the former. Among realistic dense star models developed by other authors, our models chosen here are closest to those constructed by Baym et al. (1971b). (In the medium mass region our models are also similar to those constructed by Ikeuchi et al., 1971.)

Here we wish to emphasize the importance of heavy ions for lighter stars. A sudden drop of temperature takes place at the age of about 10^7 yr for Model (III) in Figure 8 (where heavy ion energies are included). When U_{ion}^N is neglected in (2), this sudden cooling occurs near 10^6 yr (for the same model).

At the age of a few million years, which is regarded as the average age of radio pulsars (e.g., Gunn and Ostriker, 1970), the stars should be very cold. This conclusion is relatively independent of the particular models used. This is because the transparency due to strong magnetic fields, which is the main cause for the sudden cooling, is achieved by this time. For those stars the surface temperatures are from several hundred to several thousand degrees. However, the surface may be kept at somewhat above 10^5 K if we take into account the heating by accretion of interstellar matter (Ramaty, private communication). Thus, the conclusion reached in our previous work (Tsuruta et al., 1972) is still valid. Thermal ionization of atoms in the atmosphere will be negligible and only residual rotational energy may be responsible for these older pulsars. This may then explain the lack of observed pulsars with periods greater than a few seconds (Ruderman, 1971). These cold, dense stars seem to have a peculiar property, that we can see through to the center when we look down from the two magnetic poles. It may be noted that the above conclusion was reached by temporarily disregarding Model (III). This is because neutron stars with mass less than about 0.2 $M_\odot$ may not be formed (Bethe, 1972; Cohen and Borner, 1972). (This lightest model was chosen to dramatize the density dependence of cooling behavior.) The cooling behavior of a neutron star of about 0.2 $M_\odot$ is closer to our Model (II) than Model (III).

In Table II surface temperatures and internal temperatures of the three models are shown at a few interesting points during their cooling history. The first two points correspond to the approximate age of the Crab pulsar and Vela pulsar, respectively. There it seems likely that the star is hot enough so that thermal effects can be important. Earlier in this symposium it was reported that the pulsar glitch information allows some estimates of the mass of these younger pulsars. The mass of the Crab pulsar as estimated by Pines (in this volume) from the crustquake theory is about 0.5 $M_\odot$, while it is greater than 1 $M_\odot$ according to 'accretion' theory (Borner and

TABLE II

Surface temperatures T_e and internal temperatures T_i for the models described in Table I, at three interesting points during the cooling history

	Models			Age (years)	Comment
	(I)	(II)	(III)		
$\text{Log} T_e$ (K)	6.62	6.45	5.68	10^3	~ Crab
	6.40	6.30	5.56	10^4	~ Vela
	4.38	3.21	5.21	10^6	
$\text{Log} T_i$ (K)	8.62	8.74	8.54	10^3	~ Crab
	8.36	8.54	8.38	10^4	~ Vela
	4.58	3.26	7.92	10^6	

Cohen, 1972). If the core-quake theory applies, the Vela pulsar should be a heavy star (Shaham, 1972). Therefore, we may expect to be able to detect X-rays from these pulsars. It is already known that the Crab Nebula has an X-ray pulsar, though it is not of blackbody radiation. As of now I am not aware of any conclusive report about the detection of pulsed X-rays from the Vela pulsar. However, caution may be advised for these young pulsars because nucleon superfluidity, on which the above theories are based, may not yet be significant at their high temperatures. For instance, the Crab pulsar seems to too hot for the effect of the $^3P_2$ superfluidity (see Table II). On the other hand, cooling will be much faster if pions are present (Bahcall and Wolf, 1965b; Tsuruta, 1972). It was already pointed out that pions may be present in high density regions (Sawyer, 1972; Scalapino, 1972). Even though we do not know yet exactly at what density pions appear, it is possible that they are abundant for heavy hadron stars.

At this point we wish to consider the possible heating effect of crust-core rotation slippage. If neutron star interiors are superfluid, then frictional heat may be dissipated as its rotation is slowed. This process may keep the star hotter for longer periods. [Such a possibility was first suggested by Cameron (1970).] Greenstein (1971) estimated this effect and concluded that it can be significant enough so that thermal X-ray emission from certain pulsars may be detectable. The comparison shows that both for the Crab pulsar and Vela pulsar the surface temperatures in his results and those obtained in this section (Table II) are comparable if they are medium weight to heavy stars (the Models (II) and (I)). Greenstein suggests a few more older pulsars as other possible candidates for detection of X-rays due to this heating effect. Their surface temperatures estimated by Greenstein are in the vicinity of $(2 \sim 3) \times 10^5$ K. At the ages which are expected for these pulsars Model (III) is at comparable temperatures but our Models (II) and (I) are colder (with their surface temperatures at $10^3 \sim 2 \times 10^5$ K). (We may comment that the magnitude of the frictional effects depends on the strength of the coupling between the charged and the superfluid components of the stars and that quantitative estimates are not yet available.)

3.3. COOLING OF WHITE DWARFS

Before closing we wish to show typical cooling curves for white dwarfs. The properties of the particular model chosen here are listed in Table III. This is the same model as the one chosen in Tsuruta and Cameron (1970) and further details are found in that paper. A rather massive white dwarf star was chosen so that it can possess an URCA shell. However, other neutrino processes (Beaudet et al., 1967; Hansen, 1968) are also included here. The cooling curves thus obtained are shown in Figure 9. A simple ideal gas was assumed for the straight line, while the other curve includes the effect of crystallization.

TABLE III

Properties of the white dwarf model chosen in Section 3.3

Mass $= 1.373\ M_\odot$
Central density $= 10^{9.5}$ g cm^{-3}
Radius $= 1810$ km
Radius of URCA shell $= 377$ km
URCA nuclear pair ^{23}Na and ^{23}Ne
Core of $1 M_\odot$ of carbon-burning products
Envelope of ^{12}C

Composition of Core

Mass number	Mass fraction	Nuclei and electron capture thresholds (MeV)
16	0.01	^{16}O(10.4)^{16}C
20	0.41	^{20}Ne(7.03)^{20}O(21.22)^{20}C
23	0.06	^{23}Na(4.4)^{23}Ne(11.15)^{23}F(13.87)^{23}O
24	0.52	^{24}Mg(5.52)^{24}Ne(15.91)^{24}O

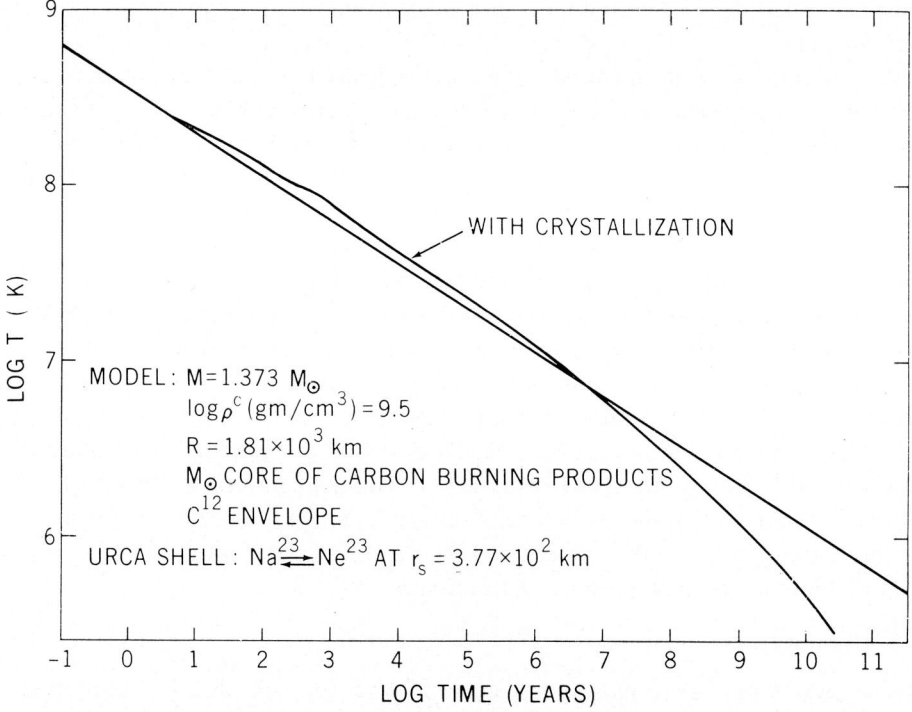

Fig. 9. Cooling curves for the white dwarf model chosen in Section 3.3. The straight line is for an ideal gas. In the other curve, the effect of crystallization is included.

Acknowledgement

I wish to thank the many persons mentioned in this paper and other participants of this symposium for valuable suggestions and discussions. The last part of this paper was completed during my visit at the Aspen Center for Physics.

References

Bahcall, J. N. and Wolf, R. A.: 1965a, *Phys. Rev.* **140**, B1452.
Bahcall, J. N. and Wolf, R. A.: 1965b, *Phys. Rev. Letters* **14**, 343.
Baym, G., Bethe, H. A., and Pethick, C. J.: 1971a, *Nucl. Phys.* **A175**, 225.
Baym, G., Pethick, C. J., and Sutherland, P.: 1971b, *Astrophys. J.* **170**, 299.
Beaudet, G., Petrosian, V., and Salpeter, E. E.: 1967, *Astrophys. J.* **150**, 979.
Bethe, H. A.: 1972 (private communication).
Borner, G. and Cohen, J. M.: 1972, Pulsar Speed-ups (preprint).
Cameron, A. G. W.: 1970 (private communication).
Canuto, V.: 1970, *Astrophys. J.* **160**, L153.
Canuto, V. and Chiu, H.-Y.: 1969, *Phys. Rev.* **188**, 2446.
Canuto, V., Lodenquai, J., and Ruderman, M.: 1971, *Phys. Rev.* **D3**, 2303.
Cohen, J. M. and Borner, G.: 1972 (private communication).
Ginzburg, V. L.: 1971, *Physica* **55**, 207.
Gold, T.: 1968, *Nature* **218**, 731.
Greenstein, G.: 1971, *Nature Phys. Sci.* **232**, 117.

Gunn, J. E. and Ostriker, J. P.: 1970, *Astrophys. J.* **160**, 979.
Hansen, C. J.: 1968, *Astrophys. Space Sci.* **1**, 499.
Hartle, J. B. and Thorne, K. S.: 1968, *Astrophys. J.* **153**, 807.
Ikeuchi, S., Nagata, S., Mizutani, T., and Nakazawa, K.: 1971, *Prog. Theor. Phys.* **46**, 95.
Itoh, N.: 1971 (private communication).
Itoh, N. and Tsuneto, T.: 1972, *Prog. Theor. Phys. Kyoto* **48**, 1849.
Landau, L. D. and Lifshitz, E. M.: 1958, *Statistical Physics*, Addison-Wesley Publ. Co., Mass.
Lodenquai, J., Canuto, V., Ruderman, M., and Tsuruta, S.: 1974 (to be published).
Mestel, L. and Ruderman, M.: 1967, *Monthly Notices Roy. Astron. Soc.* **136**, 27.
Moszkowski, S. A.: 1972, paper presented at APS Meeting, April, Washington D. C. (*Bulletin of APS* **17**, 502).
Pandharipande, V. R.: 1971, *Nucl. Phys.* **A178**, 123.
Ravenhall, D. G., Bennett, C. D., and Pethick, C. J.: 1972, *Nuclear Surface Energy and Neutron Star Matter* (preprint), (1972, *Phys. Rev. Letters* **28**, 978).
Ruderman, M.: 1971, *Phys. Rev. Letters* **27**, 1306.
Salpeter, E. E.: 1961, *Astrophys. J.* **134**, 669.
Sawyer, R. F.: 1972, *Phys. Rev. Letters* **29**, 382.
Scalapino, D. J.: 1972, *Phys. Rev. Letters* **29**, 386.
Schatzman, E.: 1958, *White Dwarfs*, North-Holland Publ. Co, Amsterdam.
Shaham, J.: 1972 (private communication).
Tamagaki, R.: 1968, *Prog. Theor. Phys. Kyoto* **39**, 91.
Tamagaki, R.: 1972 (private communication).
Takatsuka, T.: 1972, *Energy Gap in Neutron-Star Matter* (preprint) NEAP-11, May, Kyoto University, (1972, *Prog. Theor. Phys. Kyoto* **48**, 1517).
Tsuruta, S.: 1964, Thesis, Columbia University.
Tsuruta, S.: 1974 (to be published).
Tsuruta, S. and Cameron, A. G. W.: 1966a, *Can. J. Phys.* **44**, 1863.
Tsuruta, S. and Cameron, A. G. W.: 1966b, *Can. J. Phys.* **44**, 1895.
Tsuruta, S. and Cameron, A. G. W.: 1970, *Astrophys. Space Sci.* **7**, 374.
Tsuruta, S., Canuto, V., Lodenquai, J., and Ruderman, M.: 1972, *Astrophys. J.* **176**, 739.
Tsuruta, S., Truran, J. W., and Cameron, A. G. W.: 1973, *Bulletin of APS* **18**, 541.
Van Horn, H. M.: 1968, *Astrophys. J.* **151**, 227.
Wheeler, J. A.: 1966, *Ann. Rev. Astron. Astrophys.* **4**, 393.
Wolf, R. A. 1966, *Astrophys. J.* **145**, 834.

DISCUSSION

Greenstein: I notice that some of your temperatures derived for the crab pulsar can get quite high – on the order of 10^9 K in the interior. This is of the same order of the superfluid energy gap in some regions. If so, then some of the superfluid in this pulsar could contain lots of normal component. Situations of this sort have never been considered by neutron star superfluidists.

Van Horn: Your curves of $\log T_e$ vs $\log t$ show a rapid decrease of the temperature at an age of 10^4–10^6 yr. Could you explain the reason for this?

Tsuruta: The effect is due to the decrease of opacity in the presence of magnetic fields.

Itoh: I would like to comment on the opacity of the region where a neutron liquid, a proton liquid, and an electron liquid coexist. Recently, Pethick, Flowers and I have calculated the thermal conductivity of this region and find that electrons and neutrons contribute comparably to the thermal conductivity when nucleons form normal Fermi liquids.

PULSAR OBSERVATIONS AND NEUTRON STAR MODELS

GERHARD BÖRNER* and JEFFREY M. COHEN**

Laboratory for High Energy Astrophysics, NASA/Goddard Space Flight Center, Greenbelt, Md., U.S.A.

Abstract. Information about the physical parameters of neutron stars is obtained from pulsar observations. The energy balance of the Crab Nebula and the Vela X remnant allows one to derive limits for the masses of the Crab and Vela pulsars. Glitch observations provide further clues on the masses of these two pulsars. The degree of confidence with which one should believe the derived numbers is pointed out. The possibility of observing neutron stars in binary systems as pulsating X-ray sources is discussed. Finally, the importance of observing redshifted gamma ray lines from the surface of neutron stars, and thus directly measuring either individual or statistical properties of these objects, is pointed out.

1. Introduction

Besides the well-known chain of arguments that leads, by process of elimination (Maran and Cameron, 1969), to the generally accepted conclusion that pulsars are rotating neutron stars, there is another very gratifying result which stems from the comparison of neutron star models and pulsar observations. It turns out that for any reasonable equation of state the neutron star models, computed as outlined in the paper by Cohen and Börner in this volume (hereafter called Paper I), have moments of inertia which are just of the right order of magnitude to explain the observed energy input into the Crab Nebula as the loss of rotational energy from a rotating neutron star with the parameters of the Crab pulsar PSR 0531+21. This is a major triumph of the interaction of theory and experiment in this field. It has already been mentioned in the papers of H. A. Bethe and D. Pines in this volume, but this information really belongs to the subject matter of the present paper and has therefore been pointed out again. We shall discuss in detail the energy balance of the Crab and Vela Nebulae later.

Except for what was pointed out above, the interaction between theory and observation is very slim to date; so much so that no observer is participating in this conference and observations have to be discussed by theorists, one of whom has never even looked through a telescope. But this is understandable because the wealth of pulsar observations is overwhelming, and a comprehensive understanding still has not been achieved. Only the puzzling 'glitches' of the Crab and Vela pulsar have been well incorporated into various competing theories. These will be discussed at the end of this paper.

In the following we shall attempt to gather some of the pieces in the puzzle that link neutron star models and pulsar observations. We shall try very ambitiously to derive parameters for the rotating neutron stars that are represented by the names Crab and Vela in the astronomical observations.

---

* Permanent address: Max-Planck-Institut für Physik und Astrophysik München, Germany and National Academy of Science-NRC Research Associate.
** Permanent address: Physics Department, University of Pennsylvania, Philadelphia, Pennsylvania 19104. National Academy of Science-NRC Senior Research Associate.

2. Radio Pulsars

All of the 61 pulsars listed by Manchester and Taylor (1972), except for Crab and Vela, are seen only by their pulsed radio emission. It is quite obvious that the pattern, shape and polarization of these radio pulses contains quite a lot of information on the internal structure of the pulsar. We are, however, not yet able to understand the message we are getting. No satisfying quantitative description of the electromagnetic link between the rotating neutron star and the radiation pattern of the pulsar has been given so far. Indeed, not even the case of a magnetosphere of radiative particles, where the axis of the magnetic field coincides with the rotation axis of the neutron star, has been solved. Furthermore, to explain the pulse producing mechanism one would have to treat the much more complicated case of at least a slight deviation from axial symmetry.

In the absence of a detailed mechanism for pulsar emission, only energy considerations can be employed to obtain information on the physical properties of the rotating neutron star. For 22 of the 61 listed pulsars (Manchester and Taylor, 1972) both frequency Ω and change of frequency $\dot\Omega \equiv d\Omega/dt$ have been measured. Then by determining their rate of loss of energy $\dot{E}_{rot} = I\Omega\dot\Omega$ we could in principle find the moment of inertia I of these neutron stars. This in turn would precisely fix the mass and density profile of the star once an equation of state had been chosen.

The energy in radio pulses $\dot{E}_{puls}$ probably is only a small part of the total energy release, and therefore only very crude limits on the physical parameters of a neutron star may be derived. Even this modest undertaking does not look very promising, however, since neither the flux in radio pulses (a well-defined mean value for a specific pulsar) nor the distance to pulsars are known to a high accuracy.

The flux in radio pulses is known to fluctuate very strongly on a scale of weeks. This makes it very difficult to define a quantity like the mean pulse intensity at a given frequency for any pulsar. Furthermore the distance to pulsars can only be estimated (again excepting Crab and Vela) by their dispersion measure, which gives the average value of the electron density along the line of sight. To derive the distance of pulsars from the dispersion measure one would need a precise knowledge of the interstellar medium. In reality, however, we use the pulsars to obtain more information on the interstellar medium and it has been found that in a region of 100 pc around the Sun, the average electron density is almost 0.1 cm^{-3}. This is roughly comparable to the average density of the atomic hydrogen in that region, and indicates that we are surrounded neither by a classical H II region nor by an H I region (cf., Biermann, 1972).

Thus there is virtually no feedback to the neutron star models through the observations of radio pulses. If we nevertheless make some crude estimates of the energy in radio pulses for various pulsars, we find $\dot{E}_{puls} \approx 10^{28\pm 2}$ erg s^{-1}. While that result is laden with all the uncertainties discussed above, it permits one rather vague but still interesting conclusion: assuming moments of inertia between 10^{44} and 10^{45} g cm^2 we can compute the rotational energy output $(\dot{E}_{rot})$ for various pulsars. We then find that the energy in radio pulses for the Crab pulsar $\dot{E}_{puls} \approx 10^{-9} \dot{E}_{rot}$, while for many of the older pulsars $\dot{E}_{puls}$ is a much bigger fraction of $\dot{E}_{rot}$, such as PSR 0809+74, where

$\dot{E}_{\text{puls}} \approx \dot{E}_{\text{rot}}$. This seems to indicate that as pulsars grow older they spend a bigger and bigger fraction of their rotational energy output in the production of radio pulses.

3. The Crab Pulsar

The Crab pulsar PSR 0531+21 is located at the center of the Crab Nebula, the remnant of a supernova that exploded in 1054. Because of its location in the nebula a distance estimate independent of the dispersion measure can be obtained and, furthermore, the energy balance of the nebula can be used to get more information on the energy output of this pulsar. In addition, pulses from this object have been observed not only in the radio regime but also in the optical and X-ray frequencies.

The supernova of 1054 is widely considered to have been of type I, but Minkowski (1968) has cast doubt on that. Trimble (1968) has compared the radial velocities and proper motions in the thick filamentary shell, which is a projected elliptical object with semiaxes of 3' and 2', and obtained a distance of 2 kpc on the assumption that the 3-dimensional nebula is a prolate ellipsoid. Trimble and Woltjer (1971) have pointed out that the uncertainties in this value are large and that a distance as low as 1.2 kpc or as high as 2.5 kpc cannot be excluded.

Let us, however, be definite and adopt a distance of 2 kpc for the Crab Nebula. This value seems to be the most widely used. Observations of the nebula indicate that energy must be supplied to it continuously. Assuming that the Crab pulsar is the only source of energy in the nebula, one can determine limits on the energy output of the pulsar (Rees and Trimble, 1970; Börner and Cohen, 1972) by considering the energy balance.

The only well established energy loss is by synchrotron radiation which implies $\dot{E}_{\text{synch}} = 1.2 \times 10^{38}$ erg s^{-1} (Baldwin, 1971), if the distance to the Crab is 2 kpc. The pulsar has to replenish at least the electrons producing the optical and X-ray synchrotron radiation because these particles have half-lives of less than 100 yr. So a rough estimate obtained from the observed spectrum (Baldwin, 1971) indicates that the pulsar has to supply continuously at least 0.8×10^{38} erg s^{-1}. For the pulsar to replenish this energy via loss of rotational energy its moment of inertia has to be at least 1.8×10^{44} g cm^2 (cf., Cohen and Cameron, 1971), corresponding to line (a) in Figure 7 of Paper I. Neutron star models corresponding to various equations of state have been discussed in that paper and it can be seen from Figure 2 that a model with $I = 1.8 \times 10^{44}$ g cm^2 has a mass of

$$0.34 M_\odot \text{ (BPS)}, \quad 0.36 M_\odot \text{ (BJ)}, \quad 0.36 M_\odot \text{ (BBS)}, \quad 0.26 M_\odot \text{ (CCLR)}.$$

Here (BPS) corresponds to the equation of state published in Baym et al. (1971); (BJ) Bethe and Johnson (1973); (BBS) Bethe et al. (1970); (CCLR) Cohen et al. (1970). All the equations of state discussed in Paper I, except one, can easily provide neutron star models big enough to exceed this lower limit. The exception is the work of Leung and Wang (1971); the maximum moment of inertia in their equation of state, numbered (I) is less than 0.2×10^{44} g cm^2. That is, the Crab pulsar is definitely not among the stable neutron stars they compute using their equation of state which does not incorporate repulsion between baryons. Even Equation (II) of Leung and Wang (1971), which gives $I_{max} = 1.05 \times 10^{44}$ g cm^2 is too low (Leung and Wang in Figure 7 of Paper I) although some repulsion is assumed to be present in this case (II). This indicates, as has been discussed already from the nuclear physics point of view by H. A. Bethe that the repulsion between the nucleons and hyperons at short distances plays an important role which may be the dominant feature at high densities. It is interesting to see that this is also suggested by astrophysical evidence.

The evidence in favor of repulsive interactions becomes even stronger if we take into account the protons that are pulled from the surface of the rotating neutron star by strong electric fields in the model proposed by Goldreich and Julian (1969). According to this model these protons are accelerated to the same energies as the electrons producing synchrotron radiation. In this case the minimum energy loss is twice the synchrotron radiation loss. This has the consequence that the minimum moment of inertia of the Crab increases by a factor of two:

$$I_{min} \geq 4 \times 10^{44} \text{ g cm}^2.$$

The neutron star mass is then $\geq 0.5\ M_\odot$ (BBS, BJ). Thus if the model is valid we obtain the result that the Crab pulsar is a neutron star with a mass of at least $0.5\ M_\odot$.

The acceleration mechanism of Gunn and Ostriker (1969) can also be tested. They require that the protons get ten times the energy of the electrons. Assuming the fluxes of electrons and ions from the pulsar to be equal, this would require ten times the synchrotron energy for the protons, leading to a moment of inertia of 2×10^{45} g cm^2 [line (c) in Figure 7 of Paper I]. Only the CCLR equation of state has models with moments of inertia of that magnitude. But even then this condition is satisfied only over a small density range near the mass peak. It therefore seems that the model of Gunn and Ostriker (1969) should be modified quantitatively.

Conclusions based on energy losses from the Crab nebula due to the expanding supernova shell (Rees and Trimble, 1970; Börner and Cohen, 1972) are much more uncertain than the preceding considerations. Observations of the filaments in the expanding supernova shell (Woltjer, 1958; Trimble, 1968) show that the expansion velocity at present is higher than would correspond to an expansion at constant velocity since 1054. It seems that the nebula is accelerating now with an acceleration of

$$\dot{v} = 0.0014 \text{ cm s}^{-2}.$$

The nebula might, however, be decelerating now with the velocity still higher than the average if it had been accelerating rather strongly in the past. The energy of the expanding supernova shell would change due to this acceleration at a rate

$$\dot{E}_{acc} = \int \varrho v \dot{v} d(\text{Vol}), \tag{1}$$

where ϱ is the density in the nebula. It would also change by the 'snow-plow' effect wherein the change in mass of the supernova shell as interstellar material piles up along the rim implies

$$\dot{E}_{\text{plow}} = 1/2 \varrho_m v^3 A, \tag{2}$$

where A is the surface area of the nebula and ϱ_m is the density of interstellar material. Since we do not know whether the supernova shell is accelerating or decelerating at present, we investigate both cases. If the shell is decelerating, the energy gained by deceleration will be spent in the snowplow effect described by (2), and perhaps totally balance it. Thus $\dot{E}_{\text{shell}} = \dot{E}_{\text{acc}} + \dot{E}_{\text{plow}} = 0$ is a distinct possibility. No further limits on the parameters of the Crab pulsar except those derived earlier from synchrotron radiation can be found in this case. We should notice, however, that in principle one could directly measure the value of $\dot{v}$ at present, and thus decide the question of acceleration or deceleration. If the currently accepted values for acceleration and snowplow are used, we find

$$\dot{E}_{\text{acc}} = 1.6 \times 10^{38} \text{ erg s}^{-1}, \tag{3}$$

$$\dot{E}_{\text{plow}} = 1.7 \times 10^{38} \text{ erg s}^{-1}, \text{ and hence}, \tag{4}$$

$$\dot{E}_{\text{shell}} = 3.3 \times 10^{38} \text{ erg s}^{-1}. \tag{5}$$

This energy has to be supplied either directly by the pulsar *via* the low frequency waves emitted or by the adiabatic expansion of a relativistic gas (Trimble and Rees, 1970). In both cases a rotating neutron star with a moment of inertia of 10^{45} g cm^2 can continuously supply that energy [line (b) in Figure 7 of Paper I]. The equations of state that can furnish a neutron star model with a moment of inertia big enough (BBS, BJ, CCLR) give a mass of 1.2 $M_\odot$ (BBS, BJ) for this model. If little material was lost during the collapse the star would have had a mass of 1.35 $M_\odot$ prior to the collapse – a mass above the Chandrasekhar limit for typical white dwarfs. It should be remembered that these energy losses are rather uncertain, and that the pulsar has to supply the energy continuously only if the energy content of the gas of relativistic particles in the nebula is maintained at its present level. All the uncertainties can, however, be decided by future observations and thus a value of 1.2 $M_\odot$ for the Crab pulsar may be confirmed some day with a much higher degree of confidence than we have now.

If the shell is pushed out by low frequency waves from the pulsar and, if according to Gunn and Ostriker (1969), protons get ten times the energy of the electrons, then only the maximum mass (near the mass peak) neutron star models of the CCLR equation of state can fulfill this requirement, as indicated by line (d) in Figure 7 of Paper I. This particle acceleration mechanism therefore seems to be unrealistic.

If the distance of the Crab were less or more than 2 kpc, the limits derived above would have to be scaled down or up accordingly. If we go to the extreme values for the distance of 1.2 kpc and 2.5 kpc, the moment of inertia necessary to account for the short-lived synchrotron particles would vary between 0.8 and 2.4×10^{44} g cm^2 [0.2 $M_\odot$ and 0.4 $M_\odot$ respectively (BBS)].

A final remark might be of interest. It pertains to the suggestion that cosmic rays are produced in the electromagnetic field of pulsars. The spectrum of galactic cosmic rays above 300 MeV shows that about 50 times as much energy is present in protons as in electrons. It is clear from the foregoing discussion that the Crab pulsar could not produce such a ratio of proton to electron energy. This throws considerable doubt on the hypothesis that the high energy galactic cosmic rays are all produced by pulsars.

4. The Vela Pulsar

The Vela pulsar PSR 0833-45 is associated with the supernova remnant Vela X, which is about 1.1×10^4 yr old, has a radius of 10 pc, and is at a distance of 500 pc (Milne, 1970).

If the electromagnetic radiation emitted by Vela X is synchrotron, then from synchrotron theory the energy content in the gas can be estimated at $\sim 10^{49}$ (Tucker, 1971). Assuming constant velocity, if the expansion velocity of the supernova shell were constant which would mean $v = 880$ km s^{-1}, the energy loss through adiabatic expansion would be

$$\dot{E}_{ad} = 2.4 \times 10^{37} \text{ erg s}^{-1}. \tag{6}$$

On the other hand, the rotational energy lost by a neutron star with the parameters of PSR 0833-45 is between

$$\dot{E}_{rot} = 4 \times 10^{36} \text{ erg s}^{-1} \text{ to } 8.5 \times 10^{36} \text{ erg s}^{-1} \tag{7}$$

for models with mass between 0.8 $M_\odot$ and 1.7 $M_\odot$ (BBS). Although this is rather large compared to the loss of 4×10^{35} erg s^{-1} in X-rays (Tucker, 1971) and 10^{33} erg s^{-1} in radio (Milne, 1970), the pulsar cannot supply the energy given in (6). The shell must therefore have been decelerating. Let us assume that the deceleration at present is very small, and that the main contribution to $\dot{E}_{shell}$ is by snowplow (Börner and Cohen, 1972). If it is further assumed that a neutron star of 1.2 $M_\odot$ is present to balance the expansion losses, then the velocity of expansion can be determined to be 240 km s^{-1} (Börner and Cohen, 1972). It is amusing to note that at about the same time Wallerstein and Silk (1971) independently (neither group knew of the other's work until after publication) measured the expansion velocity of Vela X by an observation of Ca II lines in that direction, and they found precisely that value of 240 km s^{-1} for the expansion velocity.

5. Pulsar Glitches

The speed-ups of both Crab and Vela are discussed in this volume by D. Pines along with the various theoretical attempts to account for them. The 'starquake' theory advocated by Pines in his paper, as well as the 'accretion' model (Börner and Cohen, 1972) are the only theories of pulsar glitches that derive limits on the mass (and other parameters) of Crab and Vela. Both theories use the two-component model for the interaction between the crust-charged particle system and the neutron superfluid to account for the relaxation phenomena of the post-speed-up behavior of the pulsar.

They differ in the mechanism evoked to produce the initial glitch. Both theories are not without problems.

The starquake theory as described in Pine's paper views the smaller glitches of the Crab pulsar ($\Delta\Omega/\Omega \sim 10^{-9}$) as a sudden relaxation of elastic stresses which accumulate in the crust (i.e., the rigid outer layers composed of a lattice of nuclei) of the pulsar as it is slowing down. The magnitude of this effect depends on the magnitude of the stresses that can be built up in the crust. Baym and Pines (1971) used an equation of state developed by Baym et al. (1972), where nuclei in the lattice become very large, up to $Z \sim 200$. On the other hand, J. Negele (in this volume) came to the conclusion that $Z \sim 40$ is to be expected in the crust of neutron stars. The amount of stress that can be built up increases monotonically with Z. It therefore seems that the crustquake theory has serious problems in explaining even the Crab pulsar glitches if the neutron star crust consists only of small nuclei. Even for $Z \sim 200$ nuclei in the lattice, the requirement to have a typical Crab speed-up every two years leads to a picture of the Crab pulsar as an almost completely solidified star. The crustquake theory predicts a mass of the Crab of less than $0.15\ M_\odot$. This limit should not be taken too much at face value, but it illustrates the difficulties of this theory. A neutron star of $0.15\ M_\odot$ might not even be formed in a supernova explosion (see Paper I) because its binding energy is so low that it becomes energetically more favorable to form dispersed ^{56}Fe. Furthermore, $0.15\ M_\odot$ disagrees with all the observational limits discussed above.

The larger glitches of the Vela pulsar ($\Delta\Omega/\Omega \sim 10^{-6}$) are ascribed to corequakes by Pines; that is, sudden relaxations of stress stored in the neutron star's central core made of a hadron lattice. The question of whether or not a solid hadron core may exist in neutron stars is discussed extensively in this volume, but no definite conclusion has been reached. The work of Canuto and Chitre (reported in these proceedings) suggests that a solid core might form at densities above 1.5×10^{15} g cm^{-3}. The corequake theory therefore predicts that the Vela pulsar has a central density of at least 1.5×10^{15} g cm^{-3}, which makes Vela a rather heavy neutron star with mass greater than $0.8\ M_\odot$ (BPS), $1.2\ M_\odot$ (BBS; CCLR), and $1.5\ M_\odot$ (BJ) according to the various equations of state. The stress in the core is not built up between glitches in this model, but rather each glitch takes out a small part of a huge reservoir of elastic stress stored in the core. One has to explain why this produces glitches of ($\Delta\Omega/\Omega \sim 10^{-6}$) instead of a continuous relaxation or one extremely big jump.

In the accretion model (Börner and Cohen, 1972) the initial glitch $(\Delta\Omega/\Omega)_G$ is ascribed to the infall of material onto the neutron star, transferring angular momentum to the crust and speeding it up. After some time the initial speed-up of the crust is transferred to the interior and the pulsar settles down to a long-term frequency increase $\Delta\Omega/\Omega$. If we write

$$\left(\frac{\Delta\Omega}{\Omega}\right)_G = \frac{\Delta J_c}{J_c} - \frac{\Delta I_c}{I_c}, \qquad (8)$$

with J_c as the angular momentum of the crust and I_c the moment of inertia of the crust, then

$$\frac{\Delta\Omega}{\Omega} = \frac{\Delta J_c}{J} - \frac{\Delta I}{I}. \tag{9}$$

By choosing a definite neutron star model, everything is determined from the measured quantities $(\Delta\Omega/\Omega)_G$ and $(\Delta\Omega/\Omega)$. Even the infalling mass Δm can be found. In this simple model we find $\Delta m \sim 10^{-10}\,M_\odot$ for Crab glitches and $\Delta m \sim 10^{-6}\,M_\odot$ for Vela speed-ups. It is quite clear that initial conditions can be formulated which would exactly produce the observed behavior. A massive body flung off from the vicinity of the pulsar at some early stage in its life, not quite reaching escape velocity, but making just one loop and falling back onto a slowed down neutron star, would certainly transmit the right amount of angular momentum. It would also be able to accrete easily on the neutron star surface because it would have little excess angular momentum. The mass balance certainly is no problem, because Δm is very small compared to typical pulsar masses. The big question is, however, whether the conditions prevailing after a supernova explosion can lead (with a certain non-zero probability) to the initial conditions needed for the accretion model. There certainly is a lot of homework to do, but not withstanding these theoretical difficulties we may point out that the accretion model agrees well with the observations and it also uses the same mechanism to explain the glitches of both Crab and Vela pulsars.

Viewing $\Delta\Omega/\Omega$ and $(\Delta\Omega/\Omega)_G$ as quantities determined (with a considerable uncertainty) by the observations, one derives from (8) and (9) the condition

$$\sigma < \frac{\Delta I_c - \Delta I}{I_c} \frac{I}{I_c} \frac{\Delta\Omega}{\Omega} - \left(\frac{\Delta\Omega}{\Omega}\right)_G, \tag{10}$$

This leads to the condition that the mass of the Crab pulsar be greater than $1\,M_\odot$; for Vela (10) does not impose any restrictions on the neutron star mass.

6. Other Possibilities to Observe Neutron Stars

6.1. Pulsating X-ray sources

Recent observations from Uhuru have established the binary nature of two periodic pulsating X-ray sources: Cen X-3 (Schreier et al., 1972) with a period of 4.84 s and Her X-1 (Tananbaum et al., 1972) with a period of 1.24 s. A model may be suggested in which the X-ray source is a neutron star emitting X-rays by radiating as a black body in a number of hot spots. The rotation period of the neutron star provides the timing mechanism of the pulsation, and the X-rays are produced by the accretion of mass from a binary companion. Neutron stars accreting matter as models for X-ray sources have been proposed already by a number of authors (e.g., Shklovsky, 1967), but a reexamination of the older proposals in the light of these new observations might be worthwhile. We will not deal with the many intricate questions involved in such a model (cf., Borner et al. 1972), but just assume it to be valid and use the astronomical observation of the mass function $(M_2^3/(M_1+M_2)^2)\sin^3 i$ to get information on the mass of the neutron star. An optical identification of the main star will give its mass within rather narrow limits. The inclination $\sin i$ may be determined from an analysis of the pulse shape (Börner et al. (1972) find $\sin i \approx 1$ for Her X-1). Then the mass function will directly determine the mass M_1 of the neutron star. If, for example, for Her X-1, where $M^3/(M_1+M_2)^2 \approx 0.85$ (for $\sin i \approx 1$), M_2 is an F-type star of $\sim 2\,M_\odot$, then it follows that $M_1 \approx 1\,M_\odot$.

6.2. Gamma Ray Lines from Old Neutron Stars

Johnson et al. (1972) have reported low energy gamma ray observations from the galactic center region, showing a statistically significant spectral feature at 473 ± 30 keV. This line emission with a total photon flux of 1.8×10^{-3} cm^{-2} s^{-1} has been interpreted by Ramaty et al. (1972) as gravitationally redshifted positron annihilation radiation from the surface of old neutron stars. The production of positrons is attributed to nuclear reactions on the neutron star surface induced by the accretion of interstellar material. It is found that for an accretion rate of 10^{11} g s^{-1}, or 6×10^{34} particles s^{-1}, the redshifted positron annihilation yield of a single neutron star is 1.2×10^{33} photons s^{-1}. Thus, in order to account for the observations, a total of 1.5×10^{10} old neutron stars in the galactic center region is required. Positron annihilation radiation is normally at 511 keV, so this interpretation requires redshifts ranging from 0.016 to 0.13. Since a given redshift completely determines a specific neutron star model, as a consequence the majority of these old neutron stars have masses of less than $0.8\,M_\odot$. While the numbers are probably quite uncertain still, there appears the interesting possibility that observations of redshifted gamma ray lines provide a direct measurement of the general distribution of the physical parameters of neutron stars. The principal observational tests of the model of Ramaty et al. (1972) would be the detection of nuclear gamma ray lines from the galactic center and redshifted positron annihilation radiation from the galactic disk.

An accretion rate of 10^{17} g s^{-1} (assuming a surface composition of CNO or heavier nuclei) should produce about 10^{39} positrons s^{-1} from a single neutron star, and (if the object is at a distance of 3 kpc) a flux of redshifted positron annihilation radiation at Earth of 10^{-6} protons cm^{-2} s^{-1}. This gamma ray flux is below the presently available detector sensitivities, but in the future we might perhaps be able to see the redshifted positron annihilation radiation from binary X-ray sources or from single nearby neutron stars.

References

Baldwin, J. E.: 1971, in R. D. Davies and F. G. Smith (eds.), 'The Crab Nebula', *IAU Symp.* **46**, 22.
Baym, G. and Pines, D.: 1971, A. de Shalit Memorial Volume.
Baym, G., Bethe, H. A., and Pethick, C.: 1972, (to be published).
Baym, G., Pethick, C., and Sutherland, P.: 1971, *Astrophys. J.* **179**, 200.
Bethe, H. A. and Johnson, M.: 1973 (to be published).
Bethe, H. A., Börner, G., and Sato, K.: 1970, *Astron. Astrophys.* **7**, 279.
Biermann, L.: 1972, Boulder Lecture ('Cosmic Plasma Physics').
Börner, G. and Cohen, J. M.: 1972, *Astron. Astrophys.* **19**, 109.

Börner, G., Meyer, F., Schmidt, H. U., and Thomas, H. C.: 1972, Contribution to Fall Meeting of the DAG, Wien.
Cohen, J. M. and Cameron, A. G. W.: 1971, *Astrophys. Space Sci.* **10**, 227.
Cohen, J. M., Langer, W. D., Rosen, L. C., and Cameron, A. G. W.: 1970, *Astrophys. Space Sci.* **6**, 228.
Goldreich, P. and Julian, W. H.: 1969, *Astrophys. J.* **157**, 869.
Gunn, J. E. and Ostriker, J. P.: 1969, *Astrophys. J.* **157**, 1395.
Johnson III, W. N., Harnden, F. R., and Haymes, R. C.: 1972, *Astrophys. J. Letters* **172**, L1.
Leung, Y. C. and Wang, C. G.: 1971, *Astrophys. J.* **170**, 499.
Manchester, R. N. and Taylor, J. G.: 1972, *Astrophys. Letters* **10**, 67.
Maran, S. P. and Cameron, A. G. W.: 1969, *Earth Extraterrestr. Sci.* **1**, 3.
Milne, D. K.: 1970, *Australian J. Phys.* **23**, 425.
Minkowski, R.: 1968, *Stars and Stellar Systems* **7**, Ch. 11.
Ramaty, R., Börner, G., and Cohen, J. M.: 1972 (to be published).
Schreier, E., Levinson, R., Gursky, H., Kellogg, E., Tananbaum, H., and Giacconi, R.: 1972, *Astrophys. J. Letters* **172**, L79.
Shklovsky, I.: 1967, *Astrophys. J. Letters* **150**, L48.
Tananbaum, H., Gursky, H., Kellogg, E. M., Levinson, E., Schreier, E., and Giacconi, R.: 1972, *Astrophys. J. Letters* **174**, L143.
Trimble, V.: 1968, *Astron. J.* **73**, 535.
Trimble, V. and Rees, M. J.: 1970, *Astrophys. Letters* **5**, 93.
Trimble, V. and Woltjer, L.: 1971, *Astrophys. J. Letters* **163**, L97.
Tucker, W. H.: 1971, *Astrophys. J. Letters* **167**, L85.
Wallerstein, G. and Silk, J.: 1971, *Astrophys. J.* **170**, 289.
Woltjer, L.: 1958, *Bull. Astron. Inst. Neth.* **14**, 39.

HADRON STAR MODELS

JEFFREY M. COHEN* and GERHARD BÖRNER**

High Energy Astrophysics Laboratory, Goddard Space Flight Center, Greenbelt, Md., U.S.A.

Abstract. The properties of fully relativistic rotating hadron star models are discussed using models based on recently developed equations of state. All of these stable neutron star models are bound with binding energies as high as $\sim 25\%$. During hadron star formation, much of this energy will be released. The consequences, resulting from the release of this energy, are examined.

1. Introduction

In a number of the preceding papers of this symposium, the nuclear physics of cold dense matter was discussed and equations of state (pressure p vs. density ϱ) of the form $p = p(\varrho)$ obtained. Since the Fermi levels involved are much greater than kT for neutron stars accepted as typical pulsar models, this approximation is reasonable. Figure 1 as a flow chart of the procedure leading to observations. As can be seen from the top line (and is generally known), the use of the complete set of Einstein equations (10 non-linear simultaneous partial differential equations) have led to no neutron star models. No one has solved this complete set of equations on the computer. As the arrows indicate, all attempts to do this have led nowhere. Contemplation of Figure 1 will take the place of further discussion along this line.

A more successful approach was used by Oppenheimer and Volkoff (1939) using results of Tolman (1939). They used spherical symmetry to reduce Einstein's equations to a simple form. Generally denoted by TOV, the equations they used are obtained by eliminating ϕ from the following set

$$\partial_r m = 4\pi r^2 \varrho \tag{1a}$$

$$c^2 \partial_r \phi = G(m + 4\pi r^3 p c^{-2}) r^{-1} (r - 2GMc^{-2})^{-1} \tag{1b}$$

$$\partial_r p = -(\varrho + pc^{-2}) c^2 \partial_r \phi. \tag{1c}$$

We find the above set more useful, however, since it gives the red shift $Z = e^{-\phi} - 1$ automatically. In these equations m is the gravitational mass, ϕ is a relativistic generalization of the Newtonian gravitational potential (it is equal to the Newtonian value in the weak field limit), and r is a radial coordinate chosen in such a way that the surface area of a sphere is $4\pi r^2$ (Cohen and Cohen, 1969). A subscript r denotes differentiation with respect to r.

The equations were integrated numerically by Volkoff on a hand calculator using an equation of state for a free Fermi gas of neutrons. Neutron decay was neglected since weak interactions as well as strong interactions were neglected. These models of

---

* Permanent address: Physics Department, University of Pennsylvania, Philadelphia, Pa. 19104. National Academy of Science – NRC Senior Research Associate.
** Permanent address: Max-Planck Institut für Physik and Astrophysik, München, Germany. National Academy of Science – NRC Research Associate.

Oppenheimer and Volkoff (1939) gave rise to the name neutron star, as has been mentioned in previous papers, but the inclusion of strong and weak interactions gives rise to models composed of many varieties of hadrons. For this reason, the more precise name hadron star has been utilized. The most massive neutron star models obtained by Oppenheimer and Volkoff (1939) had a mass of $\sim 0.72\, m_\odot$. These models are non-rotating (Figure 1) and since pulsars are presently believed to be rotating neutron stars, further information is needed.

As will be seen below, typical neutron stars have a radius only a few times their gravitational radius. Thus, general relativistic effects are quite important and must be taken into account. One such general relativistic effect is the dragging along of inertial frames by rotating bodies (Brill and Cohen, 1966). Unlike Newtonian mechanics, in which a gyroscope points towards the same distant star independent of the motion of nearby masses, rotating masses in general relativity drag along the inertial frames and cause the rotation axis of a gyroscope to precess.

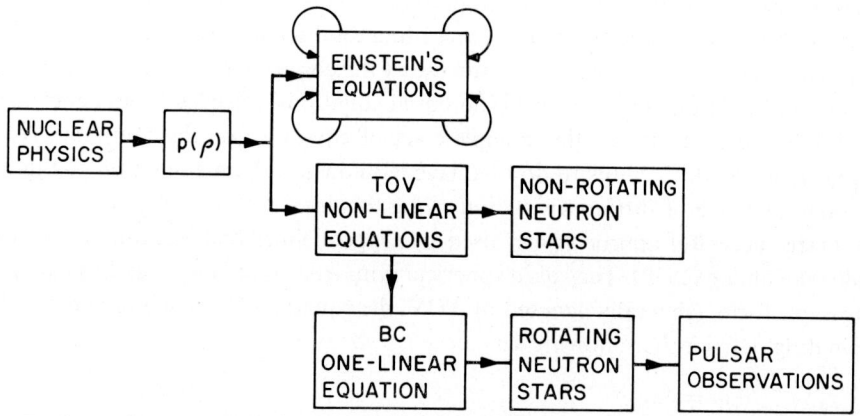

Fig. 1. Flow chart of the procedure leading from nuclear physics to astrophysical observations.

A complete general relativistic description of rotating neutron stars consists of solving the full set of Einstein equations. In order to progress further than the top line of Figure 1, another method was developed. This method (Brill and Cohen, 1966), valid for slowly rotating stars, requires the solution of only one linear equation once Equations (1) are integrated. This method is useful for treating pulsars since even the Crab pulsar can be considered a slowly rotating object in the sense that the velocity of any element of the star is small compared to the light velocity and the centrifugal force is small compared to the gravitational force.

The equation to be solved to treat rotating neutron stars is (Brill and Cohen, 1966; Cohen and Brill, 1968)

$$[A^{-1}B^{-1}r^4\Omega_r]_r = -16\pi BA^{-1}(\varrho + pc^{-2})(\omega - \Omega)\,Gc^{-2}. \qquad (2)$$

The quantity Ω is the angular velocity of inertial frames along the rotation axis where it can be measured by observing the precession rate of the axis of a gyroscope and ω is the angular velocity of the star. Except for Ω, all quantities are known from (1) or are given. To simplify the equations we have used the quantities A and B defined as

$$A = e^{\phi}, \quad B^{-2} = 1 - 2\,Gmr^{-1}c^{-2}.$$

As boundary conditions on Ω we have $\Omega \sim$ constant near the origin and $\Omega \sim r^{-3}$ outside the star.

Once Ω is determined, it is straightforward to compute quantities of astrophysical interest such as the angular momentum J and the rotational energy E_{rot}. The fully relativistic expression for the moment of inertia of a uniformly rotating body is (Cohen, 1970; Cohen, 1972; and Cohen and Cameron, 1971)

$$I = (8\pi/3) \int_{R_1}^{r} \varrho r^4 \left[(1 + p\varrho^{-1}c^{-2})\,BA^{-1}(1 - \Omega\omega^{-1}) \right] dr. \tag{3}$$

This expression differs from the corresponding Newtonian one by the quantity in brackets. Note that the pressure as well as the density contributes. Also the motion of inertial frames Ω, the red shift ($Z = e^{-\phi} - 1 = A^{-1} - 1$) and space curvature enter into this general relativistic expression. Use will be made of (2) and (3) once non-rotating models have been discussed using (1).

2. Properties of Hadron Star Models

Numerical integration of (1) gives the parameters of neutron star models (more recently called hadron star models) for various equations of state. Figure 2 shows the variation of the gravitational mass with density. As expected from their very soft equation of state, the neutron star models of Leung and Wang (1971) have very low mass. The maximum mass of their models is not only lower than any of the others shown, but it is also lower than that of Oppenheimer and Volkoff (1939) who neglected repulsion due to strong interactions. If Leung and Wang are excluded from the graph, then agreement between the remaining curves is quite reasonable. Each of these remaining curves CCLR (Cohen et al., 1970), BJ (Bethe and Johnson, 1973), BBS (Bethe et al., 1970), and BPS (Baym et al., 1971) give a maximum mass higher than that of a free Fermi gas. Such behavior is not surprising because of the repulsive core of hadron interactions.

The CCLR equation of state was obtained using the Levinger-Simons velocity dependent potentials while the later calculations BBS, BPS, BJ used the Reid soft core potential. In the high density region above $\sim 2 \times 10^{14}$ g cm^{-3}, use was made of the work of Pandharipande's pure neutron results by BBS and his hyperon results by BPS.

Figure 3 gives the density distribution of selected neutron star models. Note the kink in the curves at $\varrho_k \sim 10^{11}$ g cm^{-3}. This is the point where nuclei become unstable against the emission of neutrons; such neutron drip (Harrison et al., 1965) causes the

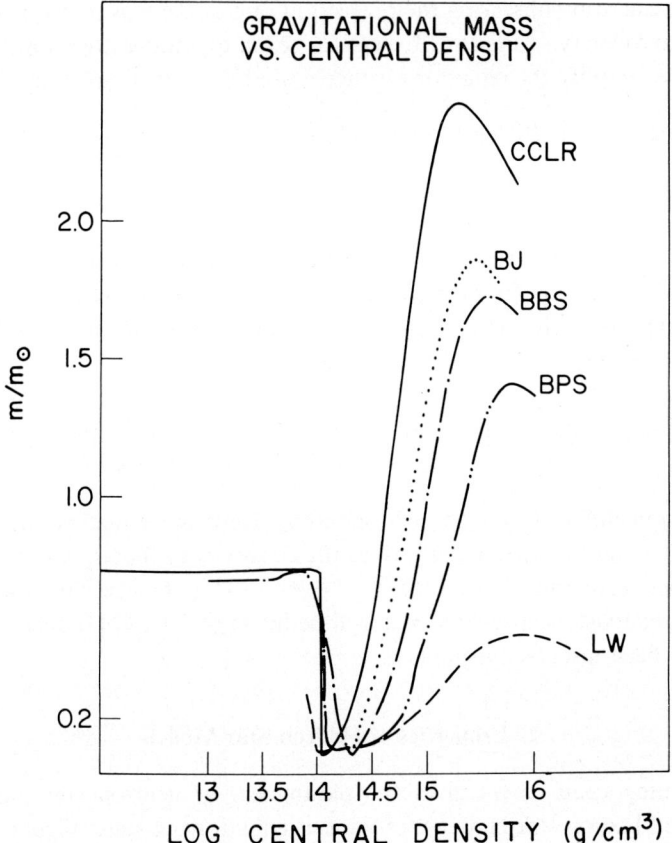

Fig. 2. Gravitational mass vs density for various equations of state.

equation of state to be quite soft in the density region from just above this density ϱ_k to the point where the neutron concentration is sufficient to give a sizeable pressure from the degenerate neutrons. From a plot of the adiabatic index Γ vs. ϱ (Cohen and Cameron, 1971), this effect manifests itself as a rapid decrease in Γ at $\sim 3 \times 10^{11}$ g cm^{-3}.

The angular velocity of inertial frames along the rotation axis as a function of radius is depicted in Figures 4, 5, and 6 for three different equations of state – CCLR, BJ, BBS. The curves are almost identical even though the underlying equations of state are based on rather different assumptions and potentials; a repulsive core is their common feature. Note that, near the center of the uniformly rotating neutron star, the inertial frames can rotate with angular velocity $\sim 70\%$ that of the star, dropping to $\sim 30\%$ near the surface.

Such large dragging of inertial frames makes a significant contribution to the moment of inertia as can be seen from (3). The moment of inertia is plotted in Figure 7 for various equations of state. An interesting property of the CCLR, BJ, and BBS curves is that the moment of inertia peaks at densities below the mass peak. Hence

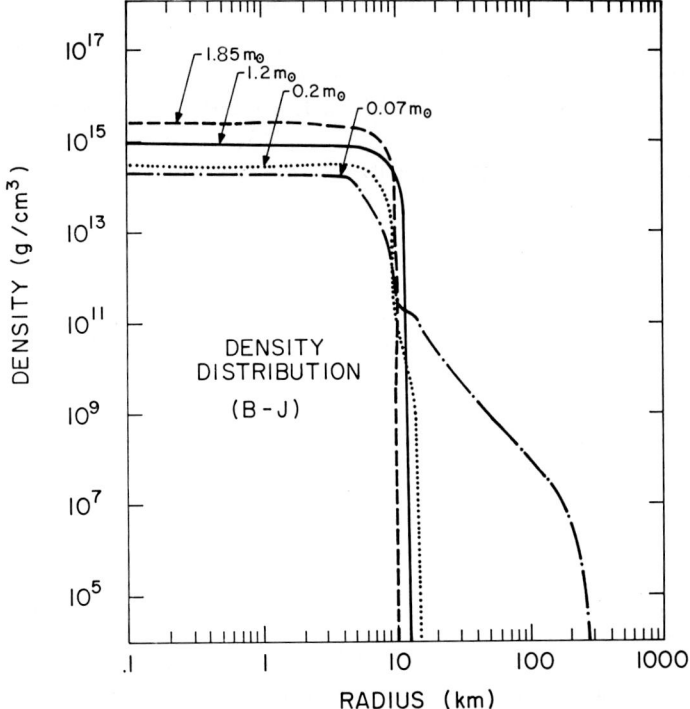

Fig. 3. Variation of density with stellar radius for a selected model based on the BBS equation of state.

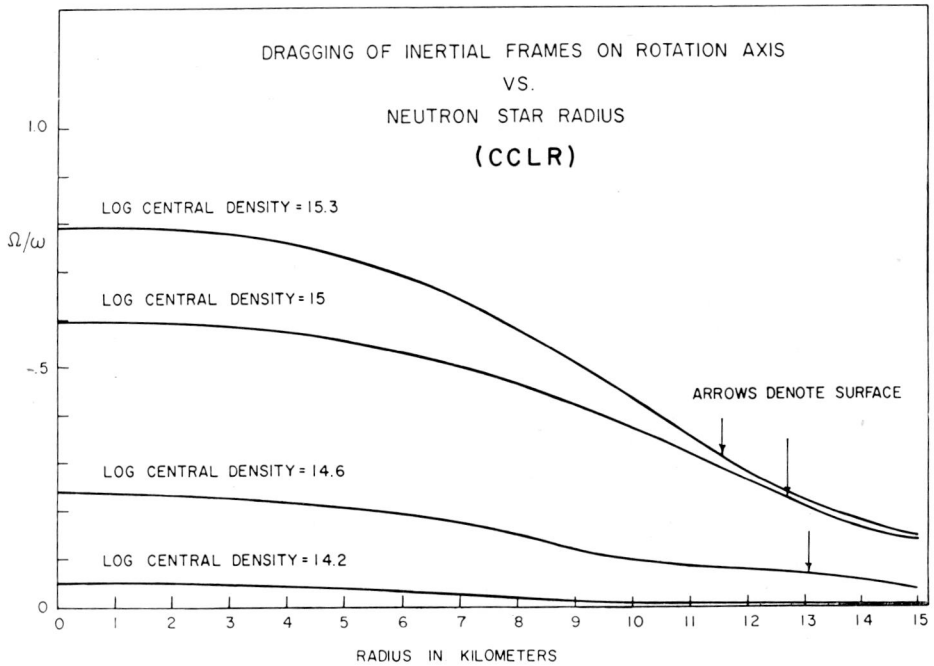

Fig. 4. Angular velocity of inertial frames as a function of radius for models based on equation of state of Cohen et al. (1970).

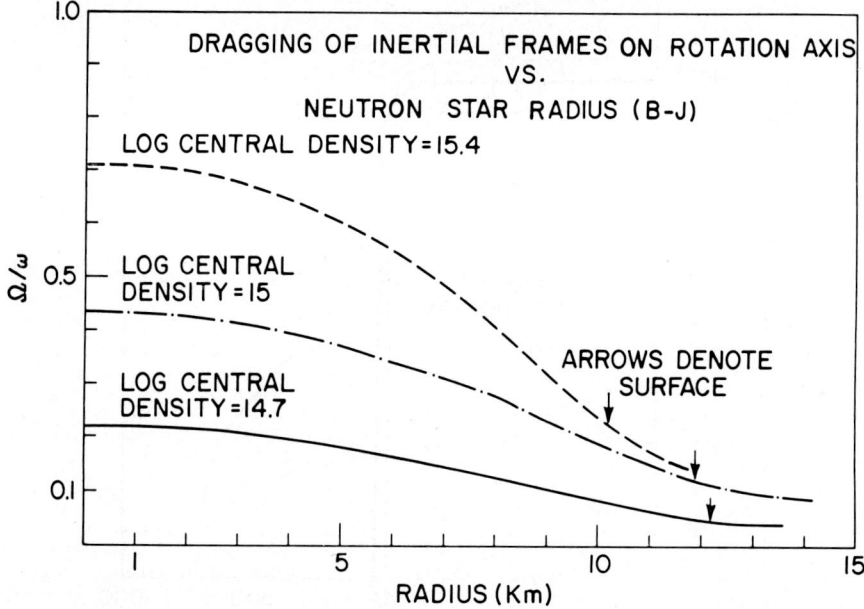

Fig. 5. Angular velocity of inertial frames as a function of radius for models based on equation of state of Bethe and Johnson.

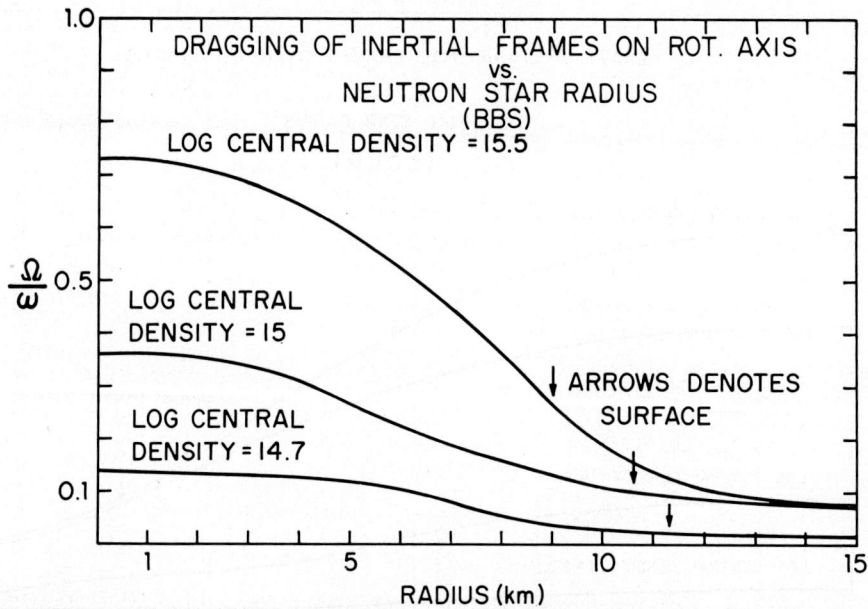

Fig. 6. Angular velocity of inertial frames as a function of radius for models based on equation of state of Bethe *et al.* (1970).

the addition of material to massive (stable) neutron stars can reduce the moment of inertia – a property which does not depend on a particular equation of state.

The horizontal lines *a* and *b* represent lower limits on the moment of inertia which can be obtained from comparison with observation. Further discussion of these lines will not be given here since a detailed discussion of all the horizontal lines appears in the next paper.

From the density distribution in Figure 3, it can be seen that the relative size of the star's outer crystalline region decreases as the star becomes more massive. Thus it is not surprising that, with increasing central density, there is a decrease in the ratio of the crust's moment of inertia to that of the entire star, as can be seen in Figure 8. We have assumed that the star's crust extends up to densities of $\sim 2 \times 10^{14}$ g cm^{-3} (Baym *et al.*, 1971).

Now that the properties of neutron star models have been determined, it is of interest to consider the question of how neutron stars are formed (Börner *et al.*, 1973).

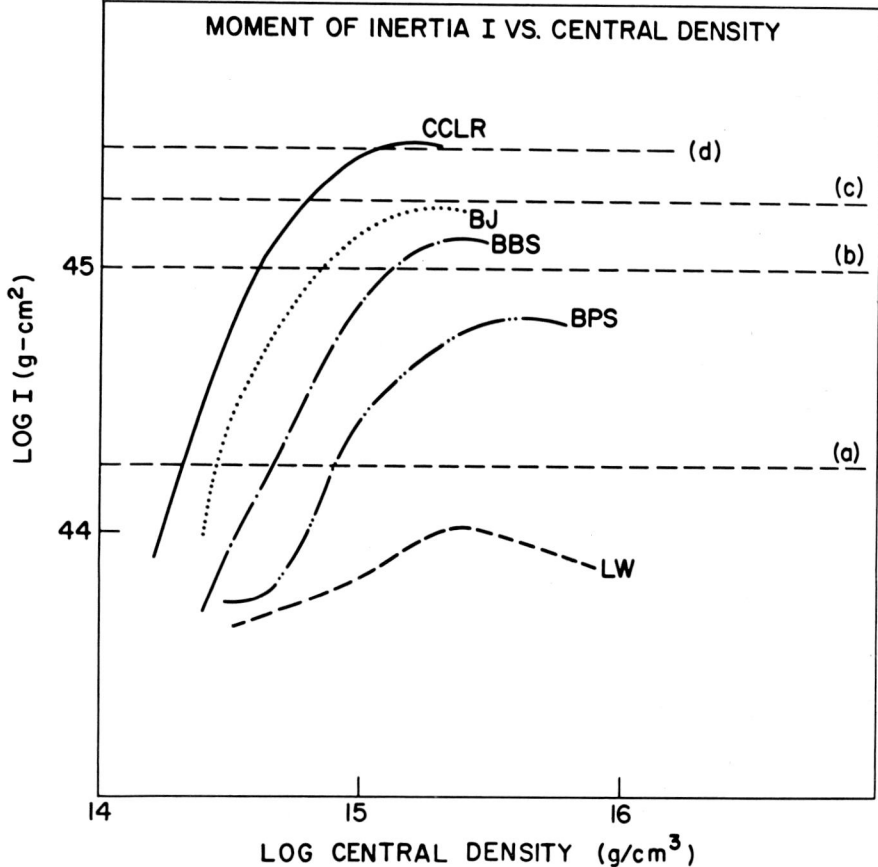

Fig. 7. Moment of inertia (curved lines) of neutron star models as a function of density for various equations of state. The value plotted includes general relativistic effects such as dragging of inertial frames, gravitational red shift, and the contribution from the pressure.

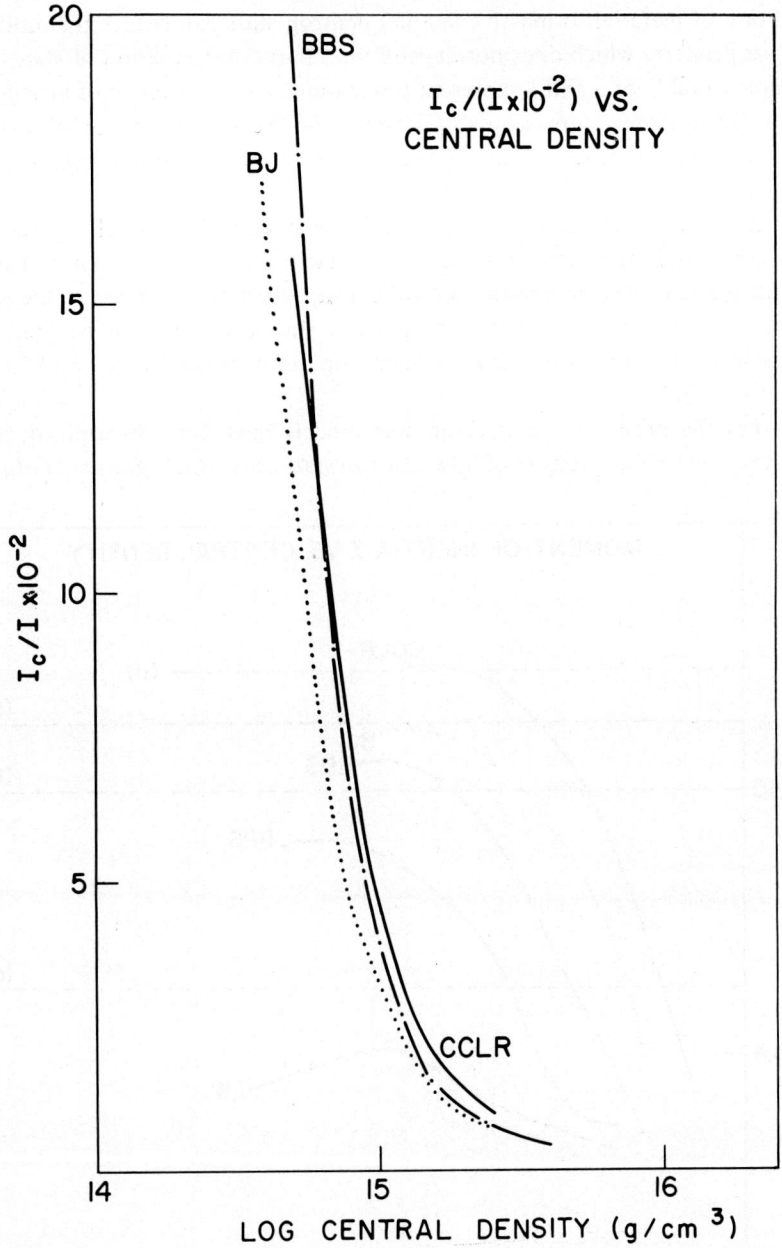

Fig. 8. Ratio of the crust moment of inertia to the moment of inertia of the entire star as a function of central density.

3. Hadron Star Formation

The endpoint of the stellar evolution of stars with mass less than $\sim 1.2\ m_\odot$ are believed to be white dwarfs (e.g., Chandrasekhar, 1939). It has been suggested (Paczynski and Ziolkowski, 1967; Lucy, 1967; Roxburgh, 1967) that stars in the mass range $\sim 1.2\ m_\odot$ and $\sim 4\ m_\odot$ develop dilute envelopes with extensive ionization regions at low gravitational potential. They predict that recombination produces an unbound envelope which is ejected leaving a white dwarf possibly surrounded by a planetary nebula (Finzi and Wolf, 1969).

In the mass range $\sim 4\ m_\odot$ to $9\ m_\odot$ evolutionary calculations have given models which develop a degenerate core with mass $\sim 1.4\ m_\odot$ (Rose, 1969; Arnett, 1969; Paczynski, 1970). Although most of the mass in these models is contained in a large dilute envelope, the core's structure is essentially independent of the model's total mass (Paczynski, 1970; Barkat, 1971). The core is degenerate and its structure is essentially that of a white dwarf. As with a white dwarf, if the core's mass exceeds the Chandrasekhar limit, gravitational collapse will ensue.

Since C^{12} has a high electron capture threshold, pure C^{12} white dwarfs undergo general relativistic instability at a central density ($\sim 3 \times 10^{10}$ g cm^{-3}) below that where electron capture sets in (Cohen *et al.*, 1969). Stars with more realistic composition become unstable at lower densities ~ 3 to 9×10^9 g cm^{-3} because of electron capture (Cohen *et al.*, 1969; Colgate, 1971; Wheeler *et al.*, 1970; Barkat *et al.*, 1970, 1971). From the above it seems reasonable to expect the formation of a collapsed remnant – neutron star or black hole (collapsar) – in the evolution of sufficiently massive stars. Since pulsars are generally believed to be rotating neutron stars (Gold, 1968) it seems likely that a neutron star remnant will be the endpoint of stellar evolution in at least some cases.

If C^{12} ignition occurs at densities lower than that where collapse sets in, Arnett (1969) has suggested that the entire core will detonate and the star will be completely disrupted. Later, Barkat *et al.* (1972) have cast serious doubt on the claim that carbon ignition in this mass range must lead to total disruption of the star. This was done by taking into account more detailed neutrino loss mechanisms. But they neglected URCA processes which, they suggest, may further reduce the possibility of disruption. Paczynski (1972) has suggested that C^{12} ignition causes the core to become convective. Such a convective core can emit a large neutrino flux *via* the URCA process. Matter from high density regions moves to regions of lower density where it β-decays emitting neutrinos in the process. Similarly, lower density matter in the core moves to higher density regions and through electron capture also emits neutrinos. The neutrino flux from such a convective region (if it exists) could remove the energy generated by carbon burning and thwart the disruptive tendencies of C^{12} ignition. Detailed convective core calculations by Barkat (1973) are in progress.

Recent measurements by Mazarakis and Stevens (1972) indicate a rapid decrease in C^{12}–C^{12} reaction cross section at low energies. Thus, a small decrease in temperature (e.g., from neutrino losses) will give a relatively large decrease in the energy generated from the C^{12}–C^{12} reaction.

Recently Gordon (Pecker-Wimel) (1972a, b) has studied the physical conditions existing in a supernova shell using observations of type I supernovae spectra. By considering emission and absorption lines from the expanding shell, she found that a continuous source of heating after the explosion is required by the observation. She suggests a neutron star remnant emitting charged particles and/or low frequency waves as a consistent explanation.

Further discussion of the situation can be found in Börner et al. (1973). A resolution of this controversy will not be attempted here. Rather we assume the formation of a neutron star remnant and examine the consequences.

4. Hadron Star Formation Energy

As mentioned above, if the degenerate core's mass exceeds the Chandrasekhar limit for white dwarfs, the core becomes unstable against gravitational collapse. If the core's collapse leads to neutron star formation and the ejection of the outer envelope, the energy released will be the difference between the binding energy of the neutron star plus that of the ejected envelope and any ejected portion of the white dwarf core minus the binding energy of the presupernova red giant. The red giant binding energy includes contributions from the envelope, nuclear binding, gravitational binding, and the kinetic energy of the degenerate electrons. Consideration of this energy balance shows that neutron stars can be formed only if there is a net positive energy release.

If a degenerate core of mass $\sim 1.4\, m_\odot$ – just above the Chandrasekhar limit – collapses to form a neutron star and little mass is lost (the idea of a small mass loss is suggested by consideration of the energy balance of the Crab and Vela nebulae by Börner and Cohen (1972)), then the resulting neutron star will have a mass $\sim 1.2\, m_\odot$. The mass difference $\sim 0.15\, m_\odot$ is released as an energy of $\sim 3 \times 10^{53}$ erg. Nowhere else in physics or astrophysics has there been proposed a mechanism which gives such a large energy release as this gravity bomb. Neutron stars with mass above $\sim 1.2\, m_\odot$ have binding energies in the range ~ 10 to 25% while thermonuclear fusion reactions release as energy only $\sim 0.7\%$ of the rest mass. Thus the efficiency of the gravity bomb is much higher than that of thermonuclear devices. Luckily, however, the problem of delivering a 2×10^{33} g device seems to be rather difficult.

A possible sequence of events leading to neutron star formation may be the following: The degenerate core collapses from the radius $\sim 10^8$ cm of a dense white dwarf near the mass peak $\sim 1.4\, m_\odot$ (Cohen et al., 1969; Barkat et al., 1971) to a neutron star of radius $\sim 10^6$ cm. The core collapses so rapidly that the neutron star is formed before much of the envelope can follow. As the core collapses, binding energy will be released as heat resulting in the cavity being filled with black-body radiation and relativistic particles in thermal equilibrium with electron-positron pairs.

As mentioned above, if little or no mass is lost from, or accreted on, the core during collapse, the resulting neutron star will have a gravitational mass of $\sim 1.2\, m_\odot$ and released binding energy $\sim 3 \times 10^{53}$ erg. In a cavity of radius 10^8 cm, this energy will give rise to a black-body temperature of $\sim 4.4 \times 10^{10}$ K. Since $kT \sim 3.8$ MeV is above the electron-positron pair production threshold, the cavity will contain relativistic positrons and electrons as well as photons. The electron-positron pair energy is 7/4 the photon energy. The above temperature includes the effect of pair production; neglect of this effect would have given a higher value.

An apparent difficulty is that the black-body curve peaks at a frequency $\sim 10^{21}$ Hz which is above the outer shell's plasma frequency of $\lesssim 10^{19}$ Hz. Because of this, one might think that the radiation will pass through the outer envelope. But this is not the case. The range of electrons and positrons is reduced by collision losses to about 1 g cm^{-2} which gives a mean free path of ~ 100 cm, much less than the envelope thickness $\sim 10^{12}$ cm. Similarly, the photon component is also stopped very efficiently at the inner edge of the envelope by Compton scattering, pair production, and free-free absorption. Some energy loss is expected from other channels (neutrinos and gravitational radiation) but this loss is expected to be small. The loss through these channels is somewhat uncertain but recent neutrino loss estimates are all of the same order ~ 1 to 4×10^{52} erg. Since this is small compared to the total binding energy released, it will have a negligible effect on the energy balance. But such a neutrino luminosity is a thousand times the optical luminosity of supernovae. It may be possible to observe these neutrinos with recently developed apparatus (Bozoki and Lande, 1973). The total energy emitted as gravitational waves cannot exceed the energy stored in pulsation ($\sim 5 \times 10^{52}$ erg) and rotation ($\sim 5 \times 10^{52}$ erg). This is an upper limit on the gravitational radiation emitted; we expect the actual value to be much less. If pulsars are rotating objects, then at least some of the rotational energy remains after the supernova explosion. Recall that there are no gravitational waves emitted during spherically symmetric collapse. Thus, we expect most of the binding energy to be deposited in black-body radiation and relativistic particles trapped within the shell.

The pressure on the outer envelope exerted by such a trapped relativistic gas will be much higher than that exerted by the white dwarf core prior to collapse. Consequently, the envelope will be pushed out quite efficiently if the gas energy is greater than the envelope's potential energy.

The outer envelope's potential energy is given by

$$PE = 4\pi G \int_{R_2}^{R_4} \varrho m(r) r dr, \qquad (4)$$

where $R_1 \geqslant 10^8$ cm is the inner radius (discussed above), and $R_2 \cong 10^{12}$ cm is the envelope's outer radius. By choosing n larger than values obtained from detailed computer models (Schwarzschild, 1957), an upper limit on the potential energy is obtainable by assuming a density profile $\varrho = \alpha r^{-n}$ with constants α and n. If $0 \leqslant n \leqslant 2.4$ (a typical value for low mass red giants is $n = 1.9$; Schwarzschild (1957)), we find that there is sufficient energy to blow off the envelope of red giants with mass in excess of 500 $m_\odot$. Here we assumed that these massive red giants develop dense cores in the late stages of their evolution as do the lower mass stars and that the core mass is less than ~ 2.5 $m_\odot$. Thus, it is energetically possible for even the most massive stars observed

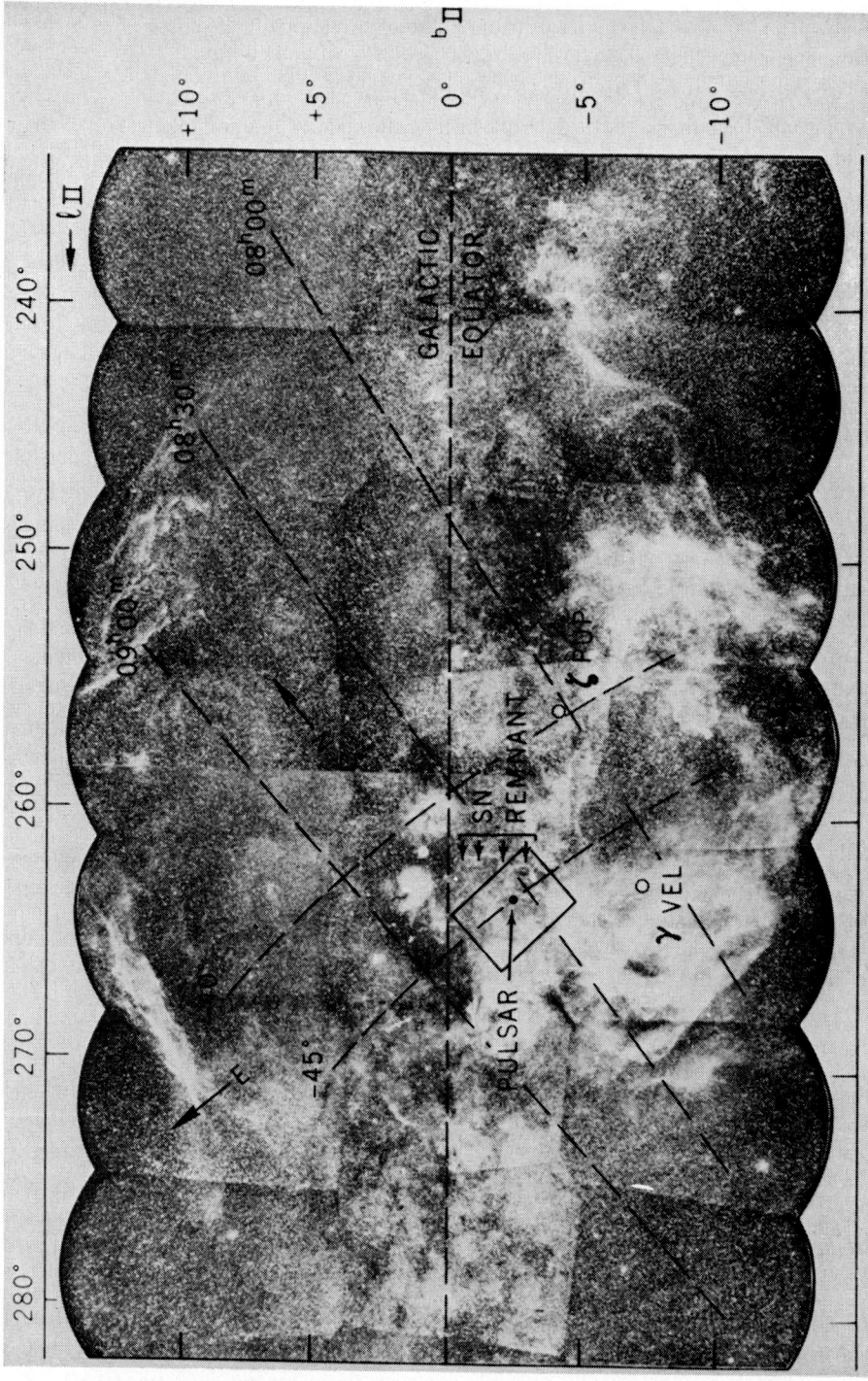

Fig. 9. The Gum Nebula. – The Vela pulsar cannot be seen in this figure since it has not been observed in the visual region of the spectrum. The arrow denotes its radio position; note the supernova shell nearby.

($m < 70\ m_\odot$, which is comparable to Chandrasekhar's limit for red giants; see also Larson and Starrfield (1971)) to leave a neutron star as a remnant of a supernova explosion.

It has been asserted (Harrison *et al.*, 1965) that once they have reached the end point of their thermonuclear evolution, stars with mass exceeding the maximum mass for stable neutron stars ($\sim 2\ m_\odot$) cannot escape from collapse into a black hole. Our results indicate that even very massive stars do not necessarily collapse into a singularity, but can leave a stable neutron star as a remnant. This suggests that black holes, if they exist, may not be formed during supernova explosion. A possible mechanism for black hole formation is discussed elsewhere (Börner *et al.*, 1973).

5. Ionization of the Gum Nebula

In Figure 9 is shown a region of the sky known as the Gum nebula. For an extensive discussion of this region see, for example, Maran *et al.* (1971). The properties of primary interest here are its size ~ 460 pc, age $\sim 10^4$ yr and the number of ionized particles it contains $\sim 2 \times 10^{62}$. Note that near the center of the Gum nebula is the Vela pulsar and supernova remnant. Here we will consider the possibility that the Vela supernova ionized the Gum nebula. To ionize interstellar hydrogen by collisions with protons with energy $\gtrsim 0.1$ MeV, an energy loss of about 36 eV per ion pair is required (Bethe and Ashkin, 1953). This implies that a total energy of $\sim 10^{52}$ erg was necessary to ionize the entire region (Ramaty and Boldt, 1971).

If the initial presupernova red giant has a mass $\sim 15\,m_\odot$ and 3×10^{53} erg was imparted to the outer envelope, then the average energy per baryon will be ~ 10 MeV. The velocity of 10 MeV protons is $\sim 4 \times 10^9$ cm s^{-1}. In 10^4 yr, these protons will cover a distance of ~ 400 pc – comparable to the observed size of the Gum nebula. Also the ionization efficiency (Ramaty and Boldt, 1971) by 10 MeV protons is such that $\sim 2 \times 10^{52}$ erg is imparted to the nebulae in this time.

References

Arnett, W. D.: 1969, *Astrophys. Space Sci.* **5**, 180.
Barkat, Z.: 1971, *Astrophys. J.* **163**, 433.
Barkat, Z.: 1973 (to be published).
Barkat, Z., Buchler, J. R., and Wheeler, J. C.: 1970, *Astrophys. Letters* **6**, 117.
Barkat, Z., Buchler, J. R., and Wheeler, J. C.: 1971, *Astrophys. J.* **8**, 21.
Barkat, Z., Wheeler, J. C., and Buchler, J. C.: 1972, *Astrophys. J.* **171**, 651.
Baym, G., Pethick, C., and Sutherland, P.: 1971, *Astrophys. J.* **170**, 299.
Bethe, H. A. and Ashkin, J.: 1953, in E. Serge (ed.), *Experimental Nuclear Physics* **1**, 233. John Wiley and Sons Inc., N.Y.
Bethe, H. A. and Johnson, M.: 1973 (to be published).
Bethe, H. A., Börner, G., and Sato, K.: 1970, *Astron. Astrophys.* **7**, 279.
Börner, G. and Cohen, J. M.: 1972, *Astron. Astrophys.* **19**, 109.
Börner, G., Cohen, J. M., and Ramaty, R.: 1973 (to be published).
Bozoki, G. and Lande, K.: 1973 (to be published).
Brill, D. R. and Cohen, J. M.: 1966, *Phys. Rev.* **143**, 1011.
Chandrasekhar, S.: 1939, *Introduction to the Study of Stellar Structure*, Dover Press, N. Y.
Cohen, J. M.: 1970, *Astrophys. Space Sci.* **6**, 263.

Cohen, J. M.: 1972, *General Relativity and Gravitation Res.* **3**, 221.
Cohen, J. M. and Brill, D. R.: 1968, *Nuovo Cimento* **50B**, 209.
Cohen, J. M. and Cameron, A. G. W.: 1971, *Astrophys. Space Sci.* **10**, 227.
Cohen, J. M. and Cohen, M. D.: 1969, *Nuovo Cimento* **60B**, 241.
Cohen, J. M., Langer, W. D., Rosen, L. C., and Cameron, A. G. W.: 1970, *Astrophys. Space Sci.* **6**, 228.
Cohen, J. M., Lapidus, A., and Cameron, A. G. W.: 1969, *Astrophys. Space Sci.* **5**, 113.
Colgate, S. A.: 1971, *Astrophys. J.* **163**, 221.
Finzi, A. and Wolf, R. A.: 1969, *Astrophys. J. Letters* **155**, L107.
Gold, T.: 1968, *Nature* **218**, 731.
Gordon (Pecker-Wimel), C.: 1972a, *Astron. Astrophys.* **20**, 79.
Gordon (Pecker-Wimel), C.: 1972b, *Astron. Astrophys.* **20**, 87.
Harrison, B., Thorne, K. S., Wakano, M., and Wheeler, J. A.: 1965, *Gravitation Theory and Gravitational Collapse*, University of Chicago Press, Ill.
Larson, R. B. and Starrfield, S.: 1971, *Astron. Astrophys.* **13**, 190.
Leung, V. C. and Wang, C. G.: 1971, *Astrophys. J.* **170**, 499.
Lucy, L.: 1967, *Astron. J.* **72**, 813.
Maran, S. P., Brandt, J. C., and Stecher, T. P.: 1971, *The Gum Nebula and Related Problems*, Goddard Space Flight Center Report X-683-71-375.
Mazarakis, M. and Stevens, W.: 1972, *Astrophys. J. Letters* **171**, L97.
Oppenheimer, J. R. and Volkoff, G.: 1939, *Phys. Rev.* **55**, 374.
Paczynski, B.: 1970, *Acta Astron.* **20**, 47.
Paczynski, B.: 1972, *Astrophys. Letters* **11**, 53.
Paczynski, B. and Ziolkowski, J.: 1967, *Acta Astron.* **17**, 7.
Ramaty, R. and Boldt, E. A.: 1971, in S. P. Maran, J. C. Brandt, and T. P. Stecher (eds.), *The Gum Nebula and Related Problems*, Goddard Space Flight Center Report X-683-71-375, p. 97.
Rose, W. K.: 1969, *Astrophys. J.* **155**, 491.
Roxburgh, I. W.: 1967, *Nature* **215**, 838.
Schwarzschild, M.: 1957, *Structure and Evolution of the Stars*, Dover Press, N.Y.
Tolman, R. C.: 1939, *Phys. Rev.* **56**, 364.
Wheeler, J. C., Barkat, Z., and Buchler, J. R.: 1970, *Astrophys. J. Letters* **162**, L129.

DISCUSSION

Van Horn: Is there any observational evidence that can be used to provide a direct estimate of the mass of the Gum nebula or the pre-supernova star?

Cohen: There are about 2×10^{62} ionized particles in the Gum nebula and its size and age are about 460 pc and 10^4 yr respectively. This distance can be traversed by 10 MeV protons. If the energy released during neutron star formation is 3×10^{53} erg then the pre-supernova red-giant mass is 15 $M_\odot$.

Wang: The minimum mass of a neutron star is accurately determined by the equation of state at subnuclear densities. The argument that for those neutron stars with mass less than 0.2 $M_\odot$, their binding energy per nucleon would be less than the parent core of the degenerate matter, and therefore can not be formed, is a dangerous one. The dynamics of stellar collapse is far from being settled. I do not know if anyone has looked into the possibility of the formation of multiple collapsed objects, due to fission, for example, and their energetics may be shared.

Regarding the energetics of the Crab Nebula: we do not know the distance to the nebula to about a factor of two. The Crab Nebula can be roughly outlined as an ellipse. The maximum Doppler velocity at the line of sight to the Nebula is 1450 km s^{-1}. Baade associated it with either the proper motion of the major or the minor axis with photographic plates taken ten years apart, he obtained 1.03 and 1.74 kpc respectively. There are many other less direct arguments to estimate the distance. I think we should bear in mind the history of astronomer's estimate of distances.

With a very stiff equation of state at supernuclear densities, that is, connect between the limit of the usual nuclear physics calculation and the causality limit with a vertical line, you get the maximum mass of a neutron star at about 1.5 $M_\odot$, the only way you can support a heavier neutron star is to make a stiff equation of state at the nuclear densities, and that region, unfortunately, is where we think we know some physics about nuclear matter.

DIFFERENTIAL ROTATION IN DEGENERATE STARS

H. M. VAN HORN

Dept. of Physics and Astronomy and C. E. Kenneth Mees Observatory,
University of Rochester, Rochester, N.Y., U.S.A.

Abstract. The problem of the development of the angular velocity distribution during the final phase of gravitational contraction of a star immediately prior to the onset of degeneracy and during the subsequent cooling phase is surveyed. Processes that may affect this distribution are discussed at some length, and estimates of the timescales for redistribution of the angular momentum are given for each process. Possible effects on the evolution and observable consequences are briefly considered.

1. Introduction

Ostriker *et al.* (1966) have shown that fully degenerate stars in differential rotation can have masses considerably greater than the Chandrasekhar limit, and Ostriker and Bodenheimer (1968) have constructed models of such stars. However, they have not discussed the problem of how such an object might be formed, and it is the aim of this paper to consider this question. The importance of an investigation of this problem has been strongly emphasized in recent years: The discovery of pulsars (Hewish *et al.*, 1968) and their subsequent interpretation as rotating neutron stars (Gold, 1968) has provided a dramatic example of a class of degenerate objects which are in rapid rotation. Still more recent has been the discovery of pulsating X-ray sources with periods in the range of 1 to 5 s (Giacconi *et al.*, 1971; Tananbaum *et al.*, 1972). The regularity of the X-ray pulsations suggests that rotation provides the basic 'clock' mechanism for these sources, in analogy with the pulsars, and the short period can again be reconciled only with a small, degenerate star.

A quantitative understanding of such objects requires the study of changes in the velocity field during the evolution of a star. For simplicity I shall consider only the final stages of evolution of an initially uniformly rotating star subsequent to the exhaustion of nuclear energy sources. The evolution then consists of two rather distinct stages: the final gravitational contraction phase immediately prior to the occurrence of complete degeneracy, and the subsequent cooling phase in which gravitational energy production is almost negligible. Models of non-rotating stars in similar phases of their evolution have been constructed, e.g., by Savedoff *et al.* (1969), and I shall make use of the qualitative features of these results for guidance in cases where rotation is important.

Most of the decrease of the stellar radius occurs in the first, gravitational contraction, stage, and it is here that the differential rotation is developed. This phase of the evolution is discussed in Section 2, where some general properties of rotating objects are reviewed, and in Section 3, where the effectiveness of meridional circulations in redistributing the angular momentum is discussed. The subsequent cooling phase is considered in Section 4. Here I discuss the effect of Ekman pumping in a white dwarf with a crystallizing, solid core, and also the question of whether a similar mechanism is operative when the core instead consists of uniformly rotating fluid. In Section 5 a schematic picture of the evolution of differentially rotating stars is sketched, and several important problem areas that require further study are pointed out.

My principal aim in this paper is to survey the general problem of the evolution of differentially rotating, degenerate stars. I shall be primarily concerned with attempting to present a qualitative physical understanding of the evolutionary process. Most of what I shall say therefore will not be new, but it has not previously been applied to the particular problem of the final phases of stellar evolution. I shall not discuss the very considerable literature dealing with uniform density objects (which is reviewed by Lebovitz (1967) and treated in detail by Chandrasekhar (1969); see also Fujimoto (1968)) or on stars in pure rotation (reviewed by Strittmatter (1969)). I shall also ignore cases where magnetic fields or general relativity are important; the fluid dynamics problems are sufficiently complex even in Newtonian mechanics, and the inclusion of these effects would complicate the situation even further.

2. General Considerations: The Necessity of Differential Rotation

The evolution of a contracting, differentially rotating star is governed by the equation of motion of a self-gravitating, viscous fluid:

$$\frac{d\mathbf{v}}{dt} \equiv \frac{\partial \mathbf{v}}{\partial t} + (\mathbf{v}\cdot\nabla)\mathbf{v} = -\frac{1}{\varrho}\nabla p - \nabla\Phi + \frac{\eta}{\varrho}\nabla^2\mathbf{v}. \tag{1}$$

In hydrostatic equilibrium the velocity $\mathbf{v}$ is zero, and (1) reduces to the familiar equation of pressure balance, $\nabla p = -\varrho\nabla\Phi$, where Φ is the gravitational potential.

From the form of (1), it follows that viscous forces act on a timescale

$$\tau_{\text{visc}} \sim \varrho L^2/\eta, \tag{2}$$

where L is the characteristic length scale of the velocity field. Even for viscosities η as large as those in white dwarfs* this timescale is $\sim 10^{10}$ to 10^{14} yr if $L \sim R$, where R is a typical dimension of the star. Thus for large scale motions the fluid can be considered to be essentially inviscid. If the evolution leads to the development of small-scale structure in the velocity field, however, the viscous forces cannot be neglected. Equation (2) then provides a lower limit to the scale of length over which appreciable differential motions can be maintained: $\Delta L_{\min} \sim (\eta \tau_{\text{ev}}/\varrho)^{1/2}$, where the timescale of the evolution is

$$\tau_{\text{ev}} \equiv (d \ln R/dt)^{-1}. \tag{3}$$

Smaller scale motions will be dissipated by viscosity in a time $\ll \tau_{\text{ev}}$. For $\tau_{\text{ev}} \sim 10^4$ yr,

---

* The viscosity $\eta \sim pn\lambda$, where p is the characteristic momentum transferred by particles of number density n over a mean free path λ. Thus for a non-relativistically degenerate electron gas, in which the Fermi energy is $\varepsilon_F = p_F^2/2m$,

$$\eta \sim p_F n_e/n_i(Ze^2/\varepsilon_F)^2 \sim \frac{\hbar^5}{Zm^2e^4}\left(\frac{Z\varrho}{AH}\right)^{5/3} \sim 10^6 Z^{-1}\varrho_6^{5/3} \text{ c.g.s.}$$

which is typical of some of the more rapid phases we shall be considering, $\Delta L_{\min} \sim 10^{-3}$ to 10^{-5} R.

A second conclusion that also follows trivially from (1) is that angular momentum is conserved during any axially symmetric motion of a ring-shaped mass element in a cylindrically symmetric, inviscid fluid. Since the angular momentum per unit mass is $\mathbf{j}=\mathbf{r}\times\mathbf{v}$, the time derivative of $\mathbf{j}$ is given by (1) as

$$\frac{d\mathbf{j}}{dt} = \mathbf{1}_\phi \left[r\left(\frac{1}{\varrho}\frac{\partial p}{\partial z} + \frac{\partial \Phi}{\partial z}\right) - z\left(\frac{1}{\varrho}\frac{\partial p}{\partial r} + \frac{\partial \Phi}{\partial r}\right) \right], \qquad (4)$$

where (r, ϕ, z) are the usual cylindrical coordinates, $\mathbf{1}_\phi$ is a unit vector in the ϕ-direction, and we have made use of the assumption of axial symmetry. Equation (4) refers to a differential mass element of volume $dr \cdot rd\phi \cdot dz$; for a ring, $d\mathbf{j}/dt$ is zero by symmetry. Since the angular momentum per unit mass of a rotating ring is $r^2\Omega\mathbf{1}_z$, where $\Omega = \Omega(r, z)$ is the angular velocity of the ring, we conclude that $r^2\Omega$ is constant in any motion of this kind.

An important consequence of this is that uniform rotation can be maintained only in a homologous contraction. Consider a case in which a mass ring with initial coordinates (r, z) undergoes an axially symmetric contraction to (r', z'). From angular momentum conservation the angular velocity Ω' at the new position is given by

$$\Omega'(r', z') = \Omega(r, z) \cdot (r/r')^2. \qquad (5)$$

If the original angular velocity distribution is uniform, the new distribution will be

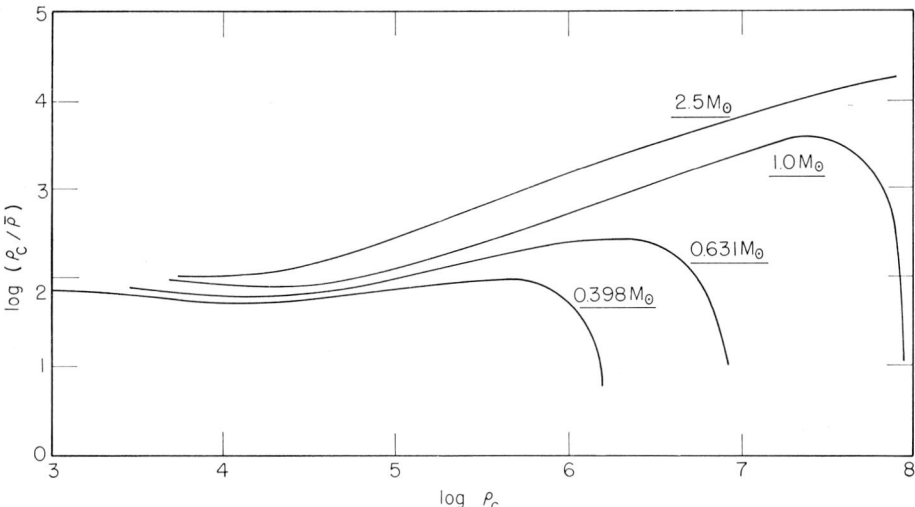

Fig. 1. The non-homologous nature of the final gravitational contraction phase, caused by copious neutrino emission from the stellar core, is indicated by the large variation of $\log(\varrho_c/\bar{\varrho})$ during the evolution. The curves shown are for the non-rotating iron star models computed by Savedoff et al. (1969); in these stars ϱ_c increases monotonically during the evolution.

uniform also only if $r' = \alpha r$, where α is a constant; i.e. the contraction must be homologous.

The final gravitational contraction phase is distinctly non-homologous, however. In a homologous contraction the ratio $\varrho_c/\bar{\varrho}$ of central to mean density is constant; but in the non-rotating iron star models of Savedoff *et al.* (1969), this ratio can vary by many orders of magnitude, as shown in Figure 1. The more rapid contraction of the central regions is caused by the copious neutrino emission from these models. Since the neutrino losses will be qualitatively the same for a rotating star, however, we may thus conclude that this phase of the evolution will in general be non-homologous and consequently that a rotating star will develop differential rotation during this phase.

There is another consequence of non-homologous contraction that considerably complicates this simple picture, however: a non-homologous contraction generates meridional circulations. Consider a cylindrical mass-shell in a star, which we assume to be initially in uniform rotation. As the star evolves (non-homologously), the central regions contract more rapidly than the rest of the star, and the cylindrical shape of the initial mass-shell, on which Ω was constant, is not preserved. The angular velocity distribution thus develops a gradient in the z-direction as well as in the r-direction.

Such a distribution is not stable, however. For an inviscid fluid which we assume to be stationary (i.e. $\partial/\partial t \to 0$), the curl of (1) gives

$$\text{curl}\,[(\mathbf{v}\cdot\nabla)\,\mathbf{v}] = -\mathbf{1}_\phi r \frac{\partial \Omega^2}{\partial z} = \nabla p \times \nabla \varrho/\varrho^2, \tag{6}$$

where for a state of pure rotation $\mathbf{v} = r\Omega \mathbf{1}_\phi$. Surfaces of constant pressure are thus not parallel to surfaces of constant ϱ. Since the net force on a stationary mass element is parallel to ∇p, as shown by (1), displacement of a mass element parallel to a surface $p = \text{constant}$ does no work. Because ∇p is not parallel to $\nabla \varrho$, however, the displaced element is of different density than its surroundings and thus experiences a net buoyancy force that accelerates it parallel to ∇p. The sense of the force is such as to restore parallelism of surfaces of constant p and constant ϱ; i.e. to drive $\partial \Omega/\partial z$ to zero. Stable configurations are thus implicit barytropes (objects in which $p = p(\varrho)$).

From (1) and (6) we estimate that z-gradients of Ω should be eliminated on a timescale of order

$$\tau_z \sim \Omega^{-1} \cdot \left(\frac{L}{\Omega}\frac{\partial\Omega}{\partial z}\right)^{-1/2}, \tag{7}$$

where $L \sim R$ is the characteristic scale of the meridional circulation currents. Thus if the evolution were so rapid as to produce $\partial\Omega/\partial z \sim \Omega/R$, we would expect very strong circulation currents that would reorganize the angular velocity distribution onto cylinders in a time $\sim \Omega^{-1}$. (However, see Section 3.) In reality, of course, this means that evolution can only lead to the development of very small gradients, of order $\partial\Omega/\partial z \sim (\Omega/R)/(\Omega\tau_{\text{ev}})^2$, as Durisen (1972) has also pointed out. Even for an evolutionary timescale as short as $\sim 10^4$ yr and a rotation period as long as $\sim$ days, this only gives $\partial\Omega/\partial z \sim 10^{-13}\Omega/R$. The final pre-white-dwarf contraction phase thus may indeed produce differential rotation on cylinders, as envisaged by Ostriker, Bodenheimer, and Lynden-Bell.

3. Meridional Circulation

We have seen that the final contraction phase of stellar evolution can be expected to generate gradients of the angular velocity distribution. However, if we are to conclude that the star maintains $\nabla\Omega \neq 0$, we must show that angular momentum transport cannot restore uniform rotation within the timescale τ_{ev} of the evolution. We must thus investigate the timescale of the meridional circulations. This problem has been discussed in considerable detail by Mestel (1965, 1970) and others, and I shall merely summarize Mestel's treatment, but in somewhat more general terms.

Perhaps the best-known and most thoroughly studied example of meridional circulation is that driven by the failure of strict radiative equilibrium in a rotating star. The timescale for this circulation, which was originally studied by Eddington (1929) and by Sweet (1950) and which now bears their names, can be crudely estimated by the following rather general argument:

The equation of energy conservation in a stellar interior is given by

$$T \frac{ds}{dt} = \varepsilon_n - \varepsilon_v - \frac{1}{\varrho} \operatorname{div} \mathbf{H}, \tag{8}$$

where $T ds/dt$ is the rate of increase of the heat content per unit mass of the stellar matter, and ε_n, ε_v, and div $\mathbf{H}/\varrho$ are, respectively, the nuclear energy production rate, the neutrino loss rate, and the net thermal energy flux from the mass element. In most calculations, this equation is applied only to the radiative zone of a stationary stellar model, in which case it reduces to the simpler form div $\mathbf{H} = 0$.

The thermal energy flux is given by

$$\mathbf{H} = -\frac{4ac}{3} \frac{T^3}{\kappa \varrho} \nabla T \tag{9}$$

and may include electron conduction processes as well as the ordinary radiative flux.

The necessity of meridional circulations derives from Eddington's resolution of von Zeipel's (1924) celebrated paradox: that div $\mathbf{H} \neq 0$ in a rotating star. This follows, after a rather lengthy argument, from the equation of hydro-stationary equilibrium,

$$-r\Omega^2 \mathbf{1}_r = -\frac{1}{\varrho} \nabla p - \nabla \Phi. \tag{10}$$

As pointed out above, the equilibrium is unstable unless Ω is independent of z. In this case the curl of the centrifugal force term vanishes, however, and it is then derivable from a scalar potential. We may thus rewrite (10) in terms of a total (gravitational plus centrifugal) potential, ψ:

$$\frac{1}{\varrho} \nabla p = -\nabla \psi. \tag{11}$$

Since (11) shows that surfaces of constant p are everywhere parallel to surfaces of constant ψ, we have $p=p(\psi)$ and consequently, by taking the gradient, $\varrho = -(\mathrm{d}p/\mathrm{d}\psi)^{-1} = \varrho(\psi)$.

Since p and ϱ are functions only of ψ, T is also, and we can therefore write

$$\mathrm{div}\,\mathbf{H} = \mathrm{div}\left[-\frac{4ac}{3}\frac{T^3}{\kappa\varrho}\frac{\mathrm{d}T}{\mathrm{d}\psi}\nabla\psi\right] =$$
$$= \frac{\mathrm{d}}{\mathrm{d}\psi}\left[-\frac{4ac}{3}\frac{T^3}{\kappa\varrho}\frac{\mathrm{d}T}{\mathrm{d}\psi}\right]|\nabla\psi|^2 - \frac{4ac}{3}\frac{T^3}{\kappa\varrho}\frac{\mathrm{d}T}{\mathrm{d}\psi}\nabla^2\psi. \tag{12}$$

This expression does not in general vanish, however. This is easy to see in the case of uniform rotation, when $\nabla^2\psi = 4\pi G\varrho - 2\Omega^2$, and the second term in (12) becomes a function only of ψ. In this case, since $|\nabla\psi|$ is not constant on an equipotential surface, it is evident that the first term leads to a variation of div $\mathbf{H}$ along an equipotential that is *not* cancelled by a corresponding variation of the second term. Evidently this must also be the case when $\Omega(r)$ is arbitrary; although the second term now varies also along an equipotential surface, cancellation of the variation of the first term would clearly occur only for very special forms of Ω.

Now in (8), ε_n and ε_v are functions only of ϱ and T and thus only of ψ. Because of the variation of div $\mathbf{H}$ along an equipotential surface, however, the net rate of change of the heat content of individual mass elements also varies along an equipotential, and it is this differential heating and cooling relative to the average that drives the Eddington-Sweet circulations. If we let $\langle\cdots\rangle$ denote the average over such a surface of constant ψ, we evidently must have

$$\left\langle T\frac{\mathrm{d}s}{\mathrm{d}t}\right\rangle = \varepsilon_n - \varepsilon_v - \frac{1}{\varrho}\langle\mathrm{div}\,\mathbf{H}\rangle.$$

Hence, subtracting from (8), we obtain

$$T\frac{\mathrm{d}s}{\mathrm{d}t} - \left\langle T\frac{\mathrm{d}s}{\mathrm{d}t}\right\rangle = -\frac{1}{\varrho}[\mathrm{div}\,\mathbf{H} - \langle\mathrm{div}\,\mathbf{H}\rangle]. \tag{13}$$

To lowest order, the difference between $T\,\mathrm{d}s/\mathrm{d}t$ and the term $\langle T\,\mathrm{d}s/\mathrm{d}t\rangle$ (which is, e.g., the negative of the average gravitational energy generation rate on an equipotential, in a contracting, non-degenerate star) is simply the convective term

$$T\mathbf{v}\cdot\nabla s = \mathbf{v}\cdot\mathbf{1}_\psi T\frac{\mathrm{d}s}{\mathrm{d}\psi}|\nabla\psi|, \tag{14}$$

where $\mathbf{1}_\psi$ is a unit vector in the direction of $\nabla\psi$, and T, s are functions of ψ only. Equations (12), (13), and (14), together with the general result $\nabla^2\psi = 4\pi G\varrho - (1/r^2)(\mathrm{d}/\mathrm{d}r)(r^2\Omega^2)$, thus give for the component of the meridional circulation velocity in the direction of $\nabla\psi$,

$$\mathbf{v} \cdot \mathbf{1}_\psi = -\left\{ \frac{d}{d\psi}\left(\frac{H}{|\nabla\psi|}\right)[|\nabla\psi|^2 - \langle|\nabla\psi|^2\rangle] \right.$$
$$\left. - \frac{H}{|\nabla\psi|}\left[\frac{1}{r}\frac{d}{dr}(r^2\Omega^2) - \left\langle\frac{1}{r}\frac{d}{dr}(r^2\Omega^2)\right\rangle\right]\right\} \bigg/ \varrho T \frac{ds}{d\psi}|\nabla\psi|. \tag{15}$$

Note that $\mathbf{H}$ is parallel to $\nabla\psi$ and, from (9), that $\mathbf{H}/|\nabla\psi|$ is a function of ψ only.

When Ω is independent of position within the star, the second term in (15) vanishes, and the first term gives the usual result for the Eddington-Sweet circulation (Mestel, 1965). Our schematic derivation indicates that this result is more general than might have been supposed, however; it appears still to be correct even when nuclear burning and gravitational contraction are included, when electron conduction is considered, and – perhaps most surprising – when neutrino energy losses are taken into account. The latter is an important point since, in the evolutionary phases we are considering, the timescale of the evolution is determined by the neutrino loss rate. Equation (15) indicates that the timescale of the meridional circulations is still determined by the variations of thermal flux, however. This is shorter for neutrino-dominated evolution than for the case in which neutrino emission is neglected, but it is still much longer than the timescale $\tau_\nu \sim E_{th}/L_\nu$ that characterizes this phase of the evolution.

Equation (15) can be used to obtain a rough approximation for the circulation timescale. If R is again a characteristic dimension of the star, $L \sim HR^2$ is the optical luminosity, $E_{th} \sim R^3\varrho T\delta s$ is the thermal energy content of the star, and if $|\nabla\psi| \sim$ $\sim GM/R^2 - R\Omega_0^2$, where Ω_0 is a characteristic rotation frequency and the centrifugal potential energy is still small compared to the gravitational term (it is at most ~ 0.14 for Ostriker and Bodenheimer's secularly stable models), we may roughly estimate

$$\tau \sim \frac{R}{v} \sim \frac{E_{th}}{L} \bigg/ \left\{ \frac{\Omega_0^2}{GM/R^3}\left[1 + \alpha\frac{\delta\Omega^2}{\Omega_0^2}\right]\right\}, \tag{16}$$

where α is a number of order unity. Apart from the (presumably small) correction term arising from the non-uniformity of the rotation this is just the usual timescale for the Eddington-Sweet circulations,

$$\tau_{ES} \sim \frac{E_{th}}{L}\frac{GM/R^3}{\Omega_0^2}. \tag{17}$$

Because Ω_0^2 is always less than GM/R^3 for an object in pressure balance, this is at best comparable to the Kelvin-Helmholtz timescale. Since this is long compared to the timescale τ_ν of the contraction, we may thus conclude that such thermally driven circulations are ineffective in eliminating the differential rotation during this evolutionary phase.

There are other processes which can also drive meridian circulations, however. For example, the instability* discussed by Goldreich and Schubert (1967) will generate

---

* The existence of this instability is due to the very large heat diffusivity in a stellar interior. This permits the growth of certain perturbations that are stable under an adiabatic displacement, by causing rapid relaxation of the isentropic constraint.

such motions if the conditions $\partial(r^2\Omega)/\partial r \geqslant 0$ and $\partial\Omega/\partial z = 0$ are violated. Since $r^2\Omega$ = constant in the contraction of an inviscid fluid, the first of these conditions is probably satisfied in the present context, although only marginally so. In cases where this criterion is violated, Goldreich and Schubert have estimated that angular momentum redistribution should occur on a timescale $\sim \Omega^{-1}(\partial \ln \Omega/\partial \ln r)^{-1/2}$. However, Colgate (1968) has pointed out that the growth of this instability is limited by non-linear processes, and he has argued that the relevant timescale for the circulation induced by this thermally driven instability is the Kelvin-Helmholtz timescale. Subsequent calculations by James and Kahn (1970, 1971) have confirmed the non-linear limiting of this instability and have shown that the relevant timescale is actually that of the Eddington-Sweet circulations, τ_{ES}. The Goldreich-Schubert instability thus appears also to be ineffective in producing a significant rearrangement of the angular momentum distribution during the timescale of the initial contraction.

Another possible driving mechanism for meridional circulation is the instability that results from the z-variation of Ω produced by the non-homologous contraction. This is not obviously identical with the Goldreich-Schubert instability in spite of the fact that the stability condition $\partial\Omega/\partial z = 0$ is the same and the timescale estimated by (7) is the same as their original estimate. In particular, it is not clear that the restriction to cases where the thermal diffusivity is large compared to the viscosity is required by the argument leading to (7), while it is essential to Goldreich and Schubert's analysis. Nevertheless, the growth of the $\partial\Omega/\partial z$ instability is probably also limited by non-linear effects. However, the timescale in which the limited-amplitude instability can produce a significant redistribution of angular momentum is probably the timescale τ in which the instability is fed by the gravitational contraction rather than the Kelvin-Helmholtz timescale that applies to the corresponding form of the Goldreich-Schubert instability. If this is correct, the $\partial\Omega/\partial z$ instability provides faster relaxation of the differential rotation than any of the other processes so far considered; but it is not enough to eliminate the differential rotation during this phase of the evolution.

4. Subsequent Evolution: Spin-up with a Crystallizing Core

If angular momentum redistribution is ineffective during the final phase of gravitational contraction immediately preceding the onset of complete degeneracy, the star is left at the end of this phase in a state of strong, almost cylindrically symmetric differential rotation. Such objects may be qualitatively very similar to the models of Ostriker and Bodenheimer. These stars are still very hot, however, and the subsequent evolution consists of gradual cooling, as in the case of non-rotating, degenerate stars. The differential rotation has several effects upon this stage of the evolution, which we consider next.

4.1. Evolution without Viscous Energy Production

Perhaps the easiest case to discuss is one in which the kinetic energy stored in differential motions, $E_{\text{diff}} \sim MR^2(\Delta\Omega)^2$, is sufficiently small that the rate of viscous dissipation of rotational kinetic energy contributes negligibly to the total luminosity of the star throughout the cooling phase. Because of the extremely long timescale for viscous processes, this does not severely restrict the angular velocity distribution; even for a maximally rotating object, in which the kinetic energy of rotation is comparable to the gravitational energy, we expect $\dot{E}_{\mathrm{diff}} \sim E_{\mathrm{diff}}/\tau_{\mathrm{visc}} \sim 10^{-4}$ to $10^0 L_\odot$. In a star where this condition is satisfied the cooling phase is thus largely unaffected by the rotation until crystallization begins at the center. With the development of a solid core of appreciable size, however, a new physical process – that of spin-up associated with the viscosity-dominated Ekman boundary layer – becomes important and produces angular momentum redistribution on a very much reduced timescale.

This process has been studied theoretically by Greenspan and Howard (1963) in the context of a simple laboratory experiment. While this case is far too idealized to be directly applicable in the context of stellar interiors, this simplification permits them to give a clear, physical exposition of the basic process, which I shall summarize here. The case which they consider is that of a homogeneous fluid initially constrained to rotate uniformly at angular velocity Ω between two parallel, semi-infinite disks located in the planes $z = -L$ and $z = +L$. At time $t = 0$ the angular velocities of the disks are both instantaneously increased to $\Omega + \Delta\Omega$, where $\Delta\Omega/\Omega \ll 1$, and the problem is to describe the subsequent process of spin-up of the fluid. This occurs in three stages. First, a viscous boundary layer (the Ekman layer) forms at each disk. The thickness δ of this layer is given to order of magnitude by $\varrho\delta^2/\eta \sim \Omega^{-1}$. This boundary layer is no longer in pressure balance, however; the pressure in the layer is that required to balance centrifugal force at the original angular velocity Ω. The excess centrifugal force thus impels fluid radially outward, away from the rotation axis, at a velocity $v_{\mathrm{E}} \sim L\Delta\Omega$. In order to conserve mass, however, this material must be replaced by fluid from the approximately inviscid region between the two boundary layers; this 'interior' fluid thus moves toward the axis, at a speed $v_{\mathrm{I}} \sim v_{\mathrm{E}}\delta/L$. Angular momentum conservation of a contracting ring of fluid spins up this material to angular velocity $\Omega + \Delta\Omega$ when the fluid ring has moved radially inward through a distance $\Delta L \sim \frac{1}{2}L\Delta\Omega/\Omega$. The timescale for spin-up of the fluid to the new angular velocity is thus

$$\tau_{\mathrm{GH}} \sim \Delta L/v_{\mathrm{I}} \sim \tfrac{1}{2}(\varrho L^2/\eta\Omega)^{1/2} \sim (\tau_{\mathrm{visc}}/\Omega)^{1/2}. \tag{18}$$

The third and final stage of the spin-up process is the gradual viscous dissipation of small residual motions about the new equilibrium state, and these ultimately decay on the timescale $\sim \tau_{\mathrm{visc}}$.

In a rapidly rotating white dwarf, the timescale given by (18) can be very short in spite of the long viscous timescale. For a star in which $\Omega^{-1} \sim 1$ s, for example, $\tau_{\mathrm{GH}} \sim 10^1$ to 10^3 yr. This is so much shorter than the cooling time, $\tau_{\mathrm{KH}} \sim 10^7$ to 10^9 yr in these phases, that it would imply virtually instantaneous spin-up of the entire star upon commencement of crystallization if this analysis were directly applicable. A star is not a homogeneous fluid, however, and this leads to an important difference from the Greenspan-Howard problem.

In particular, the stratification of density within a star restricts the effect of 'Ekman pumping' to a region of characteristic dimension $\Delta \sim R/S$, where the dimensionless 'stratification parameter' S is given by

$$S^2 \approx \xi R/H_\varrho, \quad H_\varrho = (\mathrm{d}\ln\varrho/\mathrm{d}r)^{-1}, \quad \xi = G\bar{\varrho}/\Omega^2. \tag{19}$$

This effect, first pointed out by Holton (1965), has received extensive discussion by Mestel (1970), and more recently by Sakurai *et al.* (1971), and by Clark (1972), in connection with the solar oblateness problem. In the solar case $S^2 \gg 1$ and thus $\Delta \ll R$; near the center of an object of the type we are presently considering, however, $\xi \sim 10$ to 100, $H_\varrho \sim R$ and consequently $\Delta \sim 0.1$ to $0.3\,R$. Because of the limited extent of the region in which Ekman pumping is effective in this case, spin-up now leads to a quasi-steady state of non-uniform rotation; the region of dimension Δ is brought into uniform rotation in a time $\sim \tau_\mathrm{GH}$ appropriate to this size, while the external regions not penetrated by the Ekman currents are essentially unaffected. The residual differential rotation of these outer layers then persists until the Eddington-Sweet circulations eventually complete the redistribution of angular momentum on the timescale τ_ES (Sakurai *et al.*, 1971).

It is important to note that this final redistribution may lead to ejection of matter from the star, especially for objects of larger mass. The reason is that the total angular momentum of a *uniformly* rotating, degenerate star is severely constrained by the requirement that the velocity of rotation at the equator must be less than the local velocity of escape. In a differentially rotating star, however, the outer layers can be in almost Keplerian orbits, as pointed out by Ostriker and Bodenheimer, and this restriction does not apply. This leads to a particularly interesting situation for objects more massive than the Chandrasekhar limit. As Roxburgh (1965) has shown, the maximum mass, M_U, for uniformly rotating white dwarfs is only slightly larger than for non-rotating, degenerate stars. Thus if a star of mass $M > M_U$ reaches a stable, quasi-stationary state of differential rotation at the end of the pre-degenerate contraction phase, there is no corresponding, uniformly rotating object into which it can evolve. Consequently, when the mass of the uniformly rotating core grows to exceed M_U, a catastrophic collapse must result, perhaps leading to a supernova explosion. This is essentially the mechanism suggested by Schwartz and Africk (1970). However, the time between the onset of crystallization and the collapse of the core will be considerably shorter than they found if Ekman pumping causes spin-up on the Greenspan-Howard timescale.

4.2. Evolution with Viscous Energy Production

If the energy of differential rotation is so large that $\dot{E}_\mathrm{diff}$ becomes comparable to the luminosity at some point during the radiative cooling phase, the evolution becomes qualitatively different. This is the case that Durisen (1972) has considered. Because the store of rotational energy in such objects is comparable to the total gravitational energy of the star, and because the viscous timescale that characterizes this mode of evolution is so large, such a star represents as truly a 'final' stage of evolution as does an ordinary white dwarf; the store of energy is inexhaustible in cosmological time. However, further changes in the mechanical structure of the star do occur in this case. As the core is gradually brought into uniform rotation by the viscous forces, angular momentum conservation demands an outward transfer of angular momentum to the envelope of the star. In Durisen's calculations this is implicitly assumed to take place on a timescale $\sim \tau_{\text{visc}}$; the possibility of Ekman pumping is ignored. This is probably not completely justified, however, as the following argument indicates:

It is conceivable that viscous-dominated evolution will lead to a structure consisting of a uniformly rotating core and a differentially rotating envelope containing most of the angular momentum, as found by Durisen. However, in this case there must be a layer of high shear between the core and the envelope. This may act as an Ekman layer and bring about uniform rotation on a timescale $\sim \tau_{\text{GH}}$. On the other hand, since a fluid does not have the shear resistance of a solid, it appears equally plausible *a priori* that an initial shear layer may have the opposite effect of generating differential motions within the initially uniformly rotating core. In either case Durisen's calculations – while an important first attack on this problem – needs to be extended to include an improved treatment of the fluid dynamics.

5. Summary and Conclusions

I have discussed in some detail the fluid dynamic processes associated with the final contraction and cooling phases of rotating stars near the end-points of their evolution. The strongly non-homologous nature of the neutrino-dominated, pre-degenerate contraction phase almost certainly leads to the development of strong differential rotation by the time degeneracy brings the contraction to a halt. The instability of a configuration in which $\partial \Omega / \partial z \neq 0$ further suggests that the contracting star will maintain an almost cylindrically symmetric angular velocity distribution through the action of the rapid circulations driven by this instability.* At the end of this phase the star may thus resemble the models of Ostriker and Bodenheimer.

The behavior of the star during the subsequent cooling phase depends sensitively upon the total angular momentum and the viscosity. It is furthermore uncertain because the possible effect of Ekman pumping upon the spin-up in this phase is not clear. *If* Ekman pumping is important, one can distinguish two situations, depending upon the ratio of the timescale τ_{ev} of the evolution (defined by (3)) to τ_{GH}.

In the case where $\tau_{\text{GH}} \gtrsim \tau_{\text{ev}}$, Ekman pumping is ineffective anyway, even if it must in principle be taken into account. This inequality is satisfied both for slow rotators and for stars in very rapid contraction. Defined this way, a slow rotator is one for which $\Omega^{-1} \gtrsim \tau_{\text{ev}}^2 / \tau_{\text{visc}}$. With the parameters of the non-rotating star models of Savedoff *et al.* (1969), this gives $\Omega^{-1} \gtrsim 1$ h (for a 1 $M_\odot$ white dwarf), $\gtrsim 4$ days (for 0.631 $M_\odot$), and

---

* Except perhaps near the stellar surface, where the requirement of continuity of the gravitational potential in spite of a discontinuous change in the total potential may result in the establishment of a surface boundary layer. This problem has recently been investigated by Marks and Clement (1971) and references therein.

$\gtrsim 60$ days (for $0.398\,M_\odot$). For comparison, the observational data give $\Omega^{-1} \sim 1.34$ days for the white dwarf G195-19 (Angel and Landstreet, 1971; but the mass of this object is unknown) and $\Omega^{-1} \sim 1$ h for the $0.45\,M_\odot$ white dwarf 40 Eri B (Greenstein and Trimble, 1972). Thus, if it is important at all, Ekman pumping has apparently played a role in 40 Eri B, which violates this 'slow rotator' condition.

The case of rapid contraction is illustrated by the final evolution of the $2.5\,M_\odot$ iron star model of Savedoff *et al.* This model is so much more massive than the Chandrasekhar limit that degeneracy is unable to stop the contraction, and τ_{ev} ultimately becomes orders of magnitude shorter than τ_{GH} even for maximal rotation. Strong differential rotation is therefore clearly expected for this case. Theoretical confirmation of this is provided by the calculations of LeBlanc and Wilson (1970) who studied the collapse phase of an initially rapidly and uniformly rotating iron star of $7\,M_\odot$; their calculation represents a continuation of the evolution of such a star subsequent to the phase considered by Savedoff *et al.* In LeBlanc and Wilson's fully hydrodynamic, two-dimensional (but inviscid) calculation, a strong differential rotation did indeed develop, and the angular velocity distribution was found to be approximately given by $\Omega \propto r^{-1.85}$ (except, of course, near the axis, where their finite difference equations give too coarse a representation of the physics).

In the case where $\tau_{GH} \ll \tau_{ev} \ll \tau_{ES}$, Ekman pumping may be important for the evolution. For stars in which crystallization begins at high enough luminosity so that viscous energy production can still be neglected, this process will lead to the growth of an appreciable, uniformly rotating core in the rather short timescale τ_{GH} (appropriate to a region of dimension Δ). If the stellar mass is $M \gg M_U$, this will lead to core collapse and perhaps to explosion as a supernova.

If viscous energy production cannot be ignored, growth of the crystallizing core proceeds on a much longer timescale $\sim \tau_{visc}$, and the evolution is dominated by angular momentum transfer from the core to the envelope, as Durisen has shown. The role of Ekman pumping in this case is problematical at present and is an important area for further investigation.

There are a substantial number of other problems that are also of interest in connection with the evolution of highly rotating, degenerate stars. From an observational point of view, for example, it is evidently important to establish whether such objects do in fact exist. In this regard, a determination of the rotation velocities of stars which are thought to be the progenitors of the white dwarfs – e.g. the nuclei of planetary nebulae or the hot subdwarfs – would be of considerable importance. In addition, the incontrovertible detection of a rotationally distorted object would be of still more interest. (Wegner (1972) has recently suggested that the white dwarf CD-42°14462 may be just such a star; however this remains to be confirmed.)

For the theorist, an even greater variety of problems are available: How does the instability resulting from $\partial\Omega/\partial z \neq 0$ manifest itself? How does the existence of rotation affect the evolutionary development of a star? Can the development of differential rotation halt a collapse? Is a viscosity-dominated core sufficiently similar to a solid core to permit Ekman pumping to be effective? Quantitative answers to questions of this sort will evidently require detailed numerical computations. However, the gross dissimilarity of scales of the important fluid dynamical processes (e.g., $\tau_{GH} \sim 10$ to 10^3 yr vs. $\tau_{visc} \sim 10^{10}$ to 10^{14} yr; similarly, the thickness of the Ekman layer is of order centimeters to meters, compared to a stellar radius $\sim 10^3$ km) makes it necessary also to carry out analytical studies of this problem in simpler situations. In addition, calculations of, e.g., the viscosity of relativistically degenerate electrons must be carried out, and a better understanding of the phenomenon of crystallization under conditions of high shear is also needed.

Hopefully, an attack upon some of these problems will soon be forthcoming; for the potential for increasing our understanding of the evolution of rotating, degenerate stars is truly enormous.

Acknowledgements

I am deeply indebted to the numerous people with whom I have discussed this problem for their cirticisms and assistance during the preparation of this paper. I am especially grateful to Professor H. L. Helfer for his invaluable counsel and considerable help, to Professor L. Mestel for discussions of meridional circulations and magnetic fields, to Professor J. H. Thomas for many fruitful discussions of the fluid dynamics of the spin-down problem, and to Messrs D. Q. Lamb and S. Simon for numerous discussions of the problems of evolution of rotating stars. I am grateful also to Professors A. G. W. Cameron, A. Clark, Jr., S. Colgate, J. P. Cox, A. Kovetz, and M. P. Savedoff for their various comments that have helped to effect a material improvement over an earlier version of this work. This work has been supported by the Office of Naval Research under contract number N00014-68-A-0091. Such support does not imply endorsement of the content by the Department of the Navy.

References

Angel, J. R. P. and Landstreet, J. D.: 1971, *Astrophys. J. Letters* **165**, L67.
Chandrasekhar, S.: 1969, *Ellipsoidal Figures of Equilibrium*, Yale University Press, New Haven.
Clark, A., Jr.: 1972, Lecture given at NATO Advanced Study Institute on Magnetohydrodynamic Phenomena in Rotating Fluids, Cambridge, England.
Colgate, S. A.: 1968, *Astrophys. J. Letters* **153**, L81.
Durisen, R. H.: 1972, *Astrophys. J.* **183**, 205, 215.
Eddington, A. S.: 1929, *Monthly Notices Roy. Astron. Soc.* **90**, 54.
Fujimoto, M.: 1968, *Astrophys. J.* **152**, 523.
Giacconi, R., Gursky, H., Kellogg, E., Schreier, E., and Tananbaum, H.: 1971, *Astrophys. J. Letters* **167**, L67.
Gold, T.: 1968, *Nature* **218**, 731.
Goldreich, P. and Schubert, G.: 1967, *Astrophys. J.* **150**, 571.
Greenspan, H. L. and Howard, L. N.: 1963, *J. Fluid Mech.* **17**, 385.
Greenstein, J. L. and Trimble, V.: 1972, *Astrophys. J. Letters* **175**, L1.
Hewish, A., Bell, S. J., Pilkington, J. D. H., Scott, P. F., and Collins, R. A.: 1968, *Nature* **217**, 709.
Holton, J.: 1965, *J. Atmospheric Sci.* **22**, 402.
James, R. A. and Kahn, F. D.: 1970, *Astron. Astrophys.* **5**, 232.
James, R. A. and Kahn, F. D.: 1971, *Astron. Astrophys.* **12**, 332.
LeBlanc, J. M. and Wilson, J. R.: 1970, *Astrophys. J.* **161**, 541.
Lebovitz, N. R.: 1967, *Ann. Rev. Astron. Astrophys.* **5**, 465.

Marks, D. W. and Clement, M. J.: 1971, *Astrophys. J. Letters* **166**, L27.
Mestel, L.: 1965, in *Stars and Stellar Systems* **8**, 465.
Mestel, L.: 1970, *IAU* (Commission 35) *Circular Letter* **10**.
Ostriker, J. P. and Bodenheimer, P.: 1968, *Astrophys. J.* **151**, 1089.
Ostriker, J. P., Bodenheimer, P., and Lynden-Bell, D.: 1966, *Phys. Rev. Letters* **17**, 816.
Roxburgh, I. W.: 1965, *Z. Astrophys.* **62**, 134.
Sakurai, T., Clark, A., Jr., and Clark, P. A.: 1971, *J. Fluid Mech.* **49**, 753.
Savedoff, M. P., Van Horn, H. M., and Vila, S. C.: 1969, *Astrophys. J.* **155**, 221.
Schwartz, R. and Africk, S.: 1970, *Astrophys. Letters* **5**, 141.
Strittmatter, P. A.: 1969, *Ann. Rev. Astron. Astrophys.* **7**, 665.
Sweet, P. A.: 1950, *Monthly Notices Roy. Astron. Soc.* **110**, 548.
Tananbaum, H., Gursky, H., Kellogg, E. M., Levinson, R., Schreier, E., and Giacconi, R.: 1972, *Astrophys. J. Letters* **174**, L143.
Wegner, G.: 1972, *Proc. Astron. Soc. Australia* **2**, 107.
Zeipel, H. von: 1924, *Monthly Notices Roy. Astron. Soc.* **84**, 665.

CONTINUUM POLARIZATION IN MAGNETIC WHITE DWARFS*

F. K. LAMB

Dept. of Physics, University of Illinois, Urbana, Ill. 61801, U.S.A.

and

P. G. SUTHERLAND

Dept. of Physics, Columbia University, New York, N.Y. 10027, U.S.A.

Abstract. We discuss some of the effects which magnetic fields in the range 10^4–10^8 G have on the continuum emission of white dwarfs. In order to show how atomic processes are related to the transport of radiation in an anisotropic medium we introduce the radiative transfer equation for polarized light. Using this transfer equation as a guide we develop a greatly simplified model of a white dwarf atmosphere: anisotropic absorption of unpolarized light passing through a cold, optically-thin layer. Two possible sources of continuum polarization in white dwarfs with strong magnetic fields are explored, namely bound-free transitions and cyclotron absorption. Within the hydrogenic approximation we develop a further approximation which is valid for bound-free transitions when B is in the range 10^4–10^7 G. Using this approximation it is possible to obtain simple expressions for the net circular and linear polarization in terms of the zero-field opacity. In the case of cyclotron absorption, the cross section is large and strongly peaked at the cyclotron frequency, which falls in the infrared or optical for $B \gtrsim 10^8$ G. This process can lead to net circular and linear polarization of comparable magnitude. For stars with non-uniform magnetic fields, the cyclotron absorption effects are spread over considerable wavelength ranges, with possibly quite complicated wavelength-dependent polarization.

1. Introduction

In the past three years there has been growing interest in the possible existence of white dwarfs with strong magnetic fields. A number of authors (Ostriker and Hartwick, 1968; Ginzburg *et al.*, 1969; Chow, 1969) have noted that application of the same argument put forward to explain the apparently large magnetic fields in pulsars – namely, conservation of flux in the compression of a normal star to a neutron star in a supernova – would suggest magnetic fields of the order of 10^6–10^7 G. The process by which white dwarfs are formed is unclear, however, and the extent to which this scaling argument applies is uncertain. Nevertheless, conjectures of this kind have stimulated a number of observational searches for evidence of strong magnetic fields in white dwarfs.

Three approaches have been used: examination of the wings of hydrogen lines for circular polarization due to the linear Zeeman splitting of the unperturbed line (Angel and Landstreet, 1970a), interpretation of residual displacements of hydrogen lines in terms of the quadratic Zeeman effect (Preston, 1970; Trimble, 1971; see also Lamb and Sutherland, 1971, for further discussion of this problem), and measurements of continuum circular polarization with broad pass-band filters (Angel and Landstreet,

---

* Presented by P. G. Sutherland.

1970b, 1971a, b; Landstreet and Angel, 1971; Kemp *et al.*, 1970; Kemp and Swedlund, 1970; Kemp *et al.*, 1971; and Gehrels, 1971) and multichannel photospectrometers (Angel *et al.*, 1972a, b; Angel and Landstreet, 1972). The first two approaches were confined exclusively to the DA white dwarfs and led to upper limits on the magnetic field strengths in DA stars of about 10^5 G.

The most successful search method to date has been the search for circular polarization in the continuum radiation; the data for four stars, Grw+70°8247, G99-37, G195-19, and G99-47, have been published. Grw+70°8247 is a 'peculiar' white dwarf, famous for its unassigned Minkowski bands (see Greenstein and Matthews, 1957; Greenstein, 1970; and Angel, 1972 for tentative assignments), the most notable one being at $\lambda 4135$. From the trigonometric parallax and a black-body interpretation of the $U-B$, $B-V$ colors, the temperature is near $T_e = 11\,500$ K and the radius is near $0.012\,R_\odot$ (Greenstein and Matthews, 1957). In Figure 1 we give the data for the net circular and linear polarizations, for wavelengths between $\lambda 3000$ and $\lambda 11\,000$, taken from the work of Angel *et al.* (1972b) and the work of Angel and Landstreet (1970b). The most striking features of the data are the rapid decrease and increase of the circular polarization in the $\lambda 3000$–4000 range, the slow decrease of the circular polarization with increasing wavelength above $\lambda 4000$, and the large linear polarization near $\lambda 4000$. There are also dramatic changes in the circular polarization in the vicinity of the Minkowski bands; a possible interpretation of these latter results has been put forward by Angel (1972). The second magnetic white dwarf, G99-37, is a $\lambda 4670$p star with CH band heads near $\lambda\lambda 3880$, 4300 and the Swan bands of C_2 near

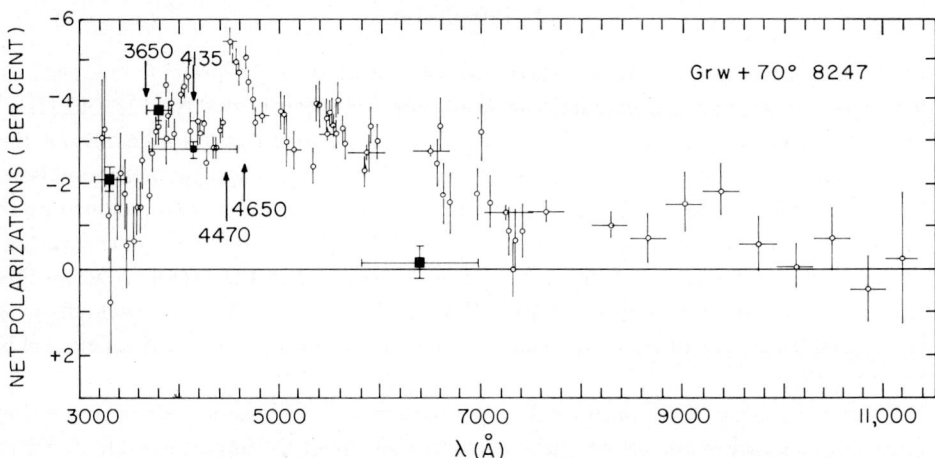

Fig. 1. The net circular and linear polarizations versus wavelength for the white dwarf Grw+70°8247. The round dots indicate circular polarization and are taken from a graph given by Angel *et al.* (1972b); the square dots at $\lambda\lambda 3300$, 3800, 4150, and 6400 indicate linear polarization and are taken from a graph given by Angel and Landstreet (1970b). The circular polarization measurements were made with a multichannel spectrophotometer and the linear polarization measurements were made with filters; the horizontal error bars correspond to the channel and filter widths, and the vertical error bars represent observational uncertainties. The vertical arrows indicate the positions of the 'Minkowski' bands that are prominent in the spectrum.

$\lambda\lambda 4380$, 4680, 5170 (Landstreet and Angel, 1971; Greenstein, 1970; Greenstein *et al.*, 1971). The polarization data for G99-37 are given in Figure 2 (the positions of the molecular absorption features are indicated by vertical arrows), taken from Landstreet and Angel (1971). These data differ markedly from the Grw + 70° 8247 data: the average circular polarization is considerably smaller (0.63% compared with 3.5%), and the ratio of linear to circular polarization is very small. The data for the third magnetic white dwarf, G195-19, a DC star, are similar to those of G99-37, with circular polarization $\sim 0.22\%$ and no detectable linear polarization (Angel and Landstreet, 1971a, b; Angel *et al.*, 1972a). In contrast to the previous two stars, the polarization of G195-19 was found to be time-varying with an amplitude of 0.25% and a period of 1.339 days, presumably due to rotation. Hence the published data, which were taken at different phases, only give a rough idea of the wavelength dependence of the polarization. The fourth magnetic white dwarf, G99-47, shows circular polarization $\sim 0.4\%$ and no detectable linear polarization (Angel and Landstreet, 1972). Recent observations of G99-37 by Angel and Landstreet, with improved sensitivity and resolution, have revealed a classical molecular linear Zeeman pattern at the CH $\lambda 4300$ band head; with 160 Å resolution the circular polarization jumps through $\sim 6\%$ across the band head, indicative of a magnetic field strength $\gtrsim 10^6$ G (Angel, private communication). This observation is perhaps the single most convincing piece of evidence that

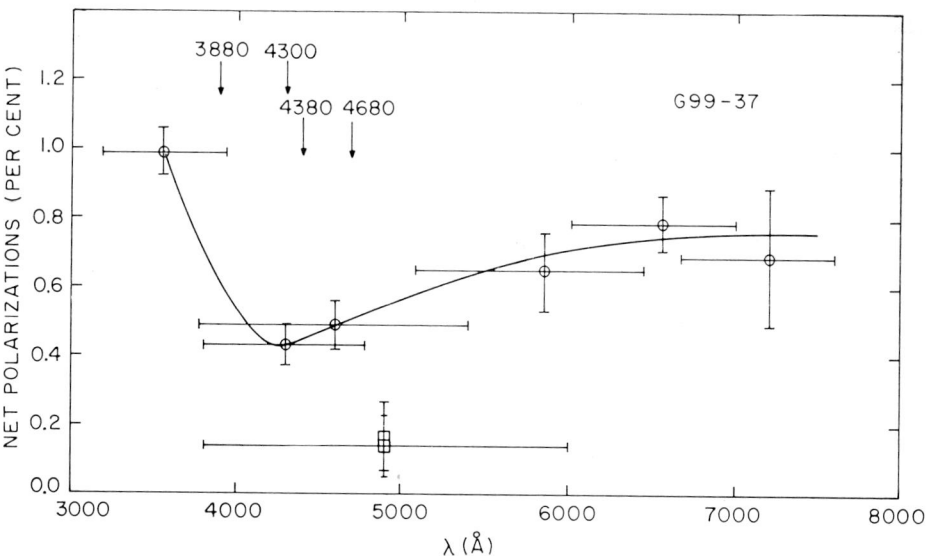

Fig. 2. The net circular and linear polarizations versus wavelength for the white dwarf G99-37, adapted from a graph given by Landstreet and Angel (1971). The round dots indicate circular polarization measurements; the curve drawn through these points is meant only to guide the eye. The two linear polarization measurements, indicated by the square dots, were made in two perpendicular directions. The horizontal error bars correspond to the filter widths and the vertical error bars represent observational uncertainties. The two vertical arrows at $\lambda\lambda 3880$, 4300 indicate the positions of band heads in CH and the two vertical arrows at $\lambda\lambda 4380$, 4680 indicate the Swan bands of C_2; these are prominent absorption features in the spectrum.

the polarizations in these four white dwarfs are in fact due to magnetic fields.

The measurements of the continuum polarizations have been interpreted using the grey-body model (Kemp, 1970a, b; see also Shipman, 1971; Chanmugam *et al.*, 1972). This model may give an order-of-magnitude estimate of the field strengths but does not agree in detail with the data. One reason for this is the assumed grey-body character of the opacity: for $\Omega_L \ll \omega$ this leads to circular polarization proportional to Ω_L/ω where $\Omega_L = eB/2mc$ is the electron Larmor frequency and ω is the angular frequency of the light. In this approximation the linear polarization is proportional to $(\Omega_L/\omega)^2$. This model almost certainly fails for Grw+70°8247, where, as mentioned above, the linear and circular polarization are comparable and $\sim 4\%$ at $\lambda 4000$. Using the grey-body model, the field strengths for G99-37 and G195-19 are estimated to be roughly 3×10^6 G and 10^6 G, respectively.

To conclude this brief survey of magnetic white dwarfs several remarks are in order. First, only four of approximately sixty white dwarfs examined were found to show detectable polarization, indicating magnetic fields greater than 10^5 G; thus the fraction of white dwarfs with magnetic fields as large as this is apparently small. Second, even among the four white dwarfs which show polarization there appears to be a considerable spread in indicated field strengths: as mentioned above, both the linear and circular polarization in Grw+70°8247 are an order of magnitude greater than the measured circular polarization in the other three, which show no detectable linear polarization. Given this context and the rather long and complex evolutionary processes which are thought to lead to white dwarf formation, use of the flux conservation model to account for magnetic fields in white dwarfs leaves unexplained at least as much as it explains.

In Section 2 we discuss the transfer equation for polarized radiation in an anisotropic medium (the anisotropy, in this case, is provided by the magnetic field). The effects of transfer on the polarization of radiation emerging from a plane-parallel semi-infinite atmosphere are briefly discussed. In cases where the characteristic states of the semi-infinite medium are orthogonal we show that Faraday rotation does not affect the polarization of the emerging radiation so that the polarization is determined by the absorption matrix alone. We then obtain the form of the absorption matrix when the absorption is due to electric dipole transitions (as is the case for those opacity sources discussed in Sections 3 and 4). To illustrate the principal polarization features that may arise, we examine a very simple model: a cold, optically thin layer through which initially unpolarized radiation passes and acquires polarization due to the anisotropic absorption.

Section 3 discusses bound-free absorption by hydrogenic atoms (specifically hydrogen and helium) in a magnetic field. Within the hydrogenic approximation we develop a further approximation which is valid for field strengths in the range 10^4–10^7 G. Using this approximation and the results of Section 2, it is possible to obtain simple expressions for the net circular and linear polarizations in terms of the zero-field opacity. We find that, away from absorption edges, the hydrogenic bound-free opacity gives the same field strength and frequency dependence as the grey-body model. However, near an absorption edge the opacity is discontinuous and the polarization undergoes sudden changes in sign and magnitude. If hydrogen-like bound-free processes form a significant source of opacity in white dwarfs with magnetic fields of order 10^7 G, characteristic polarization features should be observable near absorption edges.

In Section 4 we briefly consider polarization due to cyclotron absorption. This process is of particular interest for stars which show strong linear as well as circular polarization (such as Grw+70°8247), since it can lead to net circular and linear polarizations of comparable magnitude. The cyclotron absorption cross section is large and strongly peaked at the cyclotron frequency. The magnetic field in a typical white dwarf atmosphere is almost certainly not constant in either direction or magnitude over the surface of the star; averaging over the surface field will spread the cyclotron absorption features over broad wavelength ranges. These features occur in the infrared and optical for $B \gtrsim 10^8$ G.

2. The Transfer Equation for Polarized Radiation

In this section we formulate the radiative transfer equation for polarized light. The passage of polarized light through an anisotropic medium can be conveniently treated in terms of the two characteristic polarization states defined at each point in the medium. Light of either of these two polarization states, which are generally not orthogonal, propagates locally within the medium with its own particular complex refractive index. Providing the properties of the medium change only over distances much larger than the wavelength of the light, at each point in the medium light of arbitrary polarization can be resolved into components using as a basis the characteristic states of the medium at that point and the absorption and propagation of each component can then be followed separately. This approach is discussed in considerable detail by Lamb and ter Haar (1970); below we summarize the main results.

In order to describe the polarization properties of the light, we shall introduce a polarization matrix, analogous to the density matrix ordinarily used in quantum mechanics; rules for obtaining the corresponding values of the Stokes parameters from this polarization matrix are given in the Appendix. The radiative transfer equation can then be written in terms of a transfer matrix $\mathcal{T}$ and an emission matrix $\mathcal{E}$. The eigenvectors of $\mathcal{T}$ correspond to the characteristic polarization states of the medium; in the basis composed of these eigenvectors the explicit form of $\mathcal{T}$ is readily obtained. (We are primarily concerned here with continuum polarization, therefore we shall not discuss the effects of line emission and absorption although they may be included in a relatively straightforward way – see Lamb and ter Haar, 1970.)

Consider the local propagation of light at point **r** in an anisotropic medium. At the point **r** we establish a right-handed Cartesian triplet of unit vectors such that the light is propagating along the 3-direction. The two transverse unit vectors are $\hat{e}_1$ and $\hat{e}_2$; distance along the direction of propagation will be denoted by s. We denote by $\hat{\mathscr{E}}_\pm$ the local characteristic polarization vectors of the medium. The electric induction vector for light of one of these characteristic polarizations propagates according to

$$\mathbf{D}_\pm(\delta s) = |\mathbf{D}_\pm(0)| \hat{\mathscr{E}}_\pm e^{ik_\pm \delta s - \mathscr{K}_\pm \delta s/2 - 2\pi i v t}, \tag{1}$$

where v is the frequency of the light. We introduce the polarization matrix $\mathscr{I}$ for the light as follows: a light ray of induction vector $\mathbf{D}(s) = D_+(s)\hat{\mathscr{E}}_+ + D_-(s)\hat{\mathscr{E}}_-$ contributes to the polarization matrix a term

$$\mathscr{I}(s) \equiv \begin{pmatrix} \mathscr{I}_+(s) & \mathscr{I}_\times(s) \\ \mathscr{I}_\times^*(s) & \mathscr{I}_-(s) \end{pmatrix} = \begin{pmatrix} |D_+(s)|^2 & D_+(s) D_-^*(s) \\ D_-(s) D_+^*(s) & |D_-(s)|^2 \end{pmatrix}. \tag{2}$$

We suppose that the characteristic vectors are related to the Cartesian unit vectors by the transformation

$$\begin{aligned}\hat{\mathscr{E}}_\sigma &= \Sigma_\alpha \hat{e}_\alpha \mathscr{U}_{\alpha\sigma} \\ \hat{e}_\alpha &= \Sigma_\beta \hat{\mathscr{E}}_\beta \mathscr{U}_{\beta\alpha}^{-1} ; \end{aligned} \tag{3}$$

$\mathscr{U}$ will not be unitary unless the characteristic vectors are orthogonal. The relationship between the polarization matrices in the characteristic basis, $\mathscr{I}$, and in the Cartesian basis, $\mathscr{I}^{(c)}$, is

$$\begin{aligned}\mathscr{I}^{(c)} &= \mathscr{U} \mathscr{I} \mathscr{U}^\dagger \\ \mathscr{I} &= \mathscr{U}^{-1} \mathscr{I}^{(c)} \mathscr{U}^{-1\dagger} ; \end{aligned} \tag{4}$$

this transformation is not a similarity transformation unless $\mathscr{U}$ is unitary. The Stokes parameters of the radiation are obtained from $\mathscr{I}^{(c)}$, and rules for the determination of these are given in the Appendix. Although the approach presented in this section is quite general we shall restrict our considerations to atmospheres in which the characteristic states do not change along the direction of propagation.

The transfer equation for the polarization matrix $\mathscr{I}$ is

$$\frac{d\mathscr{I}}{ds} = -\tfrac{1}{2}(\mathscr{T}\mathscr{I} + \mathscr{I}\mathscr{T}^\dagger) + \mathscr{E}, \tag{5}$$

where $\mathscr{E}$ is the emission matrix (its form in LTE will be determined below) and $\mathscr{T}$ is the transfer matrix obtained by substituting (1) into (2) and differentiating:

$$\mathscr{T} \equiv -2i\mathscr{R} + \mathscr{K} = -2i\begin{pmatrix} k_+ & 0 \\ 0 & k_- \end{pmatrix} + \begin{pmatrix} \mathscr{K}_+ & 0 \\ 0 & \mathscr{K}_- \end{pmatrix}. \tag{6}$$

The matrix $\mathscr{K}$ represents, of course, the absorption of light; the matrix $\mathscr{R}$ contains the real parts of the refractive indices and provides for such effects as Faraday rotation.

The form of the emission matrix in LTE may be determined as follows. In a *homogeneous* (but possibly anisotropic) medium in thermodynamic equilibrium we must have $d\mathscr{I}/ds = 0$ and the light must be unpolarized black-body radiation. In the

Cartesian basis the latter condition may be written

$$\mathscr{I}^{(c)} = \tfrac{1}{2} B(T) \begin{pmatrix} 1 & 0 \\ 0 & 1 \end{pmatrix}, \quad B(T) = \frac{2h\nu^3}{c^2} \frac{1}{e^{h\nu/kT} - 1}, \tag{7}$$

or in the characteristic basis

$$\mathscr{I} = \mathscr{U}^{-1} \mathscr{I}^{(c)} \mathscr{U}^{-1\dagger} \equiv \tfrac{1}{2} B(T) \Omega, \tag{8}$$

where $\Omega = \mathscr{U}^{-1} \mathscr{U}^{-1\dagger}$. From the condition $d\mathscr{I}/ds = 0$ it follows that

$$\mathscr{E} = \tfrac{1}{2} B(T) \left[(\mathscr{K}\Omega + \Omega\mathscr{K})/2 - i(\mathscr{R}\Omega - \Omega\mathscr{R}) \right]. \tag{9}$$

But $\mathscr{E}$ is determined entirely by the local properties of the medium, and hence in LTE is given by (9) even when the medium is inhomogeneous. Note that when the characteristic states are orthogonal, Ω is the unit matrix and $\mathscr{E}$ has the familiar form $\mathscr{E} = \tfrac{1}{2} B(T) \mathscr{K}$.

Quite generally we may write $\mathscr{U}$ in the form

$$\mathscr{U} = \begin{pmatrix} a & b \\ b & a \end{pmatrix} \tag{10}$$

with a, b complex and $|a|^2 + |b|^2 = 1$. Then

$$\Omega = \frac{1}{1 - \eta^2} \begin{pmatrix} 1 & \eta \\ \eta & 1 \end{pmatrix} \tag{11}$$

with $\eta = -2 \operatorname{Re} a^* b$; η is a direct measure of the nonorthogonality of the characteristic states. In terms of the matrix components of $\mathscr{I}$, the transfer equation becomes

$$\begin{aligned}
\frac{-d\mathscr{I}_\pm}{ds} &= \mathscr{K}_\pm \left[\mathscr{I}_\pm - \tfrac{1}{2} \frac{1}{1-\eta^2} B(T) \right] \\
\frac{-d\mathscr{I}_\times}{ds} &= (\bar{\mathscr{K}} - i\varDelta) \left[\mathscr{I}_\times - \tfrac{1}{2} \frac{\eta}{1-\eta^2} B(T) \right]
\end{aligned} \tag{12}$$

with $\bar{\mathscr{K}} \equiv (\mathscr{K}_+ + \mathscr{K}_-)/2$ the average absorption coefficient and $\varDelta \equiv k_+ - k_-$.

The transfer equations for the $\mathscr{I}_+$ and $\mathscr{I}_-$ components of the polarization matrix have the standard form: there is an absorption term and an emission term, the latter being given [apart from a nonorthogonality factor $(1-\eta^2)^{-1}$] by the Planck function. On the other hand, the transfer equation for $\mathscr{I}_\times$ involves a complex absorption term, $(\bar{\mathscr{K}} - i\varDelta)\mathscr{I}_\times$, and a 'weak' source function, $\tfrac{1}{2}(\eta/1-\eta^2)B(T)$ (weak if $[1 + \varDelta^2/\bar{\mathscr{K}}^2]^{1/2} \times \eta \ll 1$).

If the polarization states are orthogonal ($\eta = 0$), (12) reduces to

$$\begin{aligned}
\frac{-d\mathscr{I}_\pm}{ds} &= \mathscr{K}_\pm \left[\mathscr{I}_\pm - \tfrac{1}{2} B(T) \right] \\
\frac{-d\mathscr{I}_\times}{ds} &= (\bar{\mathscr{K}} - i\varDelta) \mathscr{I}_\times .
\end{aligned} \tag{13}$$

In this case, the components $\mathscr{I}_\pm$ at the boundary of a semi-infinite plane-parallel atmosphere are determined only by the absorption matrix and the run of temperature in the atmosphere: the matrix $\mathscr{R}$ has no effect. Furthermore, the off-diagonal component $\mathscr{I}_x$ will be zero at the boundary because the $\mathscr{I}_x$ transfer equation has an absorption term but no source term. Faraday rotation effects are associated with Δ; in the case of a semi-infinite plane-parallel atmosphere with orthogonal characteristic states Faraday rotation does not affect the polarization of the emerging radiation because of the attenuation of the off-diagonal elements of the polarization matrix.

If the characteristic polarization states are nearly orthogonal ($[1+\Delta^2/\bar{\mathscr{K}}^2]^{1/2}\eta \ll 1$) then we may to first order in η ignore the effects of the matrix $\mathscr{R}$ in determining the polarization properties of the emergent radiation. The characteristic states are then just the eigenvectors of the absorption matrix, and in the case of electric dipole transitions these may be determined by symmetry considerations. The absorption matrix is easily expressed in the Cartesian basis; in what follows we shall work exclusively in that basis and drop the superscript c on $\mathscr{I}^{(c)}$.

Deep within the atmosphere of a magnetic white dwarf we expect the light to be essentially unpolarized black-body radiation; it is only as the light diffuses outward through the atmosphere that polarization can develop. In order to explore the qualitative features of the polarization produced by electric dipole absorption in the approximation under discussion, we consider a greatly simplified model of a white dwarf atmosphere in which the following assumptions have been made: (i) absorption takes place in a cold layer of thickness δs where $\bar{\tau}=\bar{\mathscr{K}}\delta s \ll 1$, (ii) the light incident on that absorption layer is unpolarized, and (iii) the magnetic field is homogeneous throughout the layer. Thus below the layer the polarization matrix is*

$$\mathscr{I}(0) = \tfrac{1}{2}\mathscr{I}_{\text{net}}\begin{pmatrix}1 & 0 \\ 0 & 1\end{pmatrix} \tag{14}$$

and emerging from the layer it is

$$\mathscr{I}(\delta s) = \mathscr{I}(0) - \tfrac{1}{2}[\mathscr{K}\mathscr{I}(0)+\mathscr{I}(0)\mathscr{K}]\delta s$$
$$= \tfrac{1}{2}\mathscr{I}_{\text{net}}\left[\begin{pmatrix}1 & 0 \\ 0 & 1\end{pmatrix} - \begin{pmatrix}\mathscr{K}_{11} & \mathscr{K}_{12} \\ \mathscr{K}_{12}^* & \mathscr{K}_{22}\end{pmatrix}\delta s\right]. \tag{15}$$

The Stokes parameters of the emergent radiation (see the Appendix) are given by

$$S_i = -\tfrac{1}{2}\delta s \,\text{tr}(\sigma_i \mathscr{K}) \quad i=1,2,3. \tag{16}$$

We now turn to an evaluation of the matrix elements of $\mathscr{K}$, in situations where the opacity is determined by electric dipole transitions. To be explicit, we consider that the anisotropy of the atmosphere is due to a uniform magnetic field oriented along the z-axis. If the outward normal to the atmosphere (and also the direction of light propagation) is given by the unit vector

$$\hat{k} = (\sin\theta\cos\phi, \sin\theta\sin\phi, \cos\theta), \tag{17}$$

---

* This choice of polarization matrix below the 'reversing-layer' means that Faraday rotation does not affect the polarization of the emerging radiation, just as in the case of the semi-infinite atmosphere.

then we may take the two transverse vectors $\hat{e}_1$ and $\hat{e}_2$, which together with $\hat{k}$ form our basic Cartesian triplet, to be

$$\hat{e}_1 = (\sin\phi, -\cos\phi, 0)$$
$$\hat{e}_2 = (\cos\theta\cos\phi, \cos\theta\sin\phi, -\sin\theta). \tag{18}$$

Now the absorption coefficient, in the dipole approximation, for the absorption of a photon of frequency $\nu = \omega/2\pi$ and polarization $\hat{\mathscr{E}}$ is proportional to the absorption cross section

$$\frac{4\pi^2 e^2 \omega}{c} \sum_{i,f} |\langle f | \mathbf{d} \cdot \hat{\mathscr{E}} | i \rangle|^2 \delta(E_f - E_i - \hbar\omega) \tag{19}$$

where E_i, E_f are respectively the energies of the initial and final states of the absorbing system, and $\mathbf{d}$ is the dipole operator. Because the magnetic field (along the $\hat{z}$-axis) tends to break the magnetic quantum number degeneracies in atomic states, it is convenient to express $\mathbf{d}$ as follows:

$$\mathbf{d} = \sum_q \hat{\varepsilon}_q^* d_q \qquad q = \pm 1, 0$$

with

$$\hat{\varepsilon}_\pm = \mp \frac{1}{\sqrt{2}} (\hat{x} \pm i\hat{y}) \qquad \hat{\varepsilon}_0 = \hat{z}; \tag{20}$$

d_q can only connect atomic states whose magnetic quantum numbers differ by q. We define three absorption coefficients K_q corresponding to the polarizations $\hat{\mathscr{E}} = \hat{\varepsilon}_q$ in (19); the $\hat{\varepsilon}_q$ and K_q are then the eigenvectors and eigenvalues of a 3×3 absorption matrix which can be used to describe the absorption of light of arbitrary direction of propagation. The appropriate absorption matrix in the Cartesian basis $\hat{e}_1, \hat{e}_2$ for light propagating along $\hat{k}$ [see (17) and (18)] is obtained by inserting the relevant projection factors:

$$\mathscr{K}_{ij} = \sum_{q=-1}^{+1} (\hat{e}_i \cdot \hat{\varepsilon}_q) \mathscr{K}_q (\hat{\varepsilon}_q^* \cdot \hat{e}_j) \qquad i,j = 1,2;$$
$$= \begin{pmatrix} (\mathscr{K}_+ + \mathscr{K}_-)/2 & -i/2 \cos\theta (\mathscr{K}_+ - \mathscr{K}_-) \\ i/2 \cos\theta (\mathscr{K}_+ - \mathscr{K}_-) & \frac{1}{2}[(\mathscr{K}_+ + \mathscr{K}_-) + (2\mathscr{K}_0 - \mathscr{K}_+ - \mathscr{K}_-)\sin^2\theta] \end{pmatrix}. \tag{21}$$

The values for the net circular and linear polarizations of the radiation emerging from the optically thin layer [see Equation (16)] are

$$\frac{v}{\delta s} = -\frac{1}{2}(\mathscr{K}_+ - \mathscr{K}_-)\cos\theta$$

$$\frac{p_{max}}{\delta s} = \frac{1}{4}(2\mathscr{K}_0 - \mathscr{K}_+ - \mathscr{K}_-)\sin^2\theta. \tag{22}$$

The mean optical depth of the absorption layer is

$$\bar{\tau} \equiv \delta s \tfrac{1}{2} \operatorname{tr}(\mathcal{K}) = \delta s \left[(\mathcal{K}_+ + \mathcal{K}_-) \frac{1 + \cos^2\theta}{4} + \mathcal{K}_0 \frac{\sin^2\theta}{2} \right]. \tag{23}$$

3. Bound-Free Opacities in Magnetic Fields

In the previous section we obtained the form of the absorption matrix in the dipole approximation, and developed expressions for the Stokes polarization parameters for an optically-thin absorbing layer. A major contribution to the continuum opacity in cooler white dwarfs is the bound-free opacity. The dipole approximation is always fully justified in calculating the atomic bound-free opacity in the optical. Thus in the present section we analyze, within the hydrogenic approximation, the frequency and magnetic field strength dependence of the absorption coefficients defined in (19). Besides hydrogen and helium, it may be possible to discuss the behavior of the opacities of the heavier elements within this approximation, if they are sufficiently ionized (it has been suggested by Greenstein *et al.* (1971) that heavier elements may be present in the atmospheres of the white dwarfs showing continuum polarization as a result of convection from the interior). The bound-free absorption coefficient of helium at optical wavelengths is very well represented by the hydrogenic result. This is because the optical absorption and emission spectra are due to transitions involving a single electron in an excited state. The other electron remains in the tightly bound 1s state and its only effect, to a great precision, is to reduce the effective nuclear charge to a value $\sim +e$; thus the use of hydrogenic wavefunctions in this case is fully justified. As an example of the application of this analysis, the continuum polarization produced by bound-free transitions in hydrogen is considered in detail.

The bound-free opacity is given by the photoionization cross section: a bound electron of energy $-I$ absorbs a photon of energy $\hbar\omega$ making a transition to a continuum state of energy $\hbar\omega - I$. Whenever the initial and final state wave functions are sufficiently 'rigid' with respect to perturbation by the magnetic field (this condition will be made more precise in a moment), then we may obtain simple expressions for the continuum K_q in terms of the zero-field opacity. This may be seen as follows.

Consider first the bound electron wave function. We are interested in magnetic fields strong enough ($B \gtrsim 10^4$ G) that the Paschen-Back limit has been reached. If at the same time the magnetic field is sufficiently weak and the bound electron sufficiently tightly bound so that the quadratic Zeeman effect can be ignored, then, because the linear Zeeman Hamiltonian is diagonal, the initial state wave function is unchanged by the magnetic field ('rigidity' of the initial state wave function). In this case only the energy of the bound state changes, by $m_i \hbar \Omega_L$ where m_i is the initial state magnetic quantum number and $\Omega_L = eB/2mc$ is the Larmor frequency of the bound electron. Now, far from the atom the free electron state is greatly distorted by the magnetic field: the electron must spiral along the direction of the magnetic field. On the other hand, the behavior of the free electron wave function is only needed out to an atomic distance of the order of the bound state radius, since the matrix element for photo-ionization is given by a weighted overlap of the bound electron and free electron wave functions. Thus, provided the kinetic energy of the free electron is sufficiently large, the magnetic field can again be treated in first order. If furthermore we work with eigenstates of L_z (the magnetic field being taken along the z-axis) then the free electron wave function – out to the radius of the bound state – may be approximated by the zero-field wave function, with the appropriate linear Zeeman shift $m_f \hbar \Omega_L$ in the energy ('rigidity' of the final state wave function).

The condition that the bound state quadratic Zeeman effect can be neglected in comparison to the linear Zeeman effect is

$$\frac{e^2 B^2 \langle r^2 \rangle}{8 mc^2} \ll \hbar \Omega_L, \tag{24}$$

where r is the radius of the bound state wave function. For a bound state of principal quantum number n, (24) becomes

$$n^4 B_7 \ll 10^3, \tag{25}$$

where B_7 is the magnetic field strength in units of 10^7 G. For Balmer and Paschen bound-free transitions in hydrogen ($n=2$ and $n=3$), (25) is still satisfied for $B=10^7$ G. The condition on the kinetic energy of the free electron is

$$\frac{p^2}{2m} \gg \frac{e^2 B^2}{8 mc^2} \langle r^2 \rangle \tag{26}$$

or, equivalently,

$$\langle r^2 \rangle \ll r_L^2, \tag{27}$$

where r is the radius of the bound state wave function and $r_L = v/\Omega_L$ is the Larmor radius of the electron in the magnetic field. From (24) and the condition (26) we see that for photoionization from a state of principal quantum number n, the first-order approximation will always fail in a wavelength interval

$$\Delta \lambda / \lambda = 4.53 \times 10^{-6} B_7^2 n^6, \tag{28}$$

just above threshold. For the Balmer edge ($\lambda 3650$) and the Paschen edge ($\lambda 8201$), (28) gives $\Delta \lambda = 1.04$ Å and 27.1 Å, respectively, for $B=10^7$ G. As will be shown below, this wavelength interval is negligible compared to the interval over which interesting effects occur. It is clear that in the approximation which has just been described, the absorption coefficient for photoionization in the magnetic field can be written in terms of the zero-field absorption coefficient. We turn now to the details of this program.

For bound-free processes the largest contribution to the opacity is made in the vicinity of the absorption edge. In this region the photon wavelength is much greater than the radius of the bound state wave function for the cases being considered, so that the dipole approximation is valid. Thus the bound-free absorption coefficients are given by (19):

$$\mathscr{K}_q = \text{const.} \times 4\pi^2 \frac{e^2 \omega}{c} \sum_{i,f} |\langle f| d_q |i\rangle|^2 \, \delta(E_f - E_i - \hbar\omega), \quad q = 0, \pm.\tag{29}$$

Here $\hbar\omega$ is the photon energy, and $E_i (= -\mathrm{I})$ and E_f are, respectively, the energies of the bound and ionized electrons, including the linear interaction with the external magnetic field. The matrix elements of d_q in the rigid wave function approximation are easily evaluated in terms of the zero-field matrix elements: the wave functions are unchanged by the magnetic field while the energies of the initial and final states are shifted linearly with field strength (the difference in linear Zeeman shift is the same for all pairs of states connected by a given d_q, namely, $\hbar\Omega_L q$). Thus, if in the absence of a magnetic field

$$\sum_{i,f} |\langle f| d_q |i\rangle|^2 \, \delta(E_f - E_i - \hbar\omega) = f_q(\omega),\tag{30}$$

then with a magnetic field present

$$\sum_{i,f} |\langle f| d_q |i\rangle|^2 \, \delta(E_f - E_i - \hbar\omega) = f_q(\omega - q\Omega_L).\tag{31}$$

The function f_q in (30) and (31) must be the same for all q, since only in this case will the zero-field absorption coefficient be independent of q, as required. Thus the absorption coefficients are given by

$$\mathscr{K}_q(\omega) = \text{const.} \, \omega \, f(\omega - q\Omega_L),\tag{32}$$

and these may be put directly into (22) and (23) to obtain the circular and linear polarization and mean optical depth for the optically thin absorption layer:

$$\frac{v(\theta)}{\bar{\tau}} = -\frac{[f(\omega - \Omega_L) - f(\omega + \Omega_L)] \cos\theta}{[f(\omega - \Omega_L) + f(\omega + \Omega_L)](1 + \cos^2\theta)/2 + f(\omega)\sin^2\theta}$$

$$\frac{p_{\max}(\theta)}{\bar{\tau}} = -\frac{[f(\omega) - 1/2(f(\omega - \Omega_L) + f(\omega + \Omega_L))]\sin^2\theta}{[f(\omega - \Omega_L) + f(\omega + \Omega_L)](1 + \cos^2\theta)/2 + f(\omega)\sin^2\theta}$$

$$\bar{\tau} = \frac{\delta s}{2}\left[(\mathscr{K}_+ + \mathscr{K}_-)\frac{(1 + \cos^2\theta)}{2} + \mathscr{K}_0 \sin^2\theta\right].\tag{33}$$

If $\Omega_L/\omega \ll 1$ and f is smoothly varying, then we may expand $f(\omega \pm \Omega_L)$ in a Taylor series with the result, to lowest order in Ω_L/ω,

$$\frac{v(\theta)}{\bar{\tau}} = \left(\frac{\Omega_L}{\omega}\right) \cos\theta \left[\frac{\omega}{f(\omega)} \frac{df(\omega)}{d\omega}\right]$$

$$\frac{p_{\max}}{\bar{\tau}} = \left(\frac{\Omega_L}{\omega}\right)^2 \frac{\sin^2\theta}{4}\left[\frac{\omega^2}{f(\omega)} \frac{d^2 f(\omega)}{d\omega^2}\right].\tag{34}$$

In (34) it has been assumed that $(\Omega_L^2/f) \, d^2 f/d\omega^2 \ll 1$.

Let us turn now to a specific bound-free opacity, that of hydrogen. Let $\bar{\sigma}_n$ be the absorption cross section per electron in an energy level of principal quantum number n, averaged over the orbital angular momentum quantum numbers (see Frank-Kamenetskii, 1962). For light of wavelength $\lambda \leqslant 15000$ Å we need only consider photo-ionization from levels with $n \leqslant 4$. Then the function $f(\omega)$ defined by (30) is given by

$$f(\omega) = \frac{1}{\omega}\sigma(\omega) \simeq \frac{1}{\omega}\sum_{n=1}^{4}\bar{\sigma}_n(\omega). \tag{35}$$

Figure 3(a) indicates the general behavior (although much exaggerated) of the absorption coefficients $K_q(\omega)$. The resulting net polarizations $v/\bar{\tau}$ and $p_{\max}/\bar{\tau}$ (the former calculated for propagation along the field and the latter, for propagation normal to the field) are shown in Figure 3(b) for a field of 10^7 G. As an example of the behavior of the polarization near an absorption edge, the frequency dependence of $v/\bar{\tau}$ and $p_{\max}/\bar{\tau}$ of the Balmer edge is shown in greater detail in Figure 3(c).

Except near the absorption edges, $v/\bar{\tau}$ and $p_{\max}/\bar{\tau}$ are well-represented by the functions

$$\frac{v(\theta)}{\bar{\tau}} = -4\left(\frac{\Omega_L}{\omega}\right)\cos\theta \quad \text{and} \quad \frac{p_{\max}(\theta)}{\bar{\tau}} = 5\left(\frac{\Omega_L}{\omega}\right)^2\sin^2\theta. \tag{36}$$

These results may be understood by recalling that the Kramers approximation to the bound-free opacity, which is accurate over a wide range of frequencies, gives $\bar{\sigma}_n = \text{const.}\, n^{-5}\omega^{-3}$ or $f(\omega) \sim \omega^{-4}$. In this approximation (34) gives exactly the result (36). We may remark that in the wavelength intervals where (36) is valid (away from the absorption edges), the hydrogen bound-free opacity gives, for an optically thin absorbing layer, the same field strength and frequency dependence as the grey-body model.

The relation between the bound-free case discussed here and the grey-body case discussed by Kemp can be seen from (35). In the greybody case, $\sigma(\omega) = \text{const.}$ by definition, so that $f(\omega) = (\text{const.})\,\omega^{-1}$. This gives the same behavior as that obtained here for the bound-free case, except, of course, near an absorption edge.

Near an absorption edge the cross section is discontinuous, and the polarization is given by (33). The result is that $v/\bar{\tau}$ changes sign and becomes large in the interval $\lambda_- \leqslant \lambda \leqslant \lambda_+$, where

$$\lambda_{\pm} = \lambda_0 (1 \mp \lambda_0/\lambda_L)^{-1} \tag{37}$$

in terms of the wavelength λ_0 of the absorption edge in zero field and $\lambda_L \equiv 2\pi c/\Omega_L$. For a field of 10^7 G, this interval extends for ~ 60 Å on either side of the Balmer edge ($\lambda 3646$) and for ~ 300 Å on either side of the Paschen edge ($\lambda 8201$). In these same intervals $p_{\max}/\bar{\tau}$ also becomes large and changes sign twice. The magnitudes of $v/\bar{\tau}$ and $p_{\max}/\bar{\tau}$ are independent of field strength in the intervals $\lambda_- \leqslant \lambda \leqslant \lambda_+$. The values of $v/\bar{\tau}$ and $p_{\max}/\bar{\tau}$ in these intervals may be estimated quite accurately by again using

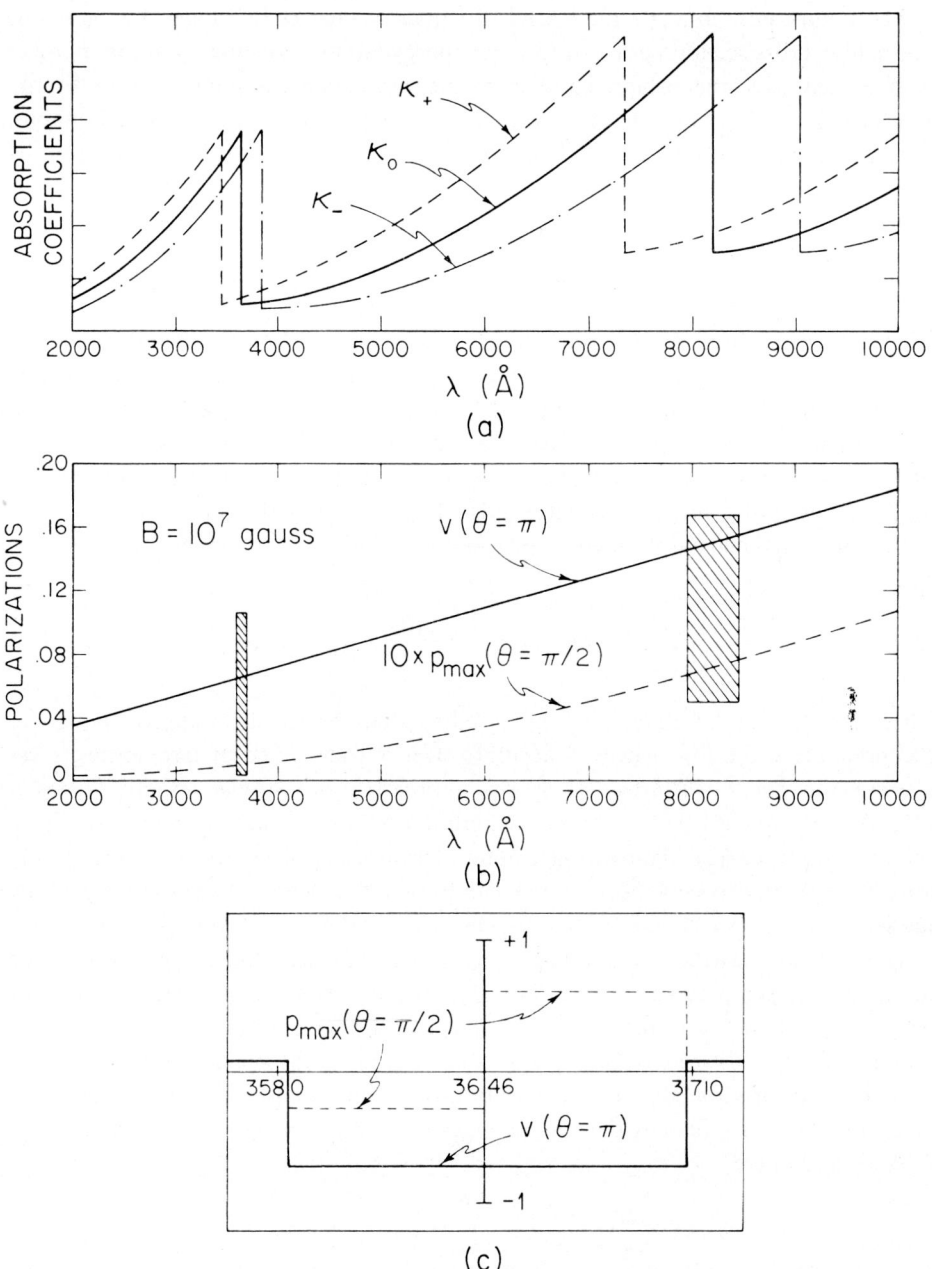

Fig. 3. The behavior of the absorption coefficients for bound-free absorption in a magnetic field and the resulting circular and linear continuum polarization. The general behavior of the absorption coefficients is shown in (a), although much exaggerated. The resulting net polarizations $v/\bar{\tau}$ and $p_{max}/\bar{\tau}$, calculated for propagation parallel and perpendicular to the field, respectively, are shown in (b) for $B = 10^7$ G. The detailed behavior of $v/\bar{\tau}$ and $p_{max}/\bar{\tau}$ at the Balmer edge is shown in (c).

the Kramers approximation. At the Balmer edge this gives

$$\frac{v(\theta=\pi)}{\bar{\tau}} \simeq -0.77 \qquad 3586 \text{ Å} \lesssim \lambda \lesssim 3706 \text{ Å}$$

and

$$\frac{p_{\max}(\theta=\pi/2)}{\bar{\tau}} \simeq \begin{cases} -0.28 & 3586 \text{ Å} \lesssim \lambda < 3646 \text{ Å} \\ +0.62 & 3646 \text{ Å} < \lambda \lesssim 3706 \text{ Å}, \end{cases}$$

while at the Paschen edge we have

$$\frac{v(\theta=\pi)}{\bar{\tau}} \simeq -0.62 \qquad 7900 \text{ Å} \lesssim \lambda \lesssim 8500 \text{ Å}$$

$$\frac{p_{\max}(\theta=\pi/2)}{\bar{\tau}} \simeq \begin{cases} -0.24 & 7900 \text{ Å} \lesssim \lambda < 8201 \text{ Å} \\ +0.45 & 8201 \text{ Å} < \lambda \lesssim 8500 \text{ Å}. \end{cases}$$

By contrast, a field of 10^7 G gives $v(\theta=\pi)/\bar{\tau}=0.073$ and $p_{\max}(\theta=\pi/2)/\bar{\tau}=0.0017$ at $\lambda 4000$, far from the edges. The Balmer edge behavior is illustrated in Figure 3(c). The behavior near the Balmer edges of helium is somewhat more complicated. Because the effective screening of the nucleus by the $1s$ electron depends somewhat upon the orbital angular momentum of the outer electron and the total spin of the two electrons, there are four separate Balmer edges (two each for orthohelium and parahelium). These edges appear at $\lambda\lambda 2600, 3420$ in orthohelium and at $\lambda\lambda 3120, 3680$ in parahelium – in both cases the separation between the edges considerably exceeds the width of the region of characteristic polarization changes, even for field strengths as great as 10^7 G. For higher wavelength edges in helium the situation simplifies because the screening of the nucleus varies less with the angular momentum states.

If some white dwarfs have magnetic fields as large as $B \sim 10^7$ G and if photoionization of hydrogen-like atoms is a significant source of opacity, then features of the type just described should be observable near absorption edges. It must be kept in mind, however, that disordered fields and the presence of other opacity sources will tend to wash out these features to a greater or lesser degree, depending on the structure of the atmosphere.

4. Free-Free Absorption in Magnetic Fields

In normal white dwarf atmospheres, an important contribution to the continuum opacity is due to free-free absorption by electrons moving in the Coulomb field of ions (inverse bremsstrahlung). In the presence of a magnetic field a new type of process becomes possible in which electrons which are otherwise free are able to absorb radiation by transferring momentum to the magnetic field (the inverse of cyclotron radiation). This process will give rise to a continuous opacity with a strong linear and circular absorption peak near the cyclotron resonance frequency, $\Omega_c = eB/mc$, corresponding to transitions between adjacent Landau levels. For magnetic field strengths $B \gtrsim 10^8$ G, this peak will fall in the infrared or optical (the wavelength in μm is $\lambda_c = 2\pi c/\Omega_c = 1.07/B_8$, where B_8 is the field strength in units of 10^8 G). For this reason, it has been suggested (see Kemp, 1970b) that the cyclotron process may be related to the polarization features seen in the white dwarf Grw + 70° 8247. In the present section we obtain a quantitative estimate for the free-free opacity in magnetic white dwarf atmospheres due to this process and compare it with the results for the bound-free opacity obtained in the previous section.

For the temperatures typical of white dwarf atmospheres ($T \sim 10^4$ K) and the magnetic field strengths of interest ($B \sim 10^8$ G), $\hbar\Omega_c \sim k_B T$ so that only the first few Landau levels will be significantly populated. Thus the cyclotron absorption cross section must be calculated quantum-mechanically. The cyclotron radius of the electron motion, for magnetic fields of these strengths and velocities determined by these temperatures, is considerably smaller than the wavelength of radiation at the cyclotron frequency. As a consequence the absorption of radiation is governed by the usual electric dipole rules, and we may thus calculate the opacities $K_q = n_e \sigma_q$ defined in (19). Here n_e is the number density of free electrons. In addition, the spiral-like motion of the electrons provides the kinematic restrictions that σ_-, σ_0 are negligible while σ_+ is peaked at the cyclotron frequency. Assuming nondegenerate electrons in LTE, the absorption cross section σ_+ is (see Lamb and Sutherland, 1971):

$$\sigma_+(\omega, \theta) \approx \frac{1}{|\cos\theta|} \frac{1}{\sqrt{\pi}} \left(\frac{e^2}{\hbar c}\right) \left(\frac{2\pi c}{\Omega_c}\right)^2 \left(\frac{mc^2}{2k_B T}\right)^{1/2} \times$$
$$\times \left(\frac{B}{B_q}\right) \frac{1}{1 - e^{-\hbar\Omega_c/k_B T}} \exp\left\{-\frac{mc^2}{2k_B T \cos^2\theta} \frac{(\omega - \Omega_c)^2}{\Omega_c^2}\right\}, \quad (38)$$

for light of angular frequency ω propagating at an angle θ with respect to the magnetic field; here $B_q = m^2 c^3/e\hbar = 4.414 \times 10^{13}$ G. From (38) it is evident that the cross section is strongly peaked about $\omega = \Omega_c$ with a width given by

$$\frac{\Delta\omega}{\Omega_c} \approx \left(\frac{2k_B T}{mc^2}\right)^{1/2} |\cos\theta|. \quad (39)$$

A measure of the strength of the absorption is provided by the frequency-integrated cross section:

$$\sigma_+ \equiv \int d\omega \, \sigma_+(\omega, \theta) \approx \frac{e^2}{\hbar c} \left(\frac{2\pi c}{\Omega_c}\right)^2 \left(\frac{B}{B_q}\right) \frac{\Omega_c}{1 - e^{-\hbar\Omega_c/k_B T}}. \quad (40)$$

For $T = 10^4$ K and $B = 10^8$ G, the peak occurs at λ 10700 with a full width at $1/e$ of roughly 39 $|\cos\theta|$ Å (cf. Figure 4); (40) gives for the integrated cross section 0.45 cm^2 s^{-1}. By comparison, the frequency-integrated cross section for hydrogen bound-free absorption from the Paschen edge to the Balmer edge is only 0.02 cm^2 s^{-1}.

Because only one K_q (that for $q = +1$) is nonzero, the light in the neighborhood of the cyclotron absorption feature should exhibit strong circular and linear polarization. For the sake of comparison with the results of Section 3, we may note that a homogeneous magnetic field in a layer of optical depth $\bar{\tau} \ll 1$ whose opacity is due only to cyclotron absorption will lead to values of $v(\theta)/\bar{\tau}$ and $p_{max}(\theta)/\bar{\tau}$ [cf. (22) and (23)] given by

$$\frac{v(\theta)}{\bar{\tau}} = -\frac{2\cos\theta}{1+\cos^2\theta},$$
$$\frac{p_{max}}{\bar{\tau}} = \frac{\sin^2\theta}{1+\cos^2\theta},$$
(41)

where θ is the angle between the direction of propagation and the magnetic field; of course these polarizations only exist in the immediate vicinity of the cyclotron frequency [cf. (39)]. Both v and p_{max} have the same sign as given by absorption due to hydrogen-like bound-free transitions at wavelengths not in the vicinity of an absorption edge.

In a typical white dwarf atmosphere, the magnetic field is expected to be far from uniform – it will vary both in magnitude and direction over the stellar surface. The effect of variations in magnitude is to shift the location of the cyclotron absorption peak. Thus at a particular frequency one observes (due to the cyclotron absorption)

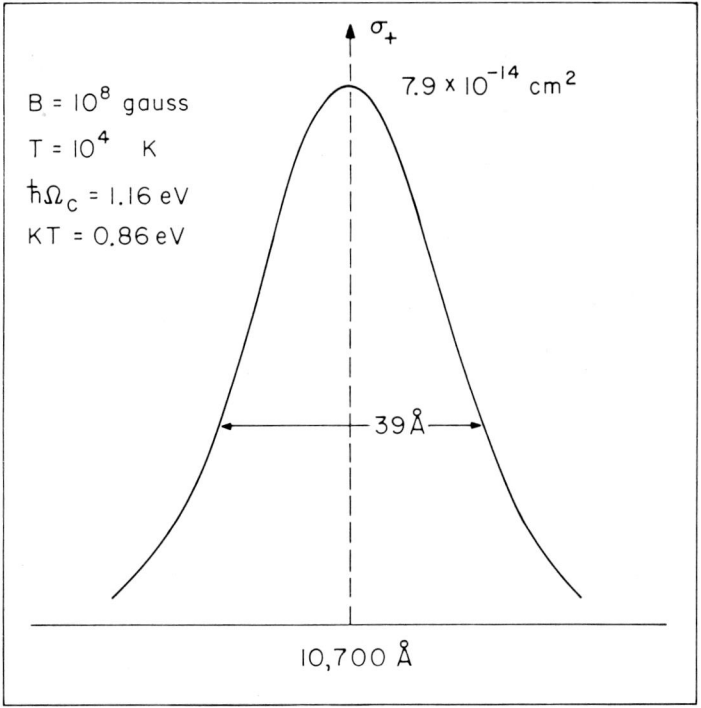

Fig. 4. Characteristics of the cyclotron absorption cross section σ_+ for the indicated temperature and magnetic field. The width shown is the full width at $\sigma_+(max)/e$.

contributions to the polarization from all points in the stellar atmosphere for which the magnetic field strength is (roughly) the same; for these points, however, the angle θ between the magnetic field and light propagating towards the observer will in general vary so that v and p_{max} as given in (41) must be suitably averaged.

5. Conclusions

In the preceding sections we have presented some of the effects magnetic fields in the range 10^4–10^8 G would have on the observed continuum emission of white dwarfs. Before examining the magnetic field effects at the atomic level, we introduced the radiative transfer equation for polarized light in an anisotropic medium. Within this framework it is possible to discuss the relative effects of the real and imaginary parts of the indices of refraction of the medium, to obtain the form of the emission matrix, and to isolate the geometrical factors related to the direction of propagation of the light relative to the magnetic field. We found that in a semi-infinite plane-parallel atmosphere, provided the characteristic states of the medium are nearly orthogonal, the polarization of the radiation is determined only by the absorption matrix and the run of temperature in the atmosphere. For the case of electric dipole absorption in a magnetic field the absorption matrix may be directly expressed in terms of the atomic absorption coefficients and the geometrical factors. To illustrate the qualitative features of the polarization induced by such anisotropic absorption, we considered absorption by a cold, optically-thin layer. In this greatly simplified model of a white dwarf atmosphere, we obtained formulae for the net circular and linear polarizations in terms of the absorption coefficients and the geometrical factors.

The continuum polarization arising from bound-free transitions in magnetic fields was calculated within the hydrogenic approximation. Away from absorption edges a hydrogenic bound-free opacity gives the same field strength and frequency dependence of the polarizations as that given by the grey-body model. However, near an absorption edge the opacity is discontinuous and the polarization undergoes sudden changes in sign and magnitude. If hydrogen or helium bound-free processes (or those of heavier atoms satisfying the hydrogenic approximation) form a significant source of opacity in white dwarfs with magnetic fields of order 10^7 G, these changes should be observable. Discovery of this type of behavior in the continuum polarization of white dwarfs would be of great interest for at least two reasons. First, it would indicate that a particular bound-free opacity makes a significant contribution to the star's opacity, thus yielding information about the composition, density, and temperature of the atmosphere. In this connection it is worth noting that fields as large as 10^7 G might make difficult the detection of an absorption edge using intensity measurements alone. Second, it would make possible a more accurate estimate of the average field strength than is possible on the basis of broad-band continuum polarization measurements. For these reasons, a search for edge behavior in the continuum polarization of white dwarfs seems desirable.

We have also discussed the opacity due to cyclotron absorption by electrons. The cross section is strongly peaked at the cyclotron frequency, which falls in the infrared or optical for $B \geqslant 10^8$ G, and is large (the integrated cyclotron cross section is $\sim 10^2$ times the hydrogen bound-free absorption cross section integrated over the optical region of the spectrum). This process can produce net linear and circular polarizations of comparable magnitude. For stars with magnetic fields that are not constant in magnitude and direction throughout the atmosphere, the cyclotron absorption effects will be spread across considerable wavelength ranges with possibly quite complicated wavelength-dependent polarizations.

Acknowledgements

We are grateful to many colleagues, both at Columbia and the University of Illinois, for many fruitful discussions, and to Roger Angel for communicating in advance of publication details of work performed by him and Landstreet. We would like to express appreciation for the kind hospitality of the Aspen Center for Physics, where part of this work was completed. This work was supported by NSF Grants GP25855 and GP32336X.

Appendix

In this appendix we discuss the polarization information contained in the polarization matrix and give rules for the determination of the Stokes parameters (further details may be found in Lamb and ter Haar (1970)). The polarization matrix is in most respects similar to the density matrix introduced in quantum mechanics; the primary difference is that the polarization matrix does not have a fixed normalization, in order that the absorption of light as it propagates through the medium may be represented within the polarization matrix formalism.

The polarization matrix in the Cartesian basis, $\mathscr{I}^{(c)}$, is Hermitian and thus its eigenvectors may be chosen to form an orthonormal basis. The congruency transformation of $\mathscr{I}^{(c)}$ under this change of orthonormal bases diagonalizes the polarization matrix:

$$\mathscr{I}' = \mathscr{U}^{-1} \mathscr{I}^{(c)} \mathscr{U}^{-1\dagger} = \begin{pmatrix} \lambda_+ & 0 \\ 0 & \lambda_- \end{pmatrix}. \tag{A1}$$

The degree of polarization p is given by

$$p = \frac{\lambda_+ - \lambda_-}{\lambda_+ + \lambda_-} \tag{A2}$$

from which it follows that p may be written in the invariant form

$$p = \left\{ 1 - \frac{4 \det(\mathscr{I})}{[\operatorname{tr}(\mathscr{I})]^2} \right\}^{1/2}. \tag{A3}$$

Because the transformation of the polarization matrix under a change of basis from the Cartesian basis [see (3) and (4)] is only a similarity transformation if the new basis is orthonormal, we must restrict ourselves to orthonormal bases so that the trace and determinant of the polarization matrix have invariant values. The trace of $\mathscr{I}$ is just the net intensity of the radiation; det $(\mathscr{I})=0$ is a necessary and sufficient condition for the radiation to be completely polarized. Note that unpolarized radiation is characterized by a polarization matrix proportional to the unit matrix, the proportionality constant being one-half the total intensity.

Various sets of Stokes parameters may be introduced to give an alternative description of polarized radiation (see, for example, Chandrasekhar, 1960, pp. 24–31); these may, in general, be obtained from the polarization matrix in the Cartesian basis. A conventional set of Stokes parameters is given by:

$$S_i = \mathrm{tr}(\sigma_i \mathscr{I}^{(c)}) \qquad i = 0, 1, 2, 3 \tag{A4}$$

where the σ_i are the Pauli matrices and the unit matrix:

$$\sigma_0 = \begin{pmatrix} 1 & 0 \\ 0 & 1 \end{pmatrix}, \quad \sigma_1 = \begin{pmatrix} 1 & 0 \\ 0 & -1 \end{pmatrix}, \quad \sigma_2 = \begin{pmatrix} 0 & 1 \\ 1 & 0 \end{pmatrix}, \quad \sigma_3 = \begin{pmatrix} 0 & -i \\ i & 0 \end{pmatrix}. \tag{A5}$$

S_0 is the net intensity and one may easily show that $v = S_3/S_0$ is the net circular polarization. The net linear polarization along an axis inclined at an angle ϕ to the 1-axis (the light propagates along the 3-axis), defined as the linear polarization measured along this axis minus the linear polarization measured along an axis at right angles to this axis, is given by

$$\begin{aligned} p(\phi) &\equiv p_\parallel - p_\perp \\ &= (S_1 \cos 2\phi + S_2 \sin 2\phi)/S_0. \end{aligned} \tag{A6}$$

The quantity $p_{\max}$ appearing in the previous sections and referred to as the 'net linear polarization' is the maximum value of $p(\phi)$ with respect to ϕ. It is conventional to choose the (as yet unspecified) orientation of the Cartesian 1-axis so that $p_{\max}$ coincides with S_1/S_0.

References

Angel, J. R. P.: 1972, *Astrophys. J. Letters* **171**, L17.
Angel, J. R. P. and Landstreet, J. D.: 1970a, *Astrophys. J. Letters* **160**, L147.
Angel, J. R. P. and Landstreet, J. D.: 1970b, *Astrophys. J. Letters* **162**, L61.
Angel, J. R. P. and Landstreet, J. D.: 1971a, *Astrophys. J. Letters* **164**, L15.
Angel, J. R. P. and Landstreet, J. D.: 1971b, *Astrophys. J. Letters* **165**, L71.
Angel, J. R. P. and Landstreet, J. D.: 1972, *Astrophys. J. Letters*, **178**, L21.
Angel, J. R. P., Illing, R. M. E., and Landstreet, J. D.: 1972a, *Astrophys. J. Letters* **175**, L85.
Angel, J. R. P., Landstreet, J. D., and Oke, J. B.: 1972b, *Astrophys. J. Letters* **171**, L11.
Chandrasekhar, S.: 1960, *Radiative Transfer*, Dover, New York, p. 24ff.
Chanmugam, G., O'Connell, R. F., and Rajagopal, A. K.: 1972, *Astrophys. J.* **175**, 157.
Chow, T. L.: 1969, *Astrophys. Letters* **3**, 85.
Frank-Kamenetskii, D. A.: 1962, *Physical Processes in Stellar Interiors* (translated from the Russian and published for the National Science Foundation by the Israel Program for Scientific Translations), p. 144.
Gehrels, T.: 1971 (unpublished).
Ginzburg, V. L., Zheleznyakov, V. V., and Zaitsev, V. V.: 1969, *Astrophys. Space Sci.* **4**, 464.
Greenstein, J. L.: 1970, *Astrophys. J. Letters* **162**, L55.

Greenstein, J. L., Gunn, J. E., and Kristian, J.: 1971, *Astrophys. J. Letters* **169**, L63.
Greenstein, J. L. and Matthews, M. S.: 1957, *Astrophys. J.* **126**, 14.
Kemp, J. C.: 1970a, *Astrophys. J. Letters* **162**, L69.
Kemp, J. C.: 1970b, *Astrophys. J.* **162**, 169.
Kemp, J. C. and Swedlund, J. B.: 1970, *Astrophys. J. Letters* **162**, L67.
Kemp, J. C., Swedlund, J. B., Landstreet, J. D., and Angel, J. R. P.: 1970, *Astrophys. J. Letters* **161**, L77.
Kemp, J. C., Swedlund, J. B., and Wolstencroft, R. D.: 1971, *Astrophys. J. Letters* **164**, L17.
Lamb, F. K. and ter Haar, D.: 1970, Oxford University, Department of Theoretical Physics, Reference No. 38/70.
Lamb, F. K. and Sutherland, P. G.: 1971, paper presented at the Conference on Line Formation in the Presence of Magnetic Fields, 30 August–2 September 1971, High Altitude Observatory, Boulder, Colorado.
Landstreet, J. D. and Angel, J. R. P.: 1971, *Astrophys. J. Letters* **165**, L67.
Ostriker, J. P. and Hartwick, F. D. A.: 1968, *Astrophys. J.* **153**, 797.
Preston, G. W.: 1970, *Astrophys. J. Letters* **160**, L143.
Shipman, H. L.: 1971, *Astrophys. J.* **167**, 165.
Trimble, V.: 1971, *Nature* **231**, 124.

POLARIZED RADIATION FROM WHITE DWARFS AND ATOMS IN STRONG MAGNETIC FIELDS

R. F. O'CONNELL

Dept. of Physics and Astronomy, Louisiana State University,
Baton Rouge, Louisiana 70803, U.S.A.

Abstract. We present the recent results of our continuing program of investigation of the behavior of matter in strong to super-strong magnetic fields ($B \sim 10^6$–10^{12} G). This work was motivated by the discovery of strong magnetic fields ($B \sim 10^7$ G) in some white dwarfs and the likely existence of super-strong fields ($B \sim 10^{12}$ G) in pulsars. Magnetic white dwarfs were discovered from observations of the continuous spectrum and one of the most intriguing challenges for the theorist is to provide an explanation for the observed wavelength dependence of the fractional circularly and linearly polarized radiation. Our initial response to this question was the determination of an exact solution of Kemp's harmonic oscillator model. These results are used as input to the ATLAS model atmosphere program and then comparison is made with observations. The disparities still existing between theory and observation convince us of the necessity for developing a new model of the continuum radiation, two likely possibilities being photoionization and free-free absorption. This leads us to present a general formulation of radiation absorption and emission processes in a magnetic field. Next we calculate the cross section for the photoionization, correct to first order in B. For the purpose of obtaining exact results for this cross section, the effect of a magnetic field on the energy spectrum and wave functions of hydrogen, helium, etc. must be obtained. The results for hydrogen are presented here. They will be useful also in determining accurate values for the displacements due to the quadratic Zeeman effect in the line spectra of DA stars, particularly for the higher excited states.

1. Introduction

The discovery of magnetic fields, of the order of 10^7 G, in some, but not all, white dwarfs must surely rank high among the many exciting astronomical discoveries of today. Such fields are orders of magnitude larger than laboratory produced fields. The magnitude is consistent with the idea that magnetic flux is conserved during the evolution preceding the formation of white dwarfs and hence lends support to the generally accepted conclusion that even higher fields, of the order of the critical magnetic field $B_c = (m^2 c^3/e\hbar) = 4.4 \times 10^{13}$ G, exist in pulsars (Gold, 1969).

Initial attempts to positively detect magnetic fields were based on observations of *line* spectra. Preston (1970) and Trimble (1971) failed to find evidence for displacement due to the quadratic Zeeman effect in the lines of the DA stars whose radial velocities had been measured by Greenstein and Trimble (1967). In addition, Angel and Landstreet (1970a) searched unsuccessfully for evidence of the Zeeman effect by looking for circular polarization in the wings of Balmer lines.

A dramatic breakthrough was achieved when Kemp (1970) predicted that the *continuous* spectrum of light from white dwarfs should exhibit a fractional circular polarization q, given by

$$q \equiv [P_+(\omega) - P_-(\omega)]/[P_+(\omega) + P_-(\omega)] \simeq -(\Omega/\omega). \tag{1}$$

Here $P_\pm(\omega)$ are the intensities of right and left circularly polarized light (RCP and

LCP, respectively) of angular frequency ω, and $\Omega = (eB/2\mu c)$ is the Larmor frequency, where B is the magnetic field. The basic feature of this idea was confirmed in the laboratory by Kemp et al. (1970a) who placed incandescent sources between the pole pieces of a 25 kG magnet, and observed q values of 10^{-5}–10^{-4}, at a mean wavelength in the near infrared.

Soon afterwards, the first discovery of circularly polarized light from a white dwarf was reported (Kemp et al., 1970b), amounting to 1–3% in visible light from the semi-DC white dwarf Grw+70°8247, from which one calculates a B field of about 10^7 G. Further work on this star (Angel and Landstreet, 1970b) led to no evidence for variation by more than 0.1% over a period of 4 days but it also gave results in conflict with Kemp's (1970) theory.

The second discovery, by Angel and Landstreet (1971a) and Kemp et al. (1971), of circular polarization was made in the DC white dwarf G195–19+GR250, with a q value of about 0.42%, implying a B value of the order of 3×10^6 G. It was later shown (Angel and Landstreet, 1971b, 1972) that the polarization is periodically variable with a period of 1.33 days, the variation in the red being different from that in the blue-green. No detectable linear polarization was found.

The third discovery (Angel and Landstreet, 1971c) of circular polarization was made in the DG p white dwarf G99-37, with a q value of about 0.63%, with the possible existence of ~ 0.2–0.3% variations. No linear polarization was found.

A search for circular polarization in about 40 other white dwarfs has produced negative results.

Motivated by the fact that Kemp's prediction that $q \sim \lambda$ is at variance with the observations, Shipman (1971) extended Kemp's theory to take into account radiative transfer in the atmosphere of Grw+70°8247 and found a λ dependence of q more in conformity with the observations. However, two important discrepancies remained (a) in the *infrared*, the observed (Kemp and Swedlund, 1970) large q values of 8.5 and 15%, at mean wavelengths of 1.15 and 1.25 respectively, are far greater than the theoretical predictions and (b) in the *ultra-violet* the observed drop in q is somewhat greater than the theoretical predictions. An updating of Shipman's work to take into account the most recent observations has led to the conclusion (Roussel and O'Connell, 1973) that the agreement in the optical region becomes worse. A similar analysis carried out for the linear polarization showed even less agreement between predictions and observations.

The work of our group at LSU has been motivated primarily by the need to develop a detailed theoretical model which would hopefully explain the many unusual features of the polarized continuum radiation, particularly the wavelength dependency of q. Our efforts have been essentially two-pronged – (a) calculation of an exact solution of Kemp's harmonic oscillator model and (b) development of a more physical model based on the behavior of atoms, ions, and electrons in a strong magnetic field.

In Section 2 we present a general development of radiation absorption and emission processes in the presence of both a magnetic field and an arbitrary central potential

$V(r)$. The results derived here form the basis for the development of the various radiation models discussed in subsequent sections.

Section 3 is devoted to Phase (a) of our program which is now essentially completed. The method used (Chanmugan et al., 1972a, b) to obtain an exact solution of Kemp's model, valid for all values of the magnetic field and the temperature, is outlined and the results discussed. Predictions relating to the fractional linear polarization, q^* say, are also obtained. The latter results have been used as input to the ATLAS model atmosphere program (Kurucz, 1969) to take account of radiative transfer, by use of Shipman's method. Finally, we discuss the results of a comparison between the output results and observations.

Kemp (1970) has remarked that "... what is now very much needed is an exact calculation of the strong B-field levels of hydrogen." In Section 4 we present the results of such a calculation (Smith et al., 1972). We also expand on our previous remark to the effect that these results are also necessary for obtaining accurate results for the quadratic Zeeman terms in strong fields, particularly for the higher excited states.

In Section 5 we discuss transition probabilities in a strong magnetic field. Bound-bound transitions (Smith et al., 1973) are useful for quadratic Zeeman calculations. Bound-free and free-free transition probabilities we believe to be the basic ingredients of a realistic physical model, which will explain the polarized radiation from magnetic white dwarfs.

2. Radiation Absorption and Emission in a Magnetic Field

The Hamiltonian for a particle of mass μ and charge $-e$ in a central potential $V(r)$ and a magnetic field $\mathbf{B}(|\mathbf{B}|=B_z)$, and interacting with radiation, is

$$H = \frac{1}{2\mu}\left(\mathbf{P} + \frac{e}{c}\mathbf{A} + \frac{e}{c}\mathbf{A}_r\right)^2 + V(\mathbf{r}), \tag{2}$$

where $\mathbf{A}$ and $\mathbf{A}_r$ are the vector potentials of the external electromagnetic field and the radiation field, respectively. We choose

$$\mathbf{A} = \tfrac{1}{2}(\mathbf{B} \times \mathbf{r}), \tag{3}$$

so that $\nabla \cdot \mathbf{A} = 0$. Then, dropping A_r^2 terms (two-photon processes), we may write

$$H = H_0 + H_1, \tag{4}$$

where

$$H_0 = \frac{P^2}{2\mu} + V(\mathbf{r}) + \Omega L_z + \tfrac{1}{2}\mu\Omega^2(x^2 + y^2), \tag{5}$$

and

$$H_1 = \frac{e}{\mu c}\mathbf{A}_r \cdot [\mathbf{P} + \mu\Omega(\hat{\mathbf{B}} \times \mathbf{r})]. \tag{6}$$

Here $\Omega = (eB/2\mu c)$ is the Larmor frequency and L_z is the z-component of the angular momentum.

Since H_0 is invariant under rotations about the z-axis and under inversion, the eigenstates can be labelled by the eigenvalues of L_z and the parity. Thus, a general form of the eigenfunctions may be written

$$\psi_m^{\pm}(\mathbf{r}) \equiv \psi_t(\mathbf{r}) = \sum_{il} a_{il}^t f_{il}^t(r) Y_{lm}(\theta, \phi). \tag{7}$$

The sum on l in (7) over all even integers leads to the state with even parity ($+$) and the sum over odd l, to the odd parity ($-$) state. Here m is the eigenvalue of L_z. The $f^t(r)$ are suitably chosen functions of r and the a^t are parameters. In special cases (as for example when $V(r)$ refers to the potential of an oscillator – see Section 3) it may be possible to solve for ψ_t in closed form. However, in general (as for example when $V(r)$ is Coulombic – see Sections 4 and 5), ψ_t must be obtained numerically. For bound states ψ_t is obtained by variational techniques, whereas for continuum states ψ_t is obtained by a numerical solution of the Schrodinger equation. Thus, the probability per unit time of a spontaneous transition from a state t of energy E_t to a state t' of energy $E_{t'}$ (where t and t', of course, refer to eigenstates of H_0), with the emission of one photon into a solid angle $d\Omega$, in the presence of an external magnetic field B is

$$A_{t't} d\Omega = \frac{e^2 \hbar \omega_{t't}}{2\pi \mu^2 c^3} |M_{t't}^{\mathbf{k}}|^2 \, d\Omega, \tag{8}$$

where

$$M_{t't}^{\mathbf{k}} = \langle t' | e^{i\mathbf{k} \cdot \mathbf{r}} \left[\nabla \cdot \hat{e}_q + \frac{i\mu}{\hbar} \Omega (\hat{\mathbf{B}} \times \mathbf{r}) \cdot \hat{e}_q \right] | t \rangle. \tag{9}$$

The propagation vector, polarization vector, and angular frequency of the photon are denoted by $\mathbf{k}$, $\hat{e}_q$ and $\omega_{t't}$, respectively, where

$$\hbar \omega_{t't} = E_t - E_{t'}, \tag{10}$$

and where the unit directions $\hat{e}_q$ are defined by

$$\hat{e}_{\pm} = \mp \frac{1}{\sqrt{2}} (\hat{e}_x \pm i \hat{e}_y); \qquad \hat{e}_0 = \hat{e}_z. \tag{11}$$

From (9) it is clear we may write

$$|M_{t't}^{\mathbf{k}}|^2 = \left| \left[D_{t't}^{\mathbf{k}} + \frac{i\mu\Omega}{\hbar} F_{t't}^{\mathbf{k}} \right] \right|^2, \tag{12}$$

where

$$D_{t't}^{\mathbf{k}} = \langle t' | e^{i\mathbf{k} \cdot \mathbf{r}} \nabla \cdot \hat{e}_q | t \rangle \tag{13}$$

is the usual momentum matrix element, and where

$$F_{t't}^{\mathbf{k}} = \langle t' | e^{i\mathbf{k} \cdot \mathbf{r}} (\hat{\mathbf{B}} \times \mathbf{r}) \cdot \hat{e}_q | t \rangle. \tag{14}$$

Now if $kr \ll 1$, which is generally true for transitions in the discrete spectrum, then we may rewrite (12) in terms of the familiar dipole length matrix element

$$\mathbf{R}_{m'm} = \langle m' | \mathbf{r} | m \rangle. \tag{15}$$

In the dipole approximation $\Delta m = \pm 1$ or 0 and the parity of the wave function must change in the transition and so with this understanding we now choose to label our initial and final states simply by m and m', respectively. Now using (5), we obtain the commutation relation

$$[\mathbf{r}, H_0]_{m'm} = \frac{\hbar^2}{\mu} \langle m'| \mathbf{V} |m\rangle - \hbar(m' - m)\Omega \langle m'| \mathbf{r} |m\rangle. \tag{16}$$

But for a transition between two eigenstates of H_0, with eigenvalues E_m and $E_{m'}$, we have

$$[\mathbf{r}, H_0]_{m'm} = (E_m - E_{m'}) \langle m'| \mathbf{r} |m\rangle. \tag{17}$$

Hence

$$\langle m'| \mathbf{V} |m\rangle = \frac{\mu}{\hbar} \omega_{m'm}(1 + f) \mathbf{R}_{m'm}, \tag{18}$$

where

$$f \equiv (m' - m)(\Omega/\omega_{m'm}) \equiv (m' - m)d. \tag{19}$$

In addition, we note that

$$\langle m'| (\hat{\mathbf{B}} \times \mathbf{r}) \cdot e_z |m\rangle = i(m' - m) \mathbf{R}_{m'm} \cdot \hat{e}_z. \tag{20}$$

Hence, we finally obtain for the transition probability

$$A_{m'm} d\Omega = \frac{e^2}{2\pi\hbar c^3} \omega_{m'm}^3 |\mathbf{R}_{m'm} \cdot \hat{e}_q|^2 d\Omega. \tag{21}$$

The corresponding expression for the intensity of emission $P_{m'm}$ is

$$P_{m'm} d\Omega = \frac{e^2}{2\pi\hbar c^3} \omega_{m'm}^4 |\mathbf{R}_{m'm} \cdot \hat{e}_q|^2 d\Omega. \tag{22}$$

Turning now to photon absorption processes in a magnetic field, we shall take as a prototype photoionization. The magnetic field affects the matrix elements in the same way as for photon emission, with the result that the cross section (Bethe and Salpeter, 1957) for photoionization σ from a bound state t' to a continuum state t is

$$\sigma d\Omega = \frac{4\pi^2 e^2 \hbar^2}{\mu^2 c \omega_{t't}} |M_{t't}^k|^2 d\Omega. \tag{23}$$

In the dipole approximation

$$\sigma d\Omega = \frac{4\pi^2 e^2}{c} \omega_{m'm} |\mathbf{R}_{m'm} \cdot \hat{e}_q|^2 d\Omega. \tag{24}$$

The results of this section constitute the basic tools we need for a consideration of various models of polarized radiation.

3. Kemp's Harmonic Oscillator Model

As a tractable model of the radiating system, Kemp (1970) chose a collection of electronic harmonic oscillators. A feature of Kemp's model is that one has to have a distribution of oscillators with essentially a continuum of natural frequencies to account for the continuous emission. Putting $V(r) = \mu \omega_0^2 r^2/2$ in (5), where ω_0^2 is the natural frequency of the oscillator, we obtain

$$H_0 = \frac{1}{2\mu}[P_x^2 + P_y^2 + \mu^2 \omega_c^2 (x^2 + y^2)] + \frac{1}{2\mu}[P_z^2 + \mu^2 \omega_0^2 z^2] + \Omega L_z. \quad (25)$$

An exact quantum-mechanical solution to this problem has been obtained. It was shown (Chanmugam et al., 1972a) that this Hamiltonian is equivalent to that of a three-dimensional anisotropic oscillator with fundamental frequencies ω_0, $\omega_c - \Omega$ and $\omega_c + \Omega$ so that the energy eigenvalues are given by (with $\omega_c^2 = \omega_0^2 + \Omega^2$)

$$E = \hbar(\omega_c + \Omega)(n_+ + 1/2) + \hbar(\omega_c - \Omega)(n_- + 1/2) + \hbar\omega_0(n_z + 1/2), \quad (26)$$

where $(n_+ n_-, n_z)$ refer to the number of quanta of different frequencies. Introducing now the radiation terms H_I, and considering transitions of frequency ω leads to the

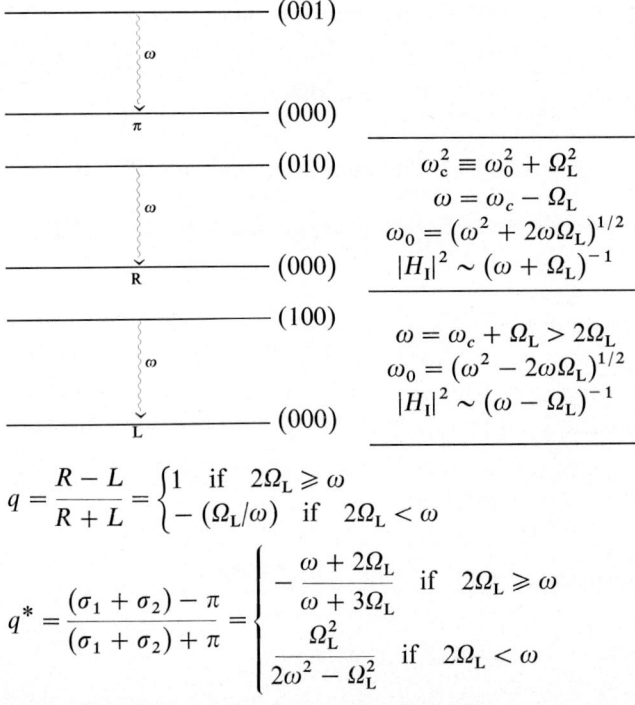

Fig. 1. Emission of linear, right-circularly, and left-circularly polarized light of frequency ω, from a system of harmonic oscillators of charge $-e$ and mass μ, with a continuous range of natural frequencies ω_0, in a magnetic field B, with associated Larmor frequency $\Omega = (eB/2\mu c)$.

results

$$|\langle 100| H_1 |000\rangle|^2 \sim (\omega - \Omega)^{-1} \tag{27}$$
$$|\langle 010| H_1 |000\rangle|^2 \sim (\omega + \Omega)^{-1} \tag{28}$$
$$|\langle 001| H_1 |000\rangle|^2 \sim \omega^{-1} \tag{29}$$

for the emission of LCP, RCP, and linearly polarized radiation, respectively. We note (see Figure 1) that LCP will not occur for $2\Omega > \omega$. Similar conclusions hold when transitions between all possible levels are considered (Chanmugan *et al.*, 1972b). Taking into account the effect of the field on the distribution of oscillators, and setting

$$s \equiv t^{-1} \equiv (\omega/2\Omega), \tag{30}$$

it follows that

$$q = (s^2 - 1)^{1/2} - s \quad \text{for} \quad s \geqslant 1 \tag{31}$$

and

$$q = 1 \quad \text{for} \quad 0 < s < 1, \tag{32}$$

in agreement with Kemp's classical results. In addition, we have also results for the

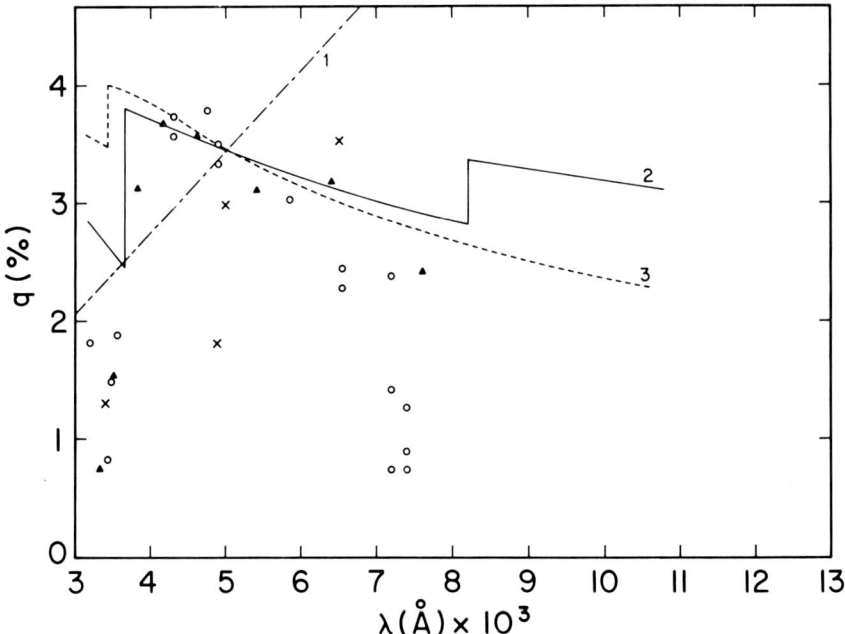

Fig. 2. The wavelength dependence of the predicted circular polarization is shown for the various models. Curve 1 corresponds to the optically thin model of Kemp (1970). Curve 2 corresponds to Model C with the parameters [$T = 12000$ K, $\log g = 8$, H $= 0.9$, He $= 0.1$, $B = 1.2 \times 10^7$ G]. Curve 3 corresponds to Model C with the parameters [$T = 14000$ K, $\log g = 8$, $H = 0.0$, He $= 1.0$, $B = 2 \times 10^7$ G]. The observed circular polarizations are indicated by the following: crosses indicate those of Kemp and Swedlund (1970), triangles those of Angel and Landstreet (1970b) and open circles those of Angel and Landstreet (1972).

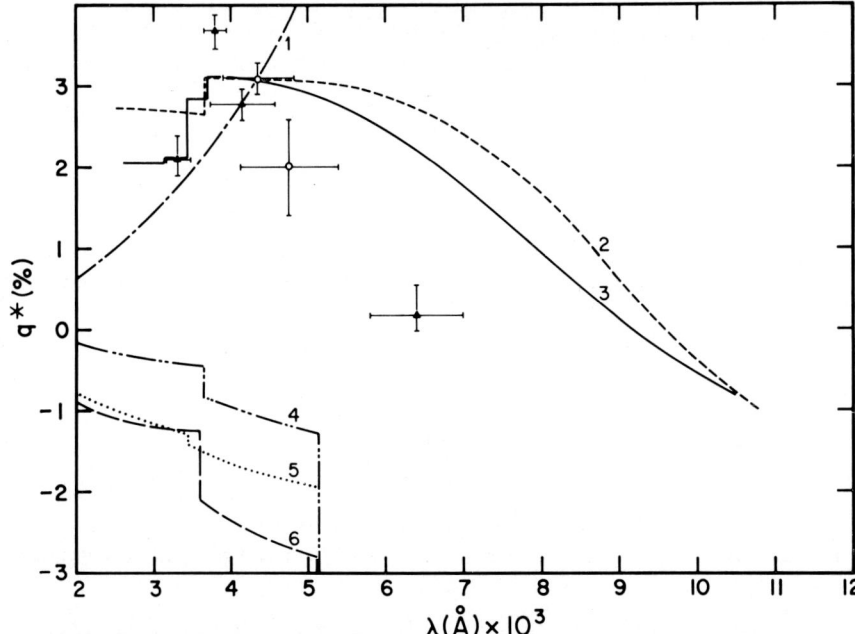

Fig. 3. The wavelength dependence of the predicted linear polarization is shown for the various models. Curve 1 corresponds to the optically thin model of Chanmugam et al. (1972a, b). Curve 2 corresponds to Model C with the parameters [$T = 12000$ K, $\log g = 8$, $H = 0.9$, He $= 0.1$, $B = 5 \times 10^7$ G]. Curve 3 corresponds to Model C with the parameters [$T = 14000$ K, $\log g = 8$, $H = 0.0$, He $= 1.0$, $B = 5.5 \times 10^7$ G]. Curve 4 corresponds to Model A with the parameters [$T = 12000$ K, $\log g = 8$, $H = 0.9$, He $= 0.1$, $B = 2.1 \times 10^8$ G]. Curve 5 corresponds to Model B with the parameters [$T = 14000$ K, $\log g = 8$, $H = 0.0$, He $= 1.0$, $B = 2.1 \times 10^8$ G]. Curve 6 corresponds to Model B with the parameters [$T = 12000$ K, $\log g = 8$, $H = 0.9$, He $= 0.1$, $B = 2.1 \times 10^8$ G]. The observed linear polarizations are indicated as follows; the triangles are those of Angel and Landstreet (1970b) and the open circles those of Angel and Landstreet (1972).

linear polarization ratio

$$q^* = [(1 + t)^{1/2} + (1 - t)^{1/2} + \\ - 2(1 - t^2)^{1/2}]/[(1 + t)^{1/2} + (1 - t)^{1/2} + 2(1 - t^2)^{1/2}] \quad \text{for} \quad t \leqslant 1$$

and

$$q^* = [s^{1/2} - 2(1 + s)^{1/2}]/[s^{1/2} + 2(1 + s)^{1/2}] \quad \text{for} \quad 1 > s > 0. \tag{33}$$

For $2\Omega \ll \omega$, these results reduce to

$$q = -(\Omega/\omega) \tag{34}$$

and

$$q^* = \tfrac{3}{4}(\Omega/\omega)^2. \tag{35}$$

If we had assumed a constant density of states even in the presence of a magnetic field, the q^* in (35) would be smaller by a factor of $\tfrac{2}{3}$ but (34) for q would be unchanged. Kemp's model of course provided the motivation for the successful search for circular polarization by Kemp et al. (1970b). However, the wavelength dependence of the observed q did not agree with the predicted $q \sim \lambda$ behavior. Now Kemp's model applies literally only to an optically thin body. Shipman (1971) was thus motivated to consider radiative transfer in Kemp's model. Actually he used Kemp's bremsstrahlung model which gives the same predictions as the harmonic oscillator model except that the q value given by (34) comes out 8 times larger. As a result Shipman was able to explain the wavelength dependence of the circular polarization, for $3000 \text{ Å} \leqslant \lambda \leqslant 9000 \text{ Å}$, with a derived B field $\sim 10^7$ G. However, discrepancies between the model predictions and the observations occurred in the ultra-violet and the infrared. In addition, it has been pointed out by Roussel and O'Connell (1973) that if the more recent observations of 1972 are included, the agreement in the optical region becomes worse, as shown in Figure 2. The observed values of q beyond 6000 Å are much smaller than predicted.

Making use of the predicted q^* values, Shipman's model, and the Atlas white-dwarf model atmosphere program, a similar analysis has been carried out for the linear polarization in Roussel and O'Connell (1973). The results are displayed in Figure 3. The agreement here is even worse than in the circular case. As a result, we feel that it is desirable to develop a new model for the polarized radiation which will predict the behavior of atoms in magnetic fields.

4. Energy Spectrum of the Hydrogen Atom in a Strong Magnetic Field

The behavior of atoms in low fields can be adequately described using hydrogenic wave functions in perturbation theory. On the other hand for super-strong magnetic fields ($B \gtrsim 3 \times 10^{10}$ G) the Coulomb interaction becomes negligible in comparison with the magnetic energy so that the wave functions are essentially oscillator-like (Cohen *et al.*, 1970). It is clearly of importance to understand at what field strengths perturbation theory using hydrogenic wave functions breaks down and to devise a scheme for analyzing the system in fields of intermediate strength. Kemp (1970) has emphasized the importance of this difficult intermediate case for magnetic white dwarfs. Thus, a program was initiated to study the behavior of atoms in fields of *any* strength, with emphasis on B values from about 10^4 G to 10^{12} G.

Our natural starting point is the hydrogen atom, for which (neglecting spin)

$$H_0 = \frac{p^2}{2\mu} - \frac{e^2}{r} + \Omega L_z + \tfrac{1}{2}\mu\Omega^2 r^2 \sin^2\theta, \tag{36}$$

where θ is the polar angle.

To assess the relative magnitudes of the various terms, it is useful to choose units $\hbar = c = \mu = 1$, so that $e^2 = \alpha = a_0^{-1}$, where α is the fine-structure constant and a_0 is the Bohr radius. Thus

$$H_0 = \frac{p^2}{2} + \frac{\alpha^2}{2}\left[-\left(\frac{r}{2a_0}\right)^{-1} + \left(\frac{B}{B_0}\right)L_z + \tfrac{1}{4}\left(\frac{B}{B_0}\right)^2\left(\frac{r}{a_0}\right)^2 \sin^2\theta\right], \tag{37}$$

where

$$B_0 \equiv \frac{\mu^2 c e^3}{\hbar^3} \equiv \alpha^2 B_c = 2.350 \times 10^9 \text{ G}. \tag{38}$$

Let
$$\psi_{nlm}(\mathbf{r}) = R_{nl}(r) P_{lm}(\theta) e^{im\phi} \tag{39}$$

be the normalized solution of the Schrödinger equation when $B=0$. Then, it is well known (Bethe and Salpeter, 1957, p. 206) that this wave function is also a solution with the B term (but not the B^2 term) included in H_0. The corresponding energy eigenvalue is

$$E = E_0 + \frac{\alpha^2}{2}\left(\frac{B}{B_0}\right) m, \tag{40}$$

where E_0 is the energy without the magnetic field.

This is the basic theory of the normal Zeeman effects and Paschen-Bach effects. The quadratic Zeeman effect (Schiff and Snyder, 1939) is obtained by treating the B^2 term using first order perturbation theory, with the result that the total energy is

$$E \equiv E_0 + E_1 + E_2 \equiv$$
$$\equiv \frac{\alpha^2}{2n^2}\left[-1 + mn^2\left(\frac{B}{B_0}\right) + F_{nlm}n^6\left(\frac{B}{B_0}\right)^2\right], \tag{41}$$

where

$$F_{nlm} = \frac{5}{4}\frac{\left[1 + \frac{1}{5n^2}(1 - 3l(l+1))\right][l(l+1) + m^2 - 1]}{(2l+3)(2l-1)}. \tag{42}$$

Thus, we expect perturbation theory to be valid only for $B \ll B_H \equiv (B_0/n^4)$. For large n and $l=1$, we have $F=(1+m^2)/4$, so that (in Rydbergs)

$$E_2 = \tfrac{1}{4}n^4(1+m^2)\left(\frac{B}{B_0}\right)^2 \text{Ry}, \tag{43}$$

which is the formula used by Preston (1970). However, since $B_H \sim n^{-4}$, we see that, for a particular B value, the perturbation results are less reliable for the higher states. For example, when $n=10$, we get $B_H = 2.350 \times 10^5$ G. It is thus clear that the deduction of B field values from an analysis of the Balmer absorption spectra from magnetic white dwarfs can be made considerably more quantitative by determining the exact energy eigenvalues of the hydrogen atom in a magnetic field. This analysis is now being carried out by Surmelian, using the results below. As already mentioned a further motivation for obtaining the spectrum is to obtain the effect of the B field on the opacities of the atmospheres of magnetic white dwarfs and, in particular, to explain the λ dependence of the circularly and linearly polarized light.

Our initial effort (Rajagopal et al., 1972) concentrated on the ionization energies of hydrogen in magnetic white dwarfs and the essence of the calculation was the use of a trial wave function which was *hydrogen*-like, in contrast to the *oscillator*-like trial wave function used by Cohen et al. (1970). Using merely a 4-parameter trial function,

$$\psi = c_1\psi_1(\beta_1 r) + c_2\psi_2(\beta_2 r), \tag{44}$$

the values of the ionization energy obtained were significantly better than those of Cohen *et al.* (1970) for fields $\geq 3 \times 10^{10}$ G (see Figure 4). A general procedure for carrying out a multi-parameter calculation was also outlined. Using standard computing techniques, it was then possible to obtain (Smith *et al.*, 1972) the energy spectrum for the 14 lowest states of the hydrogen atom. The essence of the method is to write a general trial solution of the form given in (7) with the radial function $d_{il}^t f_{il}^t$ chosen to be

$$(d_{il}^t r^l + b_{il}^t r^{l+1}) e^{-\beta_{il} t}, \tag{45}$$

where d and b are parameters.

For superstrong B fields we used a partial wave expansion with values of l up to $20+|m|$ included in summation (45). This was necessary in order to obtain convergence of the expansion of $\psi_m^{\pm}$. Up to nine Slater type orbitals were employed in the descrip-

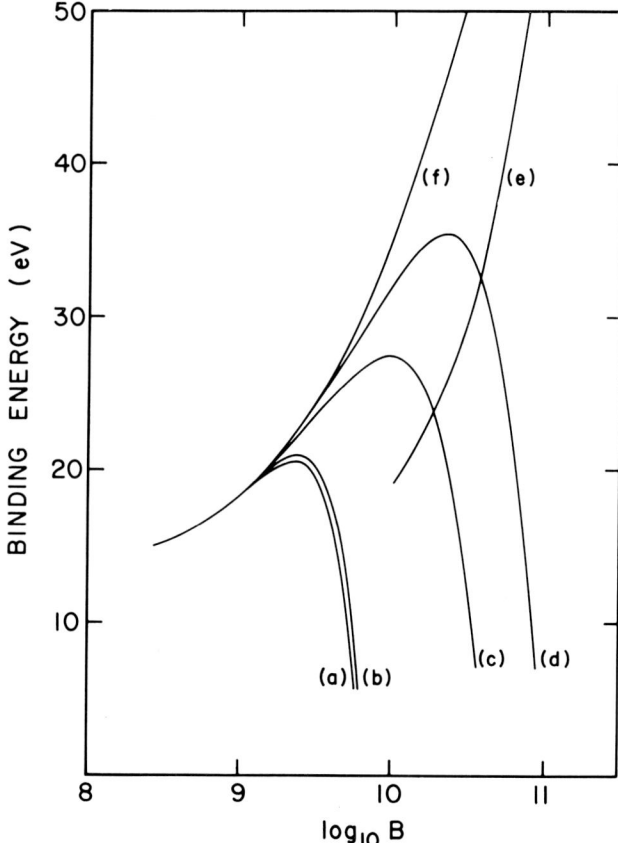

Fig. 4. The ionization energy of the ground state of hydrogen as a function of the magnetic field B calculated using (a) perturbation theory, (b) 2 linear parameter (c_1 and c_2) variational calculation, (c) 1 non-linear (β_1) parameter variational calculation, (d) 2 linear (c_1 and c_2) and 2 non-linear (β_1 and β_2) parameter variational calculation [see Equation (44)], (e) the work of Cohen *et al.* (1970), and (f) the work of Smith *et al.* (1972).

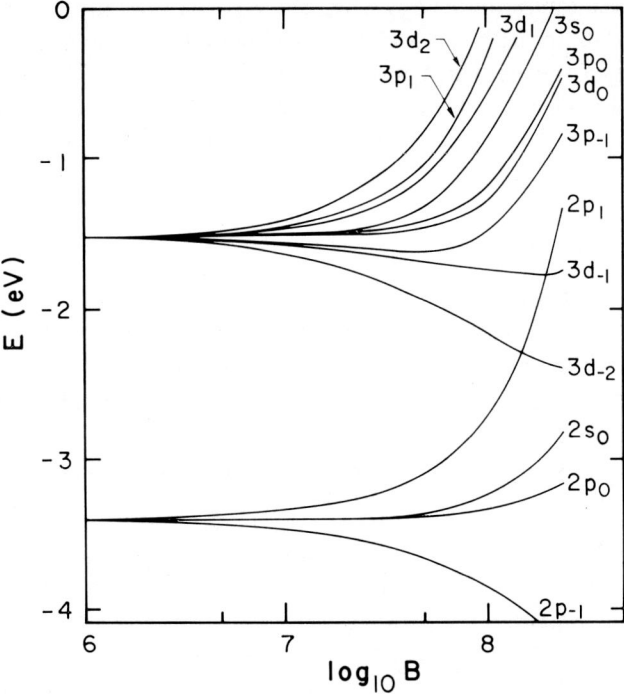

Fig. 5. The energy spectrum of hydrogen in a magnetic field for the 13 lowest states above the ground state.

tion of the radial function for the ground state and twelve for the excited states. Figure 4 gives the ionization energy E_1 (in eV) of the ground state of hydrogen as a function of the magnetic field B, calculated using (a) perturbation theory (b) two linear parameter (c_1 and c_2) variational calculation (c) one non-linear (β_1) parameter variational calculation (d) two linear (c_1 and c_2) and two non-linear (β_1 and β_2) parameter variational calculation [see Equation (44)] (e) the work of Cohen et al. (1970) (f) the work of Smith et al. (1972). The latter results are clearly superior.

In Figure 5, we present the energy spectrum for the 13 lowest states above the ground state for B values from 10^6–10^8 G. The labelling of the curves corresponds to the usual labels for the hydrogenic energy levels in the absence of a magnetic field. Thus, for example, $3d_2$ is an even-parity state with $m=2$, and, as $B \to 0$, it also has $n=3$ and $l=2$.

5. Bound-Bound and Bound-Free (Photoionization) Transition Probabilities in a Strong Magnetic Field

Bound-bound transition probabilities have been calculated (Smith et al., 1973) in the electric dipole approximation, using (21), between all states shown in Figure 5. Our results are presented in tabular form for some representative B values and are being used by Surmelian in his work on the quadratic Zeeman effect.

Henry and O'Connell are now calculating photoionization probabilities using (24), as we feel the results should provide us with an explanation of the spectral dependence of the polarized radiation. However, for magnetic fields for which $E_2 \ll E_1$ [see Equation (41)], analytic results may be obtained (Henry and O'Connell, 1972).

For $B = 0$, the cross section for ionization of the hydrogen atom in a state with principal quantum number n may be written (Karzas and Latter, 1961)

$$\sigma_0(\omega) = 2.82 \times 10^{29} n^{-5} (\omega/2\pi)^{-3} g(\omega, n), \tag{46}$$

where $g(\omega, n)$, the Gaunt factor, is of order unity and is a slowly varying function of ω and n. For our present purposes we will neglect the ω dependence in g (i.e. we essentially use the Kramers expression, while at the same time emphasizing that, apart from its complexity, there is no difficulty in principle in carrying it along).

Let $\omega_{m'm}$ be the frequency of the light absorbed for B non-zero. Then (19) and (40) tell us that

$$E_0 - E_0^1 = E - E^1 + \frac{\alpha^2}{2} \left(\frac{B}{B_0}\right)(m' - m) =$$
$$= \omega_{m'm}(1 + f) = \omega_{m'm}(1 - d\Delta m). \tag{47}$$

Thus, from (24) and (46), we obtain

$$\sigma_\pm d\Omega = (1 + f)^{-4} \sigma_0 d\Omega \tag{48}$$

which is similar to that found by Lamb and Sutherland (1972). Hence

$$(\sigma_{\pm 1}/\sigma_0) = (1 \mp d)^{-4}. \tag{49}$$

We now define the circular polarization ratio for the photoionization model,

$$q_p \equiv (\sigma_+ - \sigma_-)/(\sigma_+ + \sigma_-). \tag{50}$$

For small d, we have

$$(\sigma_{\pm 1}/\sigma_0) = (1 \pm 4d), \tag{51}$$

and

$$q_p = 4(\Omega/\omega_{m'm}). \tag{52}$$

The latter result is the same as that obtained from Kemp's model [see (34)], except for the factor of 4. We anticipate that inclusion of the dependence of the Gaunt factor, as well as B^2 terms in the energy eigenvalues may lead to more interesting predictions with respect to the λ dependence of q_p.

Acknowledgements

Most of the ideas presented here arose from continued collaboration and discussions with G. Chanmugan, R. J. W. Henry, A. K. Rajagopal, K. M. Roussel, E. R. Smith, and G. L. Surmelian, to whom the author would like to express his thanks. The author would like to thank Mrs Gladys Wiggins for excellent work with the manuscript, and Dr Henry for comments on same. It has come to our notice that many of the topics discussed above have also been investigated by Lamb and Sutherland (1972).

Note Added After the Conference

The energy spectrum of He II in a strong magnetic field, as well as bound-bound transition probabilities have now been obtained (Surmelian and O'Connell, 1973). The use of similar techniques in the study of multi-electron atoms is now being pursued. The photoionization calculation has been extended to include Gaunt factors (Roussel *et al.*, 1973). With regard to the detection of strong magnetic fields by using quadratic Zeeman displacements (see Preston, 1970), it is important to know, for the various *nlm* states, at what fields perturbation theory breaks down. This analysis has now been carried out (O'Connell, 1971). Free-free transitions in a magnetic field has also been suggested as a model for the polarized radiation (O'Connell, 1974).

References

Angel, J. R. P. and Landstreet, J. D.: 1970a, *Astrophys. J. Letters* **160**, L147.
Angel, J. R. P. and Landstreet, J. D.: 1970b, *Astrophys. J. Letters* **162**, L61.
Angel, J. R. P. and Landstreet, J. D.: 1971a, *Astrophys. J. Letters* **164**, L15.
Angel, J. R. P. and Landstreet, J. D.: 1971b, *Astrophys. J. Letters* **165**, L71.
Angel, J. R. P. and Landstreet, J. D.: 1971c, *Astrophys. J.* **165**, L67.
Angel, J. R. P. and Landstreet, J. D.: 1972, *Astrophys. J. Letters* **175**, L85.
Bethe, H. A. and Salpeter, E. E.: 1957, *Quantum Mechanics of One and Two Electron Atoms*, Springer-Verlag and Academic Press, New York.
Chanmugan, G., O'Connell, R. F., and Rajagopal, A. K.: 1972a, *Astrophys. J.* **175**, 157.
Chanmugan, G., O'Connell, R. F., and Rajagopal, A. K.: 1972b, *Astrophys. J.* **177**, 719.
Cohen, R., Lodenquai, J., and Ruderman, M. A.: 1970, *Phys. Rev. Letters* **25**, 467.
Gold, T.: 1969, *Nature* **211**, 25.
Greenstein, J. L. and Trimble, V. L.: 1967, *Astrophys. J.* **149**, 283.
Henry, R. J. W. and O'Connell, R. F.: 1972 (unpublished).
Karzas, W. J. and Latter, R.: 1961, *Astrophys. J. Suppl.* **6**, 167.
Kemp, J. C.: 1970, *Astrophys. J.* **162**, 169 and L69.
Kemp, J. C. and Swedlund, J. B.: 1970, *Astrophys. J. Letters* **162**, L67.
Kemp, J. C., Swedlund, J. B., and Evans, B. D.: 1970a, *Phys. Rev. Letters* **24**, 1211.
Kemp, J. C., Swedlund, J. B., Landstreet, J. D., and Angel, J. R. P.: 1970b, *Ap. J. Letters* **161**, L77.
Kemp, J. C., Swedlund, J. B., and Wolstencroft, R. D.: 1971, *Astrophys. J. Letters* **164**, L17.
Kurucz, R. L.: 1969, in O. Gingerich (ed.), *Theory and Observation of Normal Stellar Atmospheres*, M.I.T. Press, Cambridge, p. 375.
Lamb, F. K. and Sutherland, P. G.: 1972, in *Line Formation in the Presence of Magnetic Fields*, Manuscripts Presented at a Conference Held in Boulder, Colorado, 30 August – 2 September, 1971 (Nat'l Center for Atmospheric Research, Boulder), pp. 183–225.
O'Connell, R. F.: 1973, in T. Gehrels (ed.), 'Planets, Stars and Nebulae Studied with Photopolarimetry', *IAU Colloq.* **23** (to be published).
O'Connell, R. F.: 1974, *Phys. Letters A* (to appear).
Preston, G. W.: 1970, *Astrophys. J. Letters* **160**, L143.
Rajagopal, A. K., Chanmugan, G., O'Connell, R. F., and Surmelian, G. L.: 1972, *Astrophys. J.* **177**, 713.
Roussel, K. M. and O'Connell, R. F.: 1973, *Astrophys. J.* **182**, 277.
Roussel, K. M., Henry, R. J. W., and O'Connell, R. F.: 1973, (unpublished).
Schiff, L. I. and Snyder, H.: 1939, *Phys. Rev.* **55**, 59.
Shipman, H. L.: 1971, *Astrophys. J.* **167**, 165.
Smith, E. R., Henry, R. J. W., Surmelian, G. L., O'Connell, R. F., and Rajagopal, A. K.: 1972, *Phys. Rev.* **D6**, 3700.
Smith, E. R., Henry, R. J. W., Surmelian, G. L., and O'Connell, R. F.: 1973, *Astrophys. J.* **179**, 659 and **182**, 651 (E).
Surmelian, G. L. and O'Connell, R. F.: 1973, *Astrophys. and Space Sci.* **20**, 85.
Trimble, V. L.: 1971, *Nature Phys. Sci.* **231**, 124.

MATTER-ANTIMATTER SEPARATION AND THE ANTIMATTER PROBLEM IN COSMOLOGY

R. OMNÈS

Laboratoire de Physique Théorique et Hautes Energies, Bâtiment 211, Université de Paris-Sud, Centre d'Orsay, 91405 – Orsay France

Abstract. I shall report on some work done in France in the last few years concerning the possible existence of antimatter in the Universe. It will be shown that, by taking due account of several physical effects which have been theoretically predicted, one can propose a coherent theory for the origin of matter and galaxies that appears to be quantitatively satisfactory. In this model there should exist as many antigalaxies as galaxies. During the course of this paper, I shall try to make it clear what the basic physical processes are and the present state of their investigation. This discussion will therefore be restricted to the two aforementioned problems, *viz.*, the origin of matter and protogalaxies, leaving aside other astrophysical applications.

1. The Geometrical Background of Cosmology

As a general rule, we shall consider as a dogma our present knowledge of physics, only trying to investigate some of its consequences without allowing ourselves the freedom of changing the basic law of physics. Only if such a program were to obviously fail, could we feel justified in introducing new *ad hoc* assumptions. For that reason, general relativity will be taken as a proper overall description of space-time.

The relativistic description of space-time turns out to be particularly simple if one starts from two observed facts:

(i) The Hubble expansion of the Universe

(ii) The existence of a thermal background of radio waves with a temperature of 2.7 K. The radiation is known experimentally to be highly isotropic. We shall not discuss here any alternative interpretation for it.

Furthermore, according to the so-called cosmological principle, it is assumed that our position in the Universe is not privileged so that any other observer located on another galaxy would obtain the same picture of the Universe. Finally, a technical assumption will be to assume that Einstein's cosmological constant is zero.

Starting from these hypotheses, the structure and evolution of space-time have been well known for nearly forty years. The basic tool is provided by Einstein's equations which relate the geometry of space-time to its dynamical content, i.e. in practice to the mass density ϱ and the (radiative) pressure P. Isotropy of the background radiation implies that space itself is isotropic. Furthermore, coordinates can be chosen in such a way that there is a universal time t: this means that all co-moving observers (who are essentially located in galaxies) find the same local values of ϱ and P at a given time t and these quantities ϱ and P depend only upon t. Basically, this result expresses the fact that isotropy is incompatible with the existence of gradients for the dynamical quantities ϱ and P.

Under these conditions, it turns out that the geometry of spacetime can be characterized by two quantities, the time t and a scale factor $R(t)$ which describes expansion: an increase of $R(t)$ with time corresponds to an increase of the distances between two far-away galaxies in proportion to $R(t)$. (See Figure 1.) The Einstein equations therefore relate the geometry of space-time (R and t) to the energy-momentum (ϱ and P). The variations of ϱ and P with time are themselves simply related to the scale factor R (since, for instance, ϱ is proportional to R^{-3}, whereas, as a consequence of Einstein's equations, P can be shown to be proportional to R^{-4}: this last result is often described by saying that expansion is an isentropic process). Therefore, everything is known and one can solve the equations for $R(t)$ as a function of time.

It is found that $R(t)$ starts from a zero value at some finite time in the past which can be chosen as the origin of time and it increases up to its present value (see Figure 2). The present value of the time can be found by comparing the observed value of the Hubble constant with the theoretically predicted rate of expansion which is given by $(1/R) \times (dR/dt)$. A typical value for this time is of the order of 12×10^9 yr.

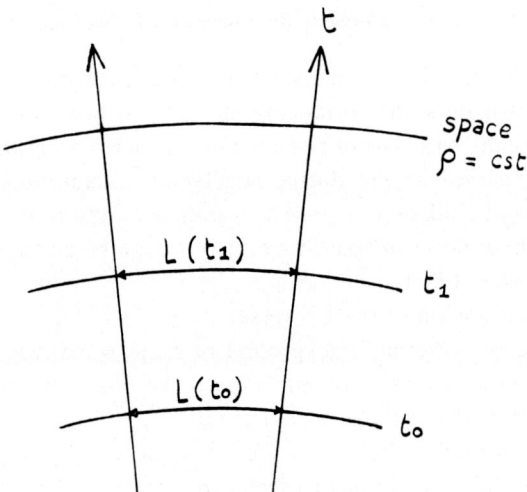

Fig. 1. The expansion of an isotropic Universe.

During the early stages of the Universe (during the first 10^6 yr or less), its evolution is governed by the thermal radiation it contains. Indeed, the conservation of entropy which is derived from the Einstein equations gives a simple relation between scale and temperature, namely

$$RT = \text{constant}. \tag{1}$$

Accordingly, for small values of $R(t)$ or small values of t, the temperature can be quite large, so that the contribution of radiation to the mass density and the pressure is dominant. Since these quantities depend only upon the temperature, the only remaining parameters are t and T which are related by

$$T(\text{MeV}) \simeq t^{-1/2}(\text{sec}). \tag{2}$$

From now on we shall stick for simplicity to this specific model. However, it is worth mentioning that most of what we shall say is essentially model-independent: we shall only exploit the existence of locally high temperatures without a real need for the global validity of the isotropic model of the Universe.

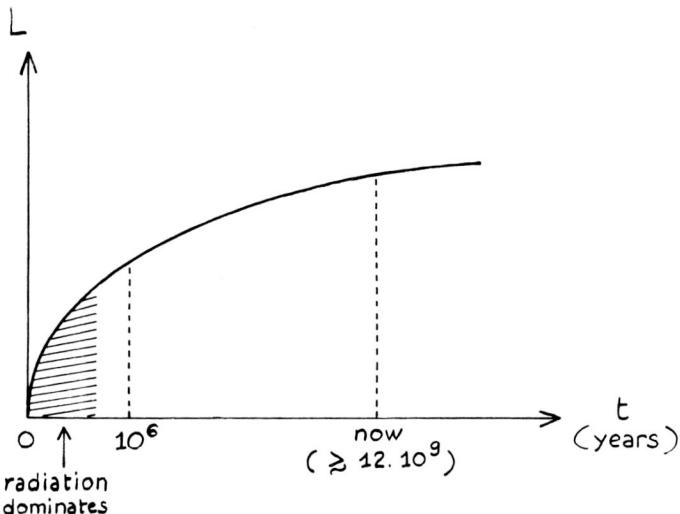

Fig. 2. The scale factor R as a function of time.

2. What is Thermal Radiation?

The conventional model of the Universe which has just been described is called the big-bang or the Gamow model. The content of the Universe is made of matter and radiation.

It has been stressed by Gamow that the nature of thermal radiation depends upon temperature: for small enough values of T, radiation consists of photons, as we all know. However, when the average thermal energy of photons is large enough to allow the production of electron-positron pairs in photon-photon collisions, one also finds positrons and electrons in radiation. In practice, this occurs when T becomes comparable with the mass energy of electrons, i.e. when T is of the order of 1 MeV or larger. At higher temperatures, other kinds of particles are also present in radiation: for instance, pions can be found when T is of the order of 100 MeV, as well as nucleons and antinucleons at higher temperatures.

From Table I let us note the important values (3000 K, 10^6 yr) which correspond to the combination of the electrons and protons in matter into atoms. Since the cross section for an interaction between a thermal photon and a neutral atom is much smaller than for an interaction with a free electron, the coupling between matter and radiation becomes very weak below 3000 K: in practice, it is found that the photons do not interact any more with matter. Their fate is thereby only to lose energy by red-shift according to (1).

It is very instructive at this stage to compare a few numbers:

(i) The ratio η between the densities of nucleons and photons

$$\eta = \frac{N \text{ nucleons}}{N \text{ photons}} = \frac{N}{N_\gamma} \qquad (3)$$

turns out to be a very fundamental quantity because it stays invariant under expansion. If the Universe were to contain antigalaxies, η would be defined as

$$\eta = \frac{N \text{ nucleons} + \bar{N} \text{ antinucleons}}{N \text{ photons}}, \qquad (4)$$

the averages being taken in large regions containing as many antigalaxies as galaxies. The observed value for η lies between 10^{-8} and 10^{-9}. This quantity can be taken as a measure of the amount of matter (and antimatter) in the Universe. Sometimes η^{-1} is also called the average entropy per baryon.

(ii) One can compare, at any given time, the quantity of nucleons which constitute matter now (their number density varies like R^{-3}) to the number of nucleons and antinucleons which are part of thermal radiation (their density varies as $T^{3/2}e^{-mc^2/T}$). By now, the second component is negligible, but it is much larger than the first one at $T \geqslant 1$ GeV (by a factor η^{-1}). Equality is obtained at $T = 30$ MeV.

TABLE I

The character of radiation at different temperatures

Temperature T	3000 K	100 keV	100 MeV
Time t	10^6 yr	10^2 s	10^{-4} s
Particles in radiation	γ	γ, e^+e^-, ν	$\gamma, e^\pm, \mu^\pm, \pi, \nu, N, \bar{N}$
Period	radiative combination	leptonic	hadronic

3. Where Does Matter Come From?

We are now ready to state the basic assumptions of the antimatter model of the Universe which we intend to investigate. We propose two very simple assumptions:

(i) There exists by now (and always existed) as much antimatter as matter. In other words, the total baryonic number of the Universe is zero.

(ii) The presently observed matter is a relic of the nucleons and antinucleons in the primordial thermal radiation.

Consideration of a few numbers will help us to envision the problems which must be solved if such a model is to make sense:

We have already noticed that the present amount of matter is comparable to what was found in radiation at a temperature of 30 MeV. Therefore, an initial separation of nucleons from antinucleons should have taken place at a higher temperature, i.e. during the hadronic period. But this period is precisely characterized by the fact that the densities of hadrons are large (of the order of the densities in atomic nuclei for temperatures above 200 MeV), so that strong interactions must play an important role in the behavior of radiation. A natural question to ask is therefore: could it be that strong interactions are responsible for a partial separation of nucleons from antinucleons during the hadronic period?

But even a separation generated by strong interactions cannot be the end of the story. Indeed, the hadronic period ends at a time $t \simeq 10^{-4}$ s. On the other hand, no physical effect can affect a region with dimensions larger than $ct \simeq 10^6$ cm. Computing the number of nucleons inside such a region at the end of the hadronic period, we find it to be much smaller than the number of nucleons in a galaxy. Since there is unquestionable evidence that normal galaxies are overwhelmingly made of only one type of matter, we must explain how an initial small-scale separation was amplified into a separation of matter from antimatter up to galactic scales.

The fundamental ingredients of our model are specific proposals for each of these mechanisms:

We propose that strong interactions produce a thermodynamic phase transition in thermal radiation at high temperatures. As a result of this transition, nucleons are partially separated from antinucleons. The corresponding effects occurred during the hadronic period.

Once nucleons and antinucleons are separated, annihilation along their boundary generates strong fluid motions. As a result of these motions, matter and antimatter are reorganized into increasingly larger regions of space. This will be called the *coalescence* effect. It is a corollary of the so-called 'Leidenfrost' mechanism which was first noticed by Alfvén and Klein (1962) and it takes place during the radiative period.

Both effects constitute physical problems that are basically independent of cosmology. The natural framework for the first one is statistical mechanics together with particle physics. For the second one, it is hydrodynamics.

In the following, we shall give our reasons for proposing these effects as well as the present status of their investigation. Later on, cosmological applications to the origin of matter and galaxies will be given.

4. Approaching Nucleon-Antinucleon Separation

We shall first describe the general behavior of nucleon-antinucleon separation for the sake of orientation.

It is a phase separation effect. To understand what that means, let us compare it with the liquid-vapor phase transition. In Figure 3, one has plotted the critical curve for liquid-vapor transition in a graph where the coordinates are density and temperature. The curve appearing in Figure 3 is the critical curve. For values of the density and temperature falling in the shaded area, liquid and vapor are separated in different regions of space. The size and shape of these regions depend upon the detailed history of the system (the time allowed since separation, for instance), but their respective volumes are completely determined by equilibrium.

In Figure 4, a similar plot has been made for the nucleon-antinucleon separation effect. The coordinates are the temperature and the baryonic density

$$B = N - \bar{N}. \tag{5}$$

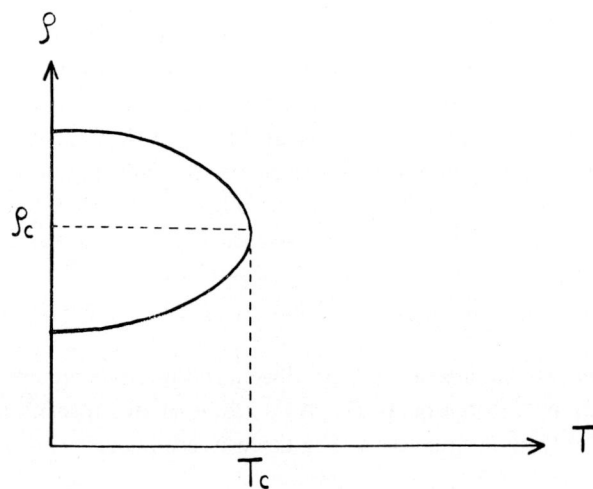

Fig. 3. Phase transition diagram for the liquid-vapor system.

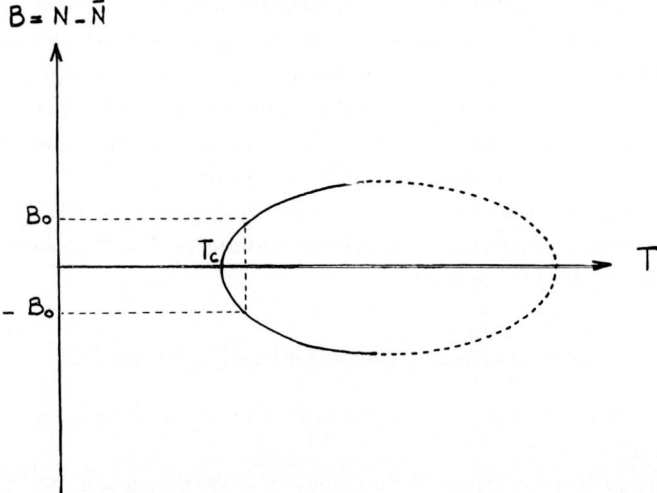

Fig. 4. Phase transition diagram for nucleon-antinucleon separation.

Since B is allowed to be different from zero, the system under consideration here is a gas of hadrons (for instance, an ordinary gas of matter) which is heated at extremely high temperatures. Here again, we have a critical curve and a critical temperature (which will be found later on to be of the order of 350 MeV). In the shaded area, the system consists of nucleons and antinucleons which are in spatially distinct regions. More exactly, there are regions with an excess B_0 of baryonic density or an excess $-B_0$ of antibaryonic density. The size of these regions depends upon the history of the system, but their volumes are completely specified by the equilibrium conditions.

The only important difference between the two cases is that, for the liquid-vapor system, the critical curve is convex towards high temperatures, whereas the reverse is true for the nucleon-antinucleon system. This can be understood as follows: separation is basically an effect which takes place at high densities and therefore at high temperatures since there are more and more nucleons and antinucleons in the system as the temperature increases. The shape of the critical curve tells us that the ordering effect due to high density is stronger than the disordering effect due to thermal motion.

It might be that the critical curve closes down at a higher temperature (see the dotted line) But this would be a region of densities and thermal energies where our knowledge of particle physics is still insufficient, so that we shall not discuss it.

Why is there a phase transition? We shall see that the basic reasons are the following:

(i) Mesons ($\pi, \varrho, \omega, \eta$) are assumed to be nucleon-antinucleon bound states, as first suggested by Fermi and Yang.

(ii) These mesonic bound states are part of thermal radiation, so that the free energy F is a stationary function of their number, i.e., for instance

$$\frac{\partial F}{\partial N_\pi} = 0. \tag{6}$$

It is interesting to note that, whereas bound states have been considered in statistical mechanics (for instance, molecules are bound states of atoms), condition (6) is never satisfied in the systems which have been met in the laboratory. In fact, the behavior of nucleons and antinucleons in thermal radiation is the simplest example where these two conditions are satisfied, so that it is worth considering with an open mind.

The task of investigating the thermodynamical behavior of nucleons and antinucleons is a rather formidable one: neither strong interaction physics nor the theory of phase transitions are easy subjects. Obviously, we must resort to highly idealized models. However, it is gratifying that there are by now two such models which both give a positive answer for the existence of the phase transition. Both together give strong support to the idea since the first one, by idealizing the particle physics, leads to a rigorous analysis of the statistical aspects while the second one is probably as good particle physics as can be done presently, approximations being made in the statistical treatment.

5. A First Model

We shall first offer a model in which statistical mechanics can be treated rigorously, while particle physics is strongly simplified. This model has the advantage of exhibiting the origin of the separation effect more clearly.

Let us treat the system of nucleons and antinucleons as a lattice gas, which means that the particles are located in the cells of a cubic lattice, the size of which is $2r_0$. (See Figure 5.) For r_0, we shall take the radius of a $N-\bar{N}$ bound state (assuming that such a notion is meaningful). Nucleons and antinucleons can be distributed among the cells but, when a nucleon is already present in a cell, the probability that an antinucleon in the same cell remains a free particle, rather than forming a bound state, is of the order of

$$P_0 = e^{-mc^2/T}. \tag{7}$$

On the other hand, according to (6), the bound states do not contribute to the free energy and may be forgotten. The model then consists in considering the small quantity P_0 as being strictly zero and passing to a new lattice with size r_0. We then have a lattice gas consisting of two species N and $\bar{N}$. Two N (or two $\bar{N}$) particles can be in neighboring cells while a N and a $\bar{N}$ cannot be first neighbors to each other. It is then intuitively clear that, when the numbers of N and $\bar{N}$ in the lattice grow as temperature increases, this statistical repulsion between N and $\bar{N}$ can only be accommodated if N and $\bar{N}$ separate into different regions of space containing many cells.

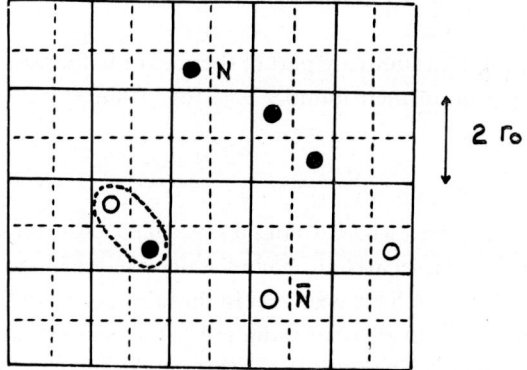

Fig. 5. A lattice space model of nucleon-antinucleon separation.

In fact, this model had been introduced independently by Widom and Rowlinson (1970) with quite different applications in mind and, recently, Ruelle (1971) has proved rigorously (in Ruelle's sense of the word) that the phase transition exists.

According to Widom and Rowlinson, the critical situation is obtained when the average number of particles N or $\bar{N}$ per cell is one, which corresponds typically, with values of r_0 less than a fermi to a temperature of the order of a few hundred MeV.

6. A Second Model

Another model has been proposed and investigated by several people (Ommes, 1969a,

1969b, 1970, 1972; Aldrovandi and Caser, 1972; Cisneros, 1972). The idea is the following: one starts from a virial expansion of the free energy F as a function of the nucleon and antinucleon densities N and $\bar{N}$. In practice, it reads

$$F(T) = F(\text{mesons}, T) + F_0(\text{free nucleons}, N, T) + F_0(\bar{N}, T) + \\ + a(N + \bar{N}) + bN\bar{N} + c(N^2 + \bar{N}^2). \tag{8}$$

The weakness of the treatment is to truncate the virial expansion to second-order terms. While this is known to preserve the critical behavior of the Van der Waals transition, it is nevertheless a strong assumption.

On the other hand, the virial coefficients a, b, c can be explicitly computed by using the scattering amplitudes for particle interactions. Indeed, it is known that a second virial coefficient can be expressed in terms of the phase shifts by means of the Beth-Uhlenbeck formula

$$2^d \text{ virial coefficient} \propto \sum_\alpha \int \frac{\mathrm{d}\delta_\alpha}{\mathrm{d}E} e^{-E/kT} \mathrm{d}E \tag{9}$$

so that the knowledge of the pion-nucleon and nucleon-nucleon phase shifts makes it possible to compute a and c. The calculation of b is more difficult but, according to Dashen et al. (1969), it can be cast into the form

$$b \propto \sum_n \int \langle N\bar{N}| S^+ |n\rangle \frac{\overleftrightarrow{\partial}}{\partial E} \langle n| S |N\bar{N}\rangle e^{-E/kT} \mathrm{d}E, \tag{10}$$

where S is the collision matrix. In this expression, terms of the type $\langle N\bar{N}|S|N\bar{N}\rangle$ have been computed by using either the mesonic theory of nuclear forces or phenomenological potentials. Annihilation matrix elements of the type $\langle N\bar{N}|S|$any number of pions$\rangle$ can be obtained from the statistical model for annihilation.

It is found in this way that b is large and negative (leading essentially to a statistical

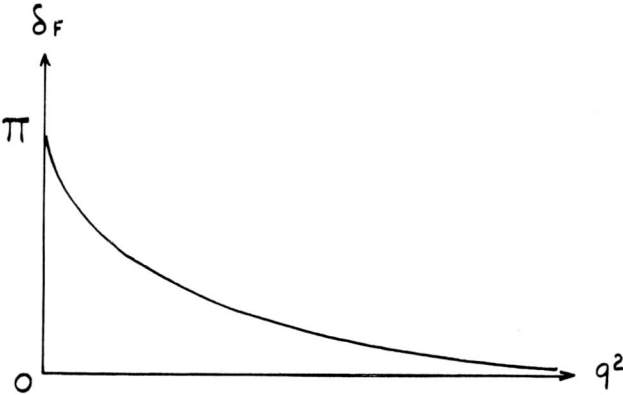

Fig. 6. A S-wave phase shift for nucleon-antinucleon scattering.

repulsion effect). This is due to the fact that, since there are bound states in all S-wave nucleon-antinucleon states, the corresponding phase shifts (or more technically, the Froissart phase shifts which take annihilation into account), decrease with the energy so that $d\delta/dE$ is negative for these states. This is the way the bound-state assumption enters into this calculation. (See Figure 6.) On the other hand, the mesons can be treated as elementary particles (as shown by Dashen *et al.* (1969), so that assumption (ii) (of Section 3) is also included in the independence of F_0(mesons) upon N and $\bar{N}$.

Imposing the thermodynamical conditions

$$\frac{\partial F}{\partial N} = 0, \quad \frac{\partial F}{\partial \bar{N}} = 0 \tag{11}$$

one finds again a phase transition with a critical temperature T_c of the order of 350 MeV.

For the sake of experimental verification, let us note that it would be important to check whether or not, as is assumed here, the real part of the S-wave $N - \bar{N}$ scattering lengths are negative. This could be observed by measuring the sign of the level shifts in the protonium (i.e., the $p - \bar{p}$) atom.

We shall conclude this part of our report by saying that, although no rigorous proof of the separation effect has been really obtained, the convergence of these approaches gives strong enough support for its existence to warrant a serious investigation of its cosmological consequences.

7. The Coalescence Effect

Now we turn to the physical situation during the radiative period: we have to consider a system where matter and antimatter occupy different regions of space more or less uniformly. The general geometry of their distribution is of an emulsion and we shall call the scale of this emulsion L, i.e., any typical length. (See Figure 7.) Because of the short annihilation mean free path, matter and antimatter penetrate each other only in a small annihilation region along the boundary with width $h \ll L$. It is important to remember that everything is embedded inside thermal radiation.

It was noticed some years ago by Alfvén and Klein (1962) that annihilation produces pressure in such a system (this is what they called a Leidenfrost effect): indeed, annihilation along the boundary produces pions which decay either into gamma rays ($\pi^0 \to \gamma + \gamma$) or into electrons and positrons ($\pi^\pm \to \mu^\pm \to e^\pm$) forgetting about the neutrinos which are unimportant here. These high energy particles (γ, $e^\pm$) with small masses carry a large momentum and therefore exert a strong pressure upon the neighboring medium.

The question we shall ask now is the following: what is the general behavior of the fluid motions which are generated by the energy release and the pressures which are due to annihilation?

A first orientation can be obtained, using a thermodynamical argument which is due to P. G. de Gennes and A. Blandin. Let us assume that the system matter + antimatter + radiation can be treated as a thermostat with temperature T. The system σ made up by the high-energy gamma rays and $e^\pm$ which are not yet thermalized is not in thermal equilibrium. According to a well-known theorem in thermodynamics, its free energy F_σ must therefore tend to decrease. When computing this free energy $F_\sigma = E_\sigma - TS_\sigma$, one finds that, because of the presence of high-energy particles, the energy part E_σ of F_σ is larger than the entropy part, so that F_σ is positive. On the other

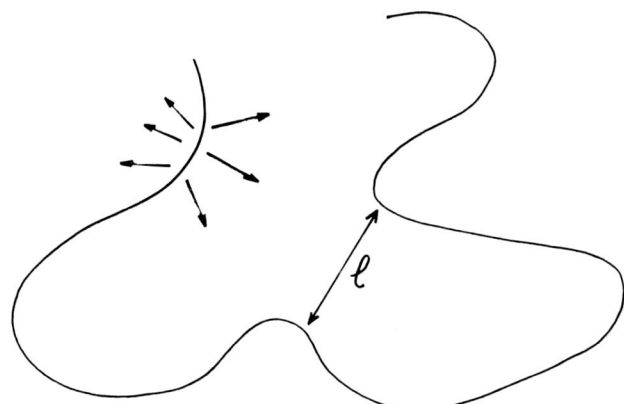

Fig. 7. An emulsion.

hand, the total number of particles in σ is obviously proportional to the area S of the matter-antimatter boundary S, so that

$$F_\sigma = AS. \tag{12}$$

Therefore S must decrease. But since matter and antimatter fill up all space in equal amounts, this is possible only if L increases. (It can be useful for comparison to notice that F_σ has exactly the same form as a surface tension free energy.)

Therefore we are led to propose the following Ansatz: *The coalescence effect*; the scale L of the emulsion increases with time as a side effect of the Leidenfrost mechan-

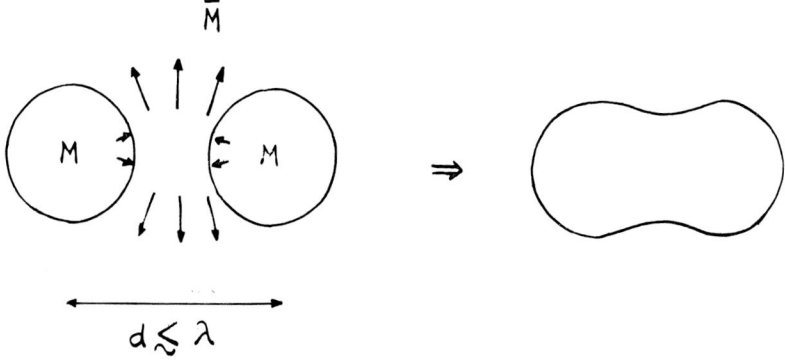

Fig. 8. The coalescence of two matter bubbles embedded in antimatter.

ism, so that matter and antimatter are coalesced into larger and larger regions of space as time increases.

What was insufficient in the Blandin – de Gennes argument? It was to consider the system matter+antimatter+radiation as a thermostat, since precisely strong temperature gradients can be expected as part of the Leidenfrost effect. Therefore, we shall now review how the problem of coalescence can be tackled using the adequate techniques of hydrodynamics and what are the results of this approach.

8. The Coalescence Effect: Hydrodynamics

Here we follow the treatment reported in Ommès (1971b, 1971c).

Let us denote the local density of nucleons (or antinucleons) by $N(x, t)$ the fluid velocity by $\mathbf{v}$ (over distances larger than the mean free path for a thermal photon, matter+radiation can be considered as making only one fluid), the mass density (of matter+radiation) by ϱ, and the energy density for thermal radiation by E. An important quantity in the problem is the energy current $c\mathbf{J}(\mathbf{x}, t)$ carried by the annihilation products to a point $\mathbf{x}$. One has

$$\mathbf{J}(\mathbf{x}) \propto \int_S dS_0 \frac{\mathbf{x} - \mathbf{x}_0}{|\mathbf{x} - \mathbf{x}_0|^3} e^{-\chi(\mathbf{x}, \mathbf{x}_0)} \tag{13}$$

where the total absorption χ is given by

$$\chi = \int_{x_0}^{x} n\sigma \, dl. \tag{14}$$

Here, σ is the cross section for interaction between an electron and high-energy photons (in practice σ is 40 times the Thomson cross section). The integration in (13) is made along the boundary.*

The basic equations are, as usual, particle conservation, energy conservation and the equation of motion. Disregarding negligible terms they read

$$(\partial N/\partial t) + \nabla (Nv) = 0, \tag{15}$$

$$(\partial E/\partial t) + \nabla (Jc - D\nabla E + \tfrac{4}{3} vE) = 0, \tag{16}$$

$$\varrho \frac{dv}{dt} = -\nabla \frac{E}{3} + \frac{J}{\lambda_\gamma^{\text{h.e.}}}. \tag{17}$$

The meaning of these equations is quite clear: in (16), the Jc term represents the heat source, the next term is conduction by diffusion and the last one conduction by convection. In (17) we notice that the radiative pressure is $E/3$ and $J/\lambda_\gamma^{\text{h.e.}}$ (with $\lambda_\gamma^{\text{h.e.}} = (N\sigma)^{-1}$) is the force exerted upon a unit volume of the fluid.

---

* When $kT > 100$ eV, χ is controlled by $\gamma + \gamma \to e^+ e^-$ (as pointed out by G. Steigman).

Of particular importance are the boundary conditions which must be added to these equations in order to describe the behavior of the particles along the matter-antimatter boundary. Here the characteristic parameter is λ_0, the mean free path for thermal photons. For distances to the boundary smaller than λ_0, the hydrodynamical approximation is not valid and one must describe the propagation and thermalization of the particles by using kinetic theory. The essential result which is found in this way is the existence of a *discontinuity pressure* δp across the boundary which is related to the curvature of the boundary by

$$\delta p = 2\alpha/R. \tag{18}$$

This relation is identical with the Laplace-Kelvin condition which holds when there is a surface tension with coefficient α. The value of α is given by

$$\alpha = \tfrac{2}{3} J(0) \lambda,$$

where

$$\lambda = \text{Max}(\lambda_0, \lambda_\gamma^{\text{h.e.}}). \tag{20}$$

Here $J(0)$ is essentially given by the rate of annihilation along the boundary which is itself controlled by diffusion.

The origin of the discontinuity δp can be easily visualized: Along the boundary, there are the same number of high-energy particles going in each direction, so that they carry the same momentum into matter and antimatter. Because of the curvature, the flux of momentum per unit area (i.e. pressure) is larger in a convex region.

Because of the formal identity of the boundary condition (18) with the effect of a surface tension and because of the large value of α, coalescence will result. Dimensional arguments show that the size of the emulsion L will increase with time according to the rule

$$L^3 = \frac{\alpha}{\varrho} t^2. \tag{21}$$

Apart from these results, which are the most important, some significant properties of (15) to (17) are also known:

(i) There is no hydrostatic solution, except for a plane or spherical boundary. The plane boundary is hydrodynamically stable.

(ii) The regime of convective motions can be investigated by freezing the motion of the boundary. It is found that, after a short transient period of the order of L/c_s (where c_s is the sound velocity), a balance is established between the force $J/\lambda_\gamma^{\text{h.e.}}$ and the radiative pressure. (The physical explanation is the following: gamma rays push electrons which drag radiation and build up a pressure gradient.) In these conditions, heat transfer, as governed by (16) determines a convective motion. The effect of this motion has been found in a special case (two bubbles of matter embedded inside antimatter) to produce further coalescence. However, these convective velocities are much smaller than the ones given by (21).

As a conclusion, the coalescence effect will most probably be well established very soon. [Though we point out that (18), a relatively new result, needs further confirmation.] We shall therefore now consider its consequences in cosmology.

9. Cosmological Consequences

Applying these ideas to the evolution of the Universe as described by the previously introduced model, we find a new succession of periods as shown in Table II. [The entries in this table have been compiled from Ommes (1971a), Schatzman (1970), Kundt (1971) but include modifications due to an effect pointed out by G. Steigman in a private communication and also some results derived from Equations (18) and (19).]

TABLE II
Evolutionary periods in our model

Temperature T	350 MeV	40 keV	3000 K
Time t	10^{-5} s	~ 15 min	10^6 yr
Emulsion size (L) (cm)	10^{-4} cm	$\sim 10^{3.5}$ cm	$\sim 10^{22}$ cm
Period	separation	annihilation	coalescence

We shall briefly describe the main features of this evolution.

(i) The separation of the two phases takes place above the critical temperature T_c. The size of the emulsion near T_c can be computed by standard techniques and is found to be given by $L = [Dt]^{1/2}$ where D is the diffusion coefficient for the baryonic number. Any of the conventional formulas for D: $D \simeq (kT/m)\mathcal{T} \simeq c\lambda$, give the same results and the value $L_c = 10^{-4}$ cm. Somewhat above T_c, the coefficient η, as computed from the baryonic density, is of the order of

$$\eta_c \simeq 7\left(\frac{T - T_c}{T_c}\right)^{1/2} \tag{22}$$

so that η_c is in fact of the order of unity 50 MeV above T_c. It tends to decrease when T approaches T_c so that, for the sake of definiteness, we shall consider that the end of the separation period takes place slightly above T_c with a value of η_c not very different from unity.

(ii) The annihilation period is characterized by a strong mixing of nucleons with antinucleons which leads to a large relative amount of annihilation. The size of the emulsion and the value of η at any given time can be known explicitly: Since particles are lost by diffusion, one has

$$\frac{dN}{dt} \simeq 2D\Delta N \simeq 2D\frac{N}{L^2}$$

(the factor 2 is for antiparticles entering). So that (taking expansion into account) one has

$$\left|\text{Log}\frac{\eta}{\eta_c}\right| \simeq |\text{Log}\,\eta| = \frac{2Dt}{L^2}. \tag{23}$$

On the other hand, this same quantity η can be computed by saying that, in a given region of size L, what baryons remain after annihilation were the fluctuations initially present in the same region. If fluctuations were as usual, one would find:

$$\eta = \eta_c \left(\frac{L_c T_c}{LT}\right)^{3/2}$$

the factor T_0/T being due to expansion. However, because of the correlations in the initial fluctuation of the baryon number (there is no volume fluctuation of the baryon number, only surface fluctuations near the boundary of the emulsion with size L), one has

$$\eta = \eta_c \left(\frac{L_c T_c}{LT}\right)^2 \tag{24}$$

Taken together (23) and (24) give a determination of η and L during the annihilation period.

As pointed out by G. Steigman, the value of D during the leptonic period is mainly given by neutron diffusion (the large neutron mean free path being controlled by electron-neutron scattering). Annihilation proceeds rapidly as long as weak interactions maintain a balance between protons and neutrons (e.g., $v+p \rightleftarrows e^+ +n$). Near $kT \simeq 3.5$ MeV, this effect of weak interactions stops and η is found to be of the order of 10^{-8}.

From 3.5 MeV to 50 keV, the system is rather quiescent. The diffusion of protons is slow because of plasma effects, and coalescence, which is beginning to take place, is not yet very effective. From 50 keV to 3000 K coalescence becomes very active and brings together large quantities of matter rapidly, the typical size of the emulsion being given by (21). It turns out that the relative amount of annihilation is small during this last period.

10. Conclusion and Perspectives

Coalescence stops at the time of recombination because of the rapid decrease in the sound velocity which would make coalescent motions supersonic.

The details of these physical processes can be rather intricate and they are presently being investigated by S. Caser, J. L. Puget, G. Valladas and myself. We cannot yet give final numbers with complete confidence until this investigation is completed, which will take some time. However I shall quote a few preliminary results which are subject to change: η is slightly smaller than 10^{-8}, the mass $M = \varrho L^3$ is of the order of 10^{11} solar masses or more (i.e., a galactic mass or a cluster mass) and there is no primeval helium because of the neutron loss during the leptonic period.

It is gratifying that the simplest model of the Universe, together with a few well-defined effects which are within the reach of theoretical investigation, leads to an evolution which can be followed in some detail. The quantitative conclusions, as to the amount of matter in the Universe and the galactic masses seem to agree rather well with observation.

The main problems will now be to link these ideas with astrophysics and we list here a few of them:

When does the gravitational collapse of galaxies take place?

How important is the role of annihilation pressure in gravitational collapse? Can it take place through the effect of a magnetic field stopping the electrons?

More generally, what are the plasma effects (e.g., magnetic instabilities) when matter and antimatter are in contact?

What are the observable effects of coalescence velocities (e.g., turbulence, angular momentum (Puget and Stecker, 1972).

Once galaxies are formed, how frequently do they collide? Are such collisions responsible for the angular momentum of galaxies, as a simple calculation suggests?

Is the clustering of galaxies associated with the correlations inside matter which are due to the geometry of the emulsion?

Does annihilation in the later stages sufficiently distort the spectrum of thermal radiation or of the X-ray background to be detectable (Stecker *et al.*, 1971; Sunyaev and Zeldovich, 1970)?

Should we revive the antimatter model for quasars: they could be made of matter (of a nearly galactic size), initially trapped within antimatter? Energy and age requirements are easily met.

What happens to a cloud of antimatter inside a galaxy? We know that annihilation leads to a strong viscosity: does the object fall into the center of the galaxy? Or is it ejected?

Where and when are helium and deuterium synthesized? (See some of the comments by A. G. W. Cameron in this volume.)

And last but not least, can we find an observational test to make sure whether antimatter does or does not exist?

References

Aldrovandi, R. and Caser, S.: 1972, *Nucl. Phys.* **B38**, 593.
Alfvén, H. and Klein, O.: 1962, *Arkiv Fysik* **23**, 187.
Cisneros, A.: 1972, California Institute of Technology Thesis (to be published).
Dashen, R., Ma, S., and Bernstein, H. J.: 1969, *Phys. Rev.* **187**, 345.
Kundt, W.: 1971, *A Survey of Cosmology,* Springer Tracts, Heidelberg.
Omnès, R.: 1969a, *Phys. Rev. Letters* **23**, 38.
Omnès, R.: 1969b, *Ann. Phys. Paris* **4**, 515.
Omnès, R.: 1970, *Phys. Rev.* **1**, 723.
Omnès, R.: 1971a, *Astron. Astrophys.* **10**, 228.
Omnès, R.: 1971b, *Astron. Astrophys.* **11**, 450.
Omnès. R.: 1971c, *Astron. Astrophys.* **15**, 275.
Omnès, R.: 1972, *Phys. Reports* **3C**, No. 1.
Puget, J. L. and Stecker, F. W.: 197? (to be published).

Ruelle, D.: 1971, *Phys. Rev. Letters* **27**, 1040.
Schatzman, E.: 1970, CERN Report.
Stecker, F. W., Morgan Jr., D. L., and Bredekamp, J.: 1971, *Phys. Rev. Letters* **27**, 1469.
Sunyaev, R. A. and Zeldovich, Ya. B.: 1970, *Comments Astrophys. Space Phys.* **11**, 66.
Widom, B. and Rowlinson, J. S.: 1970, *J. Chem. Phys.* **52**, 1670.

QUASAR COUNTS AND THE LAGGING CORE MODEL
('WHITE HOLES')

Y. NE'EMAN

University of Tel-Aviv, Tel Aviv, Israel, and University of Texas, Austin, Tex., U.S.A.

Abstract. The lagging core model for quasars (i.e., 'white holes') is shown to impose a negative exponential for the number of quasars at any given time as measured from the start of the cosmological expansion. Models in which quasars and other dense cores result from collapse yield a different behavior. A recent observational count appears to fit a negative exponential.

The morphological study of quasars lends plausibility to the following characteristics:
  (a) quasars possess very dense cores (Greenstein and Schmidt, 1964);
  (b) quasars are at cosmological distances (Bahcall and Bahcall, 1970; Gunn, 1971);
  (c) they seem to fit in a general class of objects with dense nuclei (Burbidge, 1970); and
  (d) their space distribution displays an evolutionary effect, their number appearing to follow an exponential decrease (Schmidt, 1970).

We would like to point out that the observation (d) reinforces the arguments for quasars as lagging cores of the original expansion (Ne'eman, 1965; Novikov, 1964) and would be difficult to reconcile with any other model.

Dense nuclei can be created in two ways:

(A) through the gravitational collapse of matter in a galaxy or a protogalaxy. In this case it can have the following history.
  (1) it continues to collect matter and to grow; and
  (2) at a certain point it undergoes further collapse and generates a black hole. In this case it either vanishes and thus has a finite lifetime, or it may go into a stationary state and is then equivalent to case (1).

(B) dense nuclei represent lagging – as yet unexpanded – cores ('white holes' in presently fashionable nomenclature). At some stage in their history they finally expand and turn into 'normal' distributions of matter – galaxies etc. Members of this class of objects are thus only recognized as quasars in their initial stage.

In case A1, the number of quasars (and similar objects) should grow with time. The evolutionary effect should thus be the inverse of observation (d).

In case A2, with a finite lifetime for collapsed nuclei due to the formation of black holes, the number of quasars and dense nuclei should have at least reached a steady state at some early stage in the formation of galaxies; alternatively, we would have some growth with time, though less than in case A1. Again, negative exponential behavior does not fit this picture.

Now for case B. We assume that the total amount of matter in the universe is fixed, with much of it in unexpanded cores. Assuming a fixed probability λ for any single core to start upon the final stages of its expansion in any given unit of time, we find for N the number of quasars at time t

$$N(t) = N_0 e^{-\lambda t}.$$

The time coordinate t here corresponds to some external observer, e.g., an observer linked to a universal space-averaged reference frame. We know that the expansion occurs very fast in the lagging core's own reference frame, but is extremely slow with respect to an outside observer. With the difference between the two frames becoming very large, it is natural to take a probabilistic view with respect to the occurrence of the final expansion for the outside time-coordinate.

Summing up, we see that observation (d) fits very well with the lagging core model.

References

Bahcall, J. N. and Bahcall, N. A.: 1970, *Publ. Astron. Soc. Pacific* **82**, 487.
Burbidge, G. R.: 1970, *Ann. Rev. Astron. Astrophys.* **8**, 369.
Greenstein, J. L. and Schmidt, M.: 1964, *Astrophys. J.* **140**, 1.
Gunn, J. E.: 1971, *Astrophys. J. Letters* **164**, L113.
Ne'eman, Y.: 1965, *Astrophys. J.* **141**, 1303.
Novikov, I. D.: 1964, *Astron. Zh.*, **41**, 1075.
Schmidt, M.: 1970, *Astrophys. J.* **162**, 371.

DISCUSSION

Van Horn: I am not a general relativist, but insofar as I understand the problem I believe that the total mass of the universe is directly connected with the deceleration parameter q_0. If that is so, the limitations that can be placed upon the range of possible values of q_0 from existing observational data can be used to place restrictions on the amount of mass in the universe that has not yet become observable, in the model you have described. What sort of restrictions on the amount of such 'unborn' mass does this give?

Ne'eman: I think we are as yet far from seriously influencing the overall density ϱ. The conventional value of $\varrho \sim 10^{-29}$ g cm^{-3} is indeed connected to the Hubble constant $H = \dot{Q}/Q$ for the case of Euclidean three-space $k = 0$ and deceleration $q_0 = 0.5$ by the equation

$$\varrho = \frac{3H^2}{8\pi G}.$$

This is the density which makes the universe close upon a Schwarzschild radius c/H, i.e. the universe just fills out its observational bubble. In the $k = +1$ case, at present favored by Sandage's estimates of q_0 from counts, ϱ has to be even larger than the above figure. Now to get this value of ϱ, we require some 10^{55} to 10^{56} g in the Universe, i.e. about 10^{22}–10^{23} solar masses. If the number of quasars is of the order of 10^5, this would still allow us to put in as much as 10^{16} solar masses, i.e. up to a million galaxies, without modifying the situation by more than 10%.

CONCLUDING REMARKS

A. G. W. CAMERON
Belfer Graduate School of Science, Yeshiva University, New York, N.Y., U.S.A.

Unfortunately I was unable to attend the first two days of this meeting, owing to another long-standing engagement, and hence I cannot give a fair summary of the entire symposium. Instead, I shall try to put the subject matter of this symposium into perspective, and to discuss what seem to be some of the key issues we face at the present time.

I think this is one of the most enjoyable of the IAU symposia that I have attended. The number of people working in the field of dense matter is still relatively small, and therefore this symposium has not been crowded with the large number of papers that is all too often a feature of IAU symposia. We have been able to proceed in a relaxed and orderly manner, and I have enjoyed this feature of the symposium very much.

In particular, this has been a symposium of IAU Commission 35 on Stellar Structure. It is a measure of the newness of this field of dense matter calculations that probably the majority of the people presenting papers are not members of the IAU. A large number of the participants are physicists who have entered the field very recently because of their interest in the properties of neutron stars. Probably Commission 35 should co-opt many of you in order to give dense matter a greater representation within the deliberations of the IAU.

In many respects the discussions of this symposium have run parallel to those which were held at the Aspen Workshop on Neutron Stars in August, 1971. It has been interesting to me to see to what degree the field has changed during that intervening year. There have been some important new advances, but in many respects the situation is much as it was then. There have been some aspects of this meeting which also touch upon the Aspen Workshop on The Physics of the Early Universe, held in June, 1972. I will mention some relevant results from that workshop in the course of these remarks.

There are two aspects of dense matter: hot dense matter and cold dense matter. Most of the discussions here have dealt with cold dense matter, with applications mainly to neutron stars. It is unfortunate that the other aspect, hot dense matter, has hardly been touched upon except for the exceedingly important talk by Omnès. One expects to find hot dense matter mainly in the early history of the Universe, and I would first like to speak regarding that subject.

There are two basic approaches to a discussion of the physics of the early Universe. In one approach one takes a completely symmetric Universe, in which there are equal numbers of baryons and antibaryons, and in which the baryonic number is therefore zero. We have heard Omnès describe his approach, involving a phase separation between the baryons and antibaryons at high temperature, and probably this is the only feasible scheme which may be capable of producing a large-scale separation of matter from antimatter as is required in the later stages of development of the Universe.

The alternative is to assume an asymmetric universe, in which the number of baryons exceeds the number of antibaryons. Here again we can consider two approaches. In one of these, the earlier stages of the Universe involved exceedingly high temperatures, so that the Universe is filled with baryon-antibaryon pairs, with only a very small excess of baryons over antibaryons at the earliest times. If there is no phase separation between baryons and antibaryons at high temperature, then gradually there will be a complete annihilation of the antibaryons with the baryons, leaving the small excess of baryons as ordinary matter to fill the expanding Universe in its later stages. On the other hand, if there is a separation between baryons and antibaryons at high temperature, then this picture is insignificantly different from the picture presented by Omnès, and one would expect to find large-scale separation of patches of matter and antimatter in this case as in his. Thus the whole question of the reality of the phase separation is obviously of immense importance, and deserving of much further work.

The other approach to the asymmetric universe was first proposed by Hagedorn. In this approach matter reaches a finite limiting temperature as it is squeezed to indefinitely high densities, and hence space becomes filled with baryons, but with baryons of increasingly higher masses well up on the scale of an exponential mass spectrum. We have not seen equations of state of dense matter of this type applied to the cosmological problem at this meeting, but we have seen them applied to discussions of neutron stars.

Let us follow the Hagedorn approach as we go backwards in time in the universe. The radiation background will increase in temperature, and eventually when the temperature rises to 3000 K, matter will become ionized in the Universe, and as we go still further back, when the temperature passes through 10^9 K, all of the nuclei which may exist in the Universe will be broken down by photodisintegration. Further back, we create electron pairs, then muon pairs, and eventually some pions. At that point the baryons present start to be transformed into various baryonic excited states, higher up on the mass spectrum. As the density becomes higher, the characteristic mass of the baryons present will continue to increase, but the temperature will approach an asymptotic limit which is about equal to the pion rest mass, as Craig Wheeler discussed. Hagedorn suggests large variations in baryon number are possible.

Eventually we will come to a critical era in the early Universe, at an expansion time of around 10^{-43} s, at which the typical Hubble radius would be about 3×10^{-33} cm. We expect that quantum effects will come into general relativity under these conditions. The typical mass of the baryons which will exist at this time is of the order of 10^{-5} g. Only masses of this high order will have Compton wavelengths small enough to fit into the Hubble radius of the Universe at this time. Indeed, if we multiply this mass by the expansion age of the Universe, the result is a number of the order of h; this is simply an indication that we are dealing with an epoch in the expansion of the universe at which the Heisenberg uncertainty principle is approximately fulfilled by the general relativistic quantities, and hence if there is any unity in general relativity and quantum mechanics, it does not make sense to ask what happened in the universe at an earlier time, because large fluctuations in the energy make earlier time scales lack any meaning. We would certainly like to know whether this type of picture has any validity, since it represents such an enormous extrapolation from current knowledge of the baryonic mass spectrum and of baryonic particle theory.

If we take the Omnès approach, and have a symmetric Universe, then the picture at 10^{-43} s would be rather similar, except of course that we would have equal numbers of baryons and antibaryons present. Presumably these baryons and antibaryons would have to have extremely high masses, in order that they can fit into the Universe at this time.

The reason I am stressing the epoch 10^{-43} s, is that some very interesting results concerning this epoch were presented by C. Misner and his colleagues, and by L. Parker of the University of Wisconsin (at Milwaukee), at the 1972 Aspen Workshop on The Physics of the Early Universe, which pertained to this characteristic time of 10^{-43} s. Parker, in particular, has been examining the properties of an anisotropic cosmology associated with that characteristic time. As the Universe expands, the Hubble radius encompasses larger and larger amounts of matter, and we must ask the question why different parts of the Universe should be synchronized in time when the Hubble radii centered upon those parts start to overlap and bring the two different parts into communication. If they are not synchronized, then one expects large variations in the local metric at the places where the Hubble radii start to overlap, and these large variations in the metric may be represented by anisotropic expansion rates. It is best to describe this general situation as representing gravitational chaos.

Now if we had a very strong electric field under such circumstances, then this electric field would undoubtedly be able to produce pairs of baryons and antibaryons which would be accelerated in different directions by the strong field strength, acquiring energy at the expense of the field strength. Parker has shown that strong gravitational potential gradients can perform a similar role. These can produce pairs of baryons and antibaryons, causing them to move in different directions, and acquiring energy at the expense of the gravitational potential gradient. Indeed, the greater part of the gravitational potential gradient associated with anisotropic expansion will be eliminated through the creation of rest mass in baryon-antibaryon pairs. Furthermore, his calculations show that the dimensional extent of the baryons and antibaryons created extends to about twice the distance of the associated Hubble radii. This is a hint that this pair creation process has a physical extent that extends beyond the universal horizon, and thus it is promising that this may provide a mechanism for producing a large-scale homogeneity in the structure of the Universe, thus synchronizing the rates at which different patches of the Universe join onto neighboring patches when their associated Hubble radii become large enough.

If this large-scale synchronization and homogeneity should be achieved, then this would be an argument against the appearance of Yuval Ne'eman's white holes, which he has discussed at this conference. If it is not achieved, then perhaps we should take the white holes seriously.

This work of Parker is certainly philosophically very attractive, and it produces the consequence that we should expect a symmetric Universe, with equal numbers of baryons and antibaryons. If it is correct, then we should certainly prefer the Omnès approach. I was very impressed by the progress which Omnès has achieved in working out the details of the symmetric theory, and apparently being able to separate matter and antimatter at least on a galactic scale. I hope he should eventually find, upon more precise calculation, that the separation can be done on the slightly larger scale of the mass in clusters of galaxies, because otherwise I think we would probably have trouble with too much matter-antimatter annihilation arising from interactions between galaxies in clusters. But he has certainly come sufficiently close to that point that the difference between mass separation on the scale of galaxies or of clusters of galaxies is not a very significant difference.

One of the main issues, mentioned by Omnès at the end of his talk, is the lack of any cosmological nucleosynthesis of helium, or of deuterium and tritium, in the symmetric cosmology which he has presented. The attractiveness of cosmological nucleosynthesis in a Hagedorn-type theory arises from the fact that the amount of helium produced is just of the right order of magnitude, about 25% by mass, of the amount which appears to exist in stars made both early in the Galaxy and late in the Galaxy, and also in other galaxies. This hydrogen-helium ratio appears to be a very universal function, and discussions in recent years from the observational point of view have concluded that the hydrogen to helium ratio is essentially universal. This seems to require a pre-galactic production of helium.

Recently there has also been much discussion of the amount of deuterium which we have in the solar system. The deuterium to hydrogen ratio in the solar system seems, at least in the primitive solar nebula, not to be the terrestrial ratio present in sea water but something of the order of 0.1 to 0.3 times terrestrial. This primordial deuterium would be about 50% greater than the amount of primordial ^{3}He. In a cosmology in which early nucleosynthesis occurs, these amounts of deuterium and ^{3}He would be produced in an open Universe with something like 10% of the critical closure density. This makes a very attractive picture. Some recent work by Hubert Reeves has gone further, and has suggested that in addition we can get the right amount of ^{7}Li produced in cosmological nucleosynthesis if we live in a Universe in which there is a negative lepton number, that is, an excess of antineutrinos over neutrinos in the background neutrino radiation.

Now it appears that none of this cosmological nucleosynthesis will take place in an Omnès-type symmetric cosmology. However, it does seem possible that an opportunity may exist for the production of large amounts of helium in the pre-Galactic stage. After matter recombines at a temperature of 3000 K, the Jeans length in the expanding matter encloses only about 10^6 solar masses of material. As the matter continues to expand before halting and recollapsing, the Jeans length continues to decrease, enclosing only a few thousand solar masses of material at the time of maximum expansion, where the Jeans length approach to fragmentation is probably of maximum validity. Further opportunity for fragmentation of the matter may occur during the collapse phase of the matter, so that it is entirely possible that stars in the mass range

10^2 to 10^3 solar masses may be formed as a first pre-galactic generation in space.

In a symmetric cosmology, it would appear that these stars would be entirely composed of hydrogen. In the asymmetric cosmology of the Hagedorn type, they will be composed of a mixture of hydrogen and helium. These different types of stars will behave somewhat differently, particularly during an initial collapse phase, unless angular momentum effects produce flattened disks first from which the massive stars form. In either case, a great deal of supernova explosions with accompanying formation of heavy metals can be expected to occur, and the thing that we must look into carefully is whether huge amounts of helium can be formed at the same time, particularly in the stars which would be composed of pure hydrogen. Such helium production might take place as a result of collapse, conversion of hydrogen to neutrons, production of helium, and re-explosion of the matter into space. Helium might also be produced in large quantities if the massive stars which are formed are vibrationally unstable, leading to mass shedding in which large amounts of helium are ejected. All of these processes will have to be carefully examined before we can answer the question as to whether the Omnès cosmology can acquire enough helium at the pre-galactic stage to satisfy the essential universality of the helium to hydrogen ratio which is observed throughout many galaxies.

There appears to be no opportunity to make deuterium and ^{3}He in the pre-galactic stage, at least not in the proportions appropriate to the early solar system. However, Stirling Colgate has recently suggested a supernova mechanism in which deuterium and ^{3}He might be produced during the normal course of galactic evolution. In this process a strong supernova shock wave, becoming relativistic as it approaches the outer fringes of the star, accelerates electrons forward, creating a strong electric field, which then accelerates protons and alpha-particles at different rates, leading to collisions at several tens of MeV of energy, and consequent spallation production of deuterium and ^{3}He. It is clear that this process will also require a great deal of scrutiny, because it may remove the need for cosmological production of these two light isotopes.

It is thus evident that a great deal of exciting work is in store for those working on the theory of hot matter and its applications to cosmology. Let me now turn to the question of cold dense matter.

The characteristic picture which we have for the neutron star consists of an outer crust, the lower part of that crust containing some superfluid neutrons as well as nuclei, under that a mixture of superfluid neutrons and protons, and then at the center a core which may or may not be a crystalline solid. There are a number of important basic issues to face here, among them the important question as to whether this basic picture usually assumed for the neutron star may be entirely wrong. However, let us proceed for the moment on the assumption that the general picture is right.

There now seems to be fairly general agreement that the nuclei in the crust, including those at the base of the crust which interact with the free neutron gas, tend to be on the small side rather than on the large side. There seems to be some disagreement still as to how one should go about calculating the precise character of those nuclei, but there seems definitely to be agreement that those nuclei should be rather small.

Just a year ago David Pines said at Aspen that if it turned out that the nuclei were small, that would invalidate the starquake theory, at least in terms of crustal starquakes. Not long after the 1971 Aspen workshop the second major glitch was observed in the Vela pulsar, and this caused Pines to have additional grounds for distrusting the crustquake theory, at least as applied to the Vela pulsar. Meanwhile, the work of Canuto and Chitre, and of some others, tended to indicate that probably there was a solid core at the center of neutron stars, and that led Pines and others to suggest that maybe we have glitches in the form of corequakes. I think the whole problem of glitches is still very much uncertain.

People have suggested that the infall of matter can produce such glitches. Personally I find this to be a very implausible suggestion because of the difficulty of arranging to have a suitable supply of matter to do the infalling. Crust-quakes seem not to produce glitches. Corequakes may or may not work. I am a little concerned about the ability of cores to maintain distortions as their rotation decelerates, in the presence of density-sensitive reactions which can convert solids to liquids as the density changes. George Greenstein has mentioned that we have in principle another way to get glitches, although we do not know how to make use of it yet. In any model of a neutron star in which there is a fair amount of superfluid, the superfluid must slow down by frictional processes, and if one can arrange for variations in the friction, one can certainly have what Pines calls the 'restless' behavior of the Crab nebula. If one can have sudden transfers of angular momentum from the superfluid neutron reservoir to the charged particle system in the pulsar, then we may get major glitches as well. Greenstein was suggesting the 'boiling pot theory' of glitches, which he did not indicate that he took very seriously, so I suppose the rest of us will not take it very seriously either. At any rate, some such mechanism for a sudden angular momentum transfer always remains a possibility if we can think of the appropriate physics that could produce it.

Now let me come to the question of the solid core. I think there are some extremely challenging aspects that are associated with the calculations which Canuto presented here, in particular concerning the hyperonic components of the crystalline core in his calculations. The basic problem is that the $^3P_1$ interaction between neutrons and protons is repulsive; it plays a big role in the stability of nuclei, but when Canuto and Chitre computed the interactions that should exist between hyperons using SU(3) techniques, they found that the interaction is as attractive as those of the other interactions, at least until very short distances are reached. The result is that if one makes a plot of the energy per nucleon vs. density, then by far the lowest energies that emerge from the calculations are those in which the crystalline lattice contains only hyperons, such as a Λ^0 lattice or a mixed lattice with Λ^0 and Σ hyperons. Furthermore, the calculations even indicate that such purely hyperonic lattices may form highly bound systems, with negative energies, at densities very much greater than that of ordinary nuclear matter. Such lattices are unstable against shear motions, and the implication is that the liquid state of a mixture of pure hyperons may be even more tightly bound than indicated for the lattice calculations of Canuto and Chitre.

If one takes these calculations at face value, we must realize that we have a fundamental uncertainty concerning our standard picture of the neutron star. A purely hyperonic star would be a more stable configuration. The calculations of Canuto and Chitre indicate that such a star may have a mass as high as 1.1 $M_\odot$, and its moment of inertia is certainly great enough to provide the requirements of the Crab nebula pulsar. If the mixture of pure hyperons has a large negative binding energy at a density of a few times 10^{15} g cm^{-3}, then it is even possible that this hyperonic star has an abrupt sharp surface at this density, without even a fringe of ordinary neutrons and protons If Σ hyperons are included within the mixture, then this configuration can maintain a strong magnetic field, so that all of the basic characteristics of the pulsar could be provided by such a model. Even restless behavior in glitches might occur in the lower density regime, near the surface, where the material has fluid properties, if in fact the properties are superfluid, and there is only a small amount of friction between the Λ^0 hyperons and the charged Σ hyperons.

I have tried very hard to think of pulsar observations which might exclude this type of model which emerges if we take the work of Canuto and Chitre seriously, but I have been unable to do so. This indicates the enormous importance for the future development of neutron star physics for particle physicists to determine the character of the $^3P_1$ interaction between hyperons, and between hyperons and neutrons or protons. If this interaction should be attractive between hyperons, as indicated by the SU(3) calculations of Canuto and Chitre, then this pure compact hyperonic star would seem to be a leading candidate for the explanation of pulsars.

This suggests also that further investigations of the physics of cold dense matter are likely to be exceedingly exciting. This is a very challenging field in which to work, and I am sure that there will be many exciting future conferences on the physics of dense matter.

It remains for me to express great thanks on behalf of myself and also on behalf of everyone here, to the organizers of this conference for the very excellent meeting which they have hosted and for the great hospitality which they have shown us. I propose a vote of thanks to the local organizing committee.

This report has been supported in part by a grant from the National Science Foundation.